Broadband Network & Device Security

Broadband Network & Device Security

Benjamin M. Lail

McGraw-Hill/Osborne
New York Chicago San Francisco
Lisbon London Madrid Mexico City
Milan New Delhi San Juan
Seoul Singapore Sydney Toronto

McGraw-Hill/Osborne
2600 Tenth Street
Berkeley, California 94710
U.S.A.

To arrange bulk purchase discounts for sales promotions, premiums, or fund-raisers, please contact **McGraw-Hill**/Osborne at the above address. For information on translations or book distributors outside the U.S.A., please see the International Contact Information page immediately following the index of this book.

Broadband Network & Device Security

1234567890 CUS CUS 0198765432

ISBN 0-07-219424-3

Publisher
Brandon A. Nordin

Vice President & Associate Publisher
Scott Rogers

Senior Acquisitions Editor
Jane K. Brownlow

Project Editor
Julie M. Smith

Acquisitions Coordinator
Emma Acker

Technical Editor
Eric Rosenfeld

Copy Editor
Nancy McLaughlin

Proofreader
Linda Medoff

Indexer
Jack Lewis

Computer Designers
Elizabeth Jang
John Patrus
Kelly Stanton-Scott

Illustrators
Michael Mueller
Lyssa Wald

Series Design
Peter F. Hancik

This book was composed with Corel VENTURA™ Publisher.

To my grandparents Bill and Sue,
whose eternal love, support,
and encouragement have touched my life
in more ways than they could ever imagine.

About the Author

Benjamin M. Lail is a Senior Systems Engineer within the Developer Solutions Group of RSA Security, Inc. specializing in broadband network security infrastructures and solutions. He is an MCSE and CCNA, and has experience with many aspects of information security including cryptography, Public Key Infrastructure, network and router security, secure Internet and e-commerce transactions, and numerous security protocols and specifications. Before joining RSA, Ben worked as a systems administrator for Advance Micro Devices, and as a technical manager for Dell Computer Corporation. Ben is also a frequent speaker at industry conferences and seminars.

Contents at a Glance

Part I Broadband Network Security Fundamentals

Chapter 1 An Overview of Broadband Communication 3

Chapter 2 Choosing the Right Tools: Security Services and Cryptography 17

Chapter 3 The Need for Security: Network Threats and Countermeasures 61

Chapter 4 Broadband Networking Technologies 103

Chapter 5 A Survey of Existing Broadband Security Standards and Specifications 145

Part II Broadband Security Design Considerations

Chapter 6 Existing Network Security Protocols 171

Chapter 7 Placing Security Services and Mechanisms 223

Chapter 8 Security Side Effects 247

Part III Case Studies

Chapter 9 Securing Broadband Internet Access: DOCSIS BPI+ 273

Chapter 10 Securing Real-Time Multimedia: PacketCable Security 311

Chapter 11 Securing Interactive Television: DVB MHP Security 359

Chapter 12 Design Scenarios 383

Appendix A TCP/IP Primer 415

Appendix B Digital Certificates and Public-Key Infrastructure 437

Index 457

Contents

Foreword . . . xvii
Acknowledgments . . . xix
Preface . . . xxi

Part I Broadband Network Security Fundamentals

Chapter 1 An Overview of Broadband Communication . . . 3

A Brief History of Telecommunication . . . 4
- That Was Then . . . 6
- This Is Now . . . 6

What is Broadband Access? . . . 7
Existing Broadband Access Technologies . . . 7
- Cable . . . 8
- DSL . . . 9
- Fixed Wireless . . . 9
- Two-Way Satellite . . . 10

The Future of Broadband . . . 10
- Fiber Optics . . . 11

The Importance of Security in Broadband Networks . . . 12
- Security and the Average User . . . 12
- Securing the Network Infrastructure . . . 15

References . . . 16

Chapter 2 Choosing the Right Tools: Security Services and Cryptography . . . 17

Security Services and Mechanisms . . . 17
- Confidentiality . . . 18
- Integrity . . . 19
- Authentication . . . 19
- Nonrepudiation . . . 19
- Authorization and Access Control . . . 20
- Availability . . . 21

The Basics of Cryptography . . . 21
- Random Number Generation . . . 23
- Symmetric-Key Cryptography . . . 25
- Message Digests . . . 36
- Public-Key Cryptography . . . 40
- Public-Key Cryptography Standards . . . 48
- Federal Information Processing Standards and Certification . . . 51

Store-and-Forward vs. Session-Based Encryption . . . 52

Choosing the Appropriate Cryptographic Tools 53
  Using Stream Ciphers 53
  Using Block Ciphers 54
  Using Message Digests 56
  Using Public-Key Algorithms 56
  Interoperability Notes 57
  How Secure Is Too Secure? 57
References 58

Chapter 3 The Need for Security: Network Threats and Countermeasures 61

Who, What, and Why? Attackers and Their Motivations 62
When? "The Network Administrator Went Home Hours Ago..." 66
Where? The Internet's a Big Place! 67
  Broadband Access vs. Dial-up Access 67
Categorizing Common Attacks 68
  Passive Attacks vs. Active Attacks 69
  Eavesdropping 69
  Impersonation 73
  Denial of Service 75
  Data Modification 77
  Packet Replay 79
  Routing Attacks 80
TCP/IP-Specific Attacks 83
  Address Spoofing 83
  Session Hijacking 86
  Countermeasures for Address-Spoofing and Session-Hijacking Attacks 89
  TCP/IP Denial of Service 90
  IP and ICMP Fragmentation 92
Attacks on Cryptography 95
  Cryptanalysis 95
  Testing for Weak Keys 97
  Block Replay 97
  Man-in-the-Middle Attacks 97
  Countermeasures for Attacks Against Cryptographic Mechanisms 98
  Social Engineering and Dumpster Diving 100
References 100

Chapter 4 Broadband Networking Technologies 103

The Origins of Broadband 104
The ISO/OSI Reference Model 106
  Layer 7—Application 107

Layer 6—Presentation . . . 107
Layer 5—Session . . . 107
Layer 4—Transport . . . 107
Layer 3—Network . . . 108
Layer 2—Data Link . . . 108
Layer 1—Physical . . . 109
The TCP/IP Reference Model . . . 110
Data Encapsulation . . . 111
Communication Protocol Characteristics . . . 113
Service Provider Networks . . . 114
Cable . . . 115
Digital Subscriber Line . . . 120
Fixed Wireless Technology . . . 123
Two-Way Satellite Communication . . . 126
Quality of Service . . . 129
QoS Parameters . . . 130
Degrees of QoS . . . 136
The Great Debate: Cell-Relay vs. Standard Packet Switching . . 136
Models for QoS over IP Networks . . . 139
References . . . 143

Chapter 5 A Survey of Existing Broadband Security Standards and Specifications . . . 145

Standards Bodies and the Role of Standardization . . . 146
ANSI (American National Standards Institute) . . . 146
The BWIF (Broadband Wireless Internet Forum) . . . 146
Cable Television Laboratories . . . 147
The DVB (Digital Video Broadcasting) Project . . . 147
The DSL Forum . . . 147
ETSI (European Telecommunications Standards Institute) . . . 147
The IETF (Internet Engineering Task Force) . . . 148
The ITU (International Telecommunication Union) . . . 148
The IEEE (Institute of Electrical and Electronics Engineers) . . . 148
The ISO (International Standards Organization) . . . 148
Current Broadband Security Standards and Specifications . . . 149
The DOCSIS 1.0 Baseline Privacy Interface . . . 149
The DOCSIS 1.1 Baseline Privacy Plus Interface . . . 151
The PacketCable Security Specification . . . 154
The H.235 Security Standard . . . 154
The DVB Multimedia Home Platform . . . 160
The OpenCable Copy Protection System . . . 161
Security Gone Wrong—A Case Study of 802.11 WEP Encryption . . . 165
References . . . 168

Part II Broadband Security Design Considerations

Chapter 6 Existing Network Security Protocols 171
IPSec 172
Transport and Tunnel Modes 174
Security Associations 177
Security Policy Database 179
Security Associations Database 180
Authentication Header 181
Encapsulating Security Payload 186
Internet Key Exchange 191
SSL and TLS 197
A Brief History of SSL 198
SSL in Detail 198
Application Layer—Kerberos 216
Kerberos Authentication 217
Cross-Realm Authentication 219
Public-Key Authentication with Kerberos 220
References 220

Chapter 7 Placing Security Services and Mechanisms 223
Binding Security Services and Mechanisms to Data 223
Which Network Layer? 225
Application Transparency 225
Extent of Coverage 230
Performance 232
Comparing Existing Security Protocols 233
Security Protocol Implementation 234
Host-Based Security vs. Security Gateways 237
Extent of Coverage 238
Implementation, Configuration, and Maintenance 241
Securing Traffic Between a Large Number of Hosts or Applications 242
Distinct Traffic Flows 243
User Contexts 243
Coordination with Existing Security Policy 244
A Final Word on Encryption and Protocol Headers 245
References 245

Chapter 8 Security Side Effects 247
Network Performance and QoS 248
Embedded Device Constraints 249
Cryptography and Performance 250

General Considerations for Choosing Cryptographic Algorithms . . . 251
Dedicated Cryptographic Hardware . . . 262
Encryption and Compression . . . 263
Security Protocol Tuning . . . 264
Additional Tips for Improving Security in Real-Time Multimedia Applications . . . 265
Manageability . . . 266
References . . . 269

Part III Case Studies

Chapter 9 Securing Broadband Internet Access: DOCSIS BPI+ . . . 273

An Overview of the Baseline Privacy Plus Interface . . . 275
DOCSIS MAC Layer Frame Formats . . . 277
Baseline Privacy Key Management Protocol . . . 279
Authorization State Machine . . . 280
TEK State Machine . . . 285
BPI+ Key Encryption, Traffic Encryption, and Authentication Algorithms . . . 289
DOCSIS 1.1 BPI+ X.509
Certificate Usage and PKI Hierarchies . . . 292
BPI+ Cable Modem Certificate Hierarchy . . . 292
BPI+ Certificate Formats . . . 297
Certificate Validation on the CMTS . . . 301
Certificate Revocation and Hot Lists . . . 303
TFTP Configuration Files . . . 303
Signed Software Upgrade Verification . . . 304
Generation and Verification of Signed Software Upgrade Files . . . 307
References . . . 309

Chapter 10 Securing Real-Time Multimedia: PacketCable Security . . . 311

Overview of PacketCable Security . . . 319
IPSec . . . 322
Internet Key Exchange . . . 325
SNMPv3 Security . . . 326
PacketCable's Use of Kerberos . . . 327
Kerberized Key Management for IPSec and SNMPv3 . . . 330
Cross-Realm Operation . . . 336
Securing RTP and RTCP . . . 337
Key Management for RTP and RTCP . . . 341
PacketCable Security Certificate Usage and PKI Hierarchies . . . 345
PacketCable Certificate Validation . . . 356

Physical Protection of Keying Material 357
Secure Software Upgrades 358
References 358

Chapter 11 Securing Interactive Television: DVB MHP Security 359

The Multimedia Home Platform 360
MHP Security Overview 362
Authentication Messages 363
- Hash Files 365
- Signature Files 366
- Certificate Files 368
- The Object Authentication Process 369

MHP X.509 Certificate Usage and PKI Hierarchy 372
- Storage and Management of Root Certificates 373
- Certificate Revocation 374

Application Security Policy 375
- Permission Request File 375

Return Channel Security 379
Supported Java Security Classes 380
References 381

Chapter 12 Design Scenarios 383

Initial Design Steps 383
- Step 1: Identify Your Assets and Assess Their Value 384
- Step 2: Identifying the Threats 385
- Step 3: Selecting the Appropriate Security Services 386
- Step 4: Choosing Suitable Security Mechanisms 388
- Step 5: Identifying the Need for Persistent Security Services and Mechanisms 390
- Step 6: Choosing a Network Layer 391
- Step 7: Choosing Between Host-Based Security and Security Gateways 391
- Step 8: Identifying Existing Security Protocols That Meet Your Needs 392
- Step 9: Designing a New Protocol 393

Sample Design Scenarios 394
- Scenario 1: A Flawed Design 394
- Scenario 2: Designing Security from the Ground Up 403

Appendix A TCP/IP Primer 415

Encapsulation 416
Internet Protocol 417
- IP Headers 419
- IP Routing 423

Address Resolution Protocol and Reverse Address Resolution Protocol 425
Internet Control Message Protocol 426
Transmission Control Protocol 429
TCP Headers 430
Windowing 432
User Datagram Protocol 433
UDP Headers 434
Resources 435

Appendix B Digital Certificates and Public-Key Infrastructure 437
Digital Certificates 437
Certificate Types and Classes 439
Contents of a Digital Certificate 440
Validating a Digital Certificate 443
Certificate Revocation 445
Public-Key Infrastructure 448
CA Operations 448
Trust Models 450
Path Discovery and Validation 454
References 455

Index 457

Foreword

Welcome to *Broadband Network & Device Security*. This book, the latest in a series of e-security titles published by RSA Press, covers one of the newest and most critical frontiers of network security. While broadband dates back to the introduction of cable television in the 1960s, it wasn't until the 1980s that broadband technology began to be utilized within computer networking. In the late 1990s, broadband-based DSL and cable modems helped drive the meteoric growth of the Internet and the increased utilization of audio and video over the Web.

Today, there are four main broadband access technologies: DSL, cable, fixed wireless, and two-way satellite. For all their incredible capabilities, they – and the networks that deliver their various services—are vulnerable to security breaches, intrusions, and theft of service. This poses considerable risks to both organizations and individuals—risks that range from unauthorized data capture to identity theft.

Sophisticated broadband users may already understand the importance of installing personal firewall and antivirus software on their PCs. But unless the underlying broadband network infrastructure is secure, no one is safe. This book, then, is intended to help software developers, hardware designers, and corporate and institutional IT professionals to design, build, and implement secure broadband networks. In doing so, organizations can assure the availability and integrity of their services, while protecting their customers' privacy.

Written by RSA Security Senior Systems Engineer Benjamin Lail, *Broadband Network & Device Security* is divided into three parts. Part 1 introduces readers to the history and growth of broadband networks, the various security threats to which broadband networks are vulnerable, the primary security services and mechanisms for combating these threats, and the current security standards employed in both wired and wireless environments. This section provides valuable background information that will prepare readers for the comprehensive discussion of how to design a security system for broadband networks that appears in Part 2.

The heart of the book, Part 2 begins with a discussion of communication protocol characteristics, and introduces readers to a number of popular security protocols, including IPSec, SSL/TLS, and Kerberos. The ensuing chapters address the physical placement of security solutions within a broadband network, the impact that the implementation of these components may have on your network (including factors related to performance, cost, interoperability, and manageability), and the actions that can be undertaken to offset them.

In Part 3, case studies are used to demonstrate the actual steps involved in making broadband Internet access, voice-over-IP applications, and on-demand multimedia services secure. Step-by-step design scenarios walk readers through analyzing and building a security infrastructure from the ground up. *Broadband Network & Device Security* closes with two valuable appendixes—a "TCP/IP Primer" and an introduction to "Digital Certificates and Public-Key Infrastructure" that present the concepts of the TCP/IP protocol suite, digital certificates, and PKI as they pertain to the security issues presented in this book.

We hope that readers will benefit from this and other titles in the RSA Press catalog. We always welcome your comments and your suggestions for future books. For more information on RSA Security, please visit our Website at www.rsasecurity.com; to learn more about RSA Press, please visit www.rsapress.com.

Victor Chang
Vice President, Engineering
RSA Security Inc.

Acknowledgments

First of all, I would like to thank my family and friends for their patience and support while I toiled over this book. Writing this book was at times stressful, and I realize that I sometimes forgot about how it affected those close to me. I would like to thank my parents, Mike and Debi, for their enthusiasm in both my education and career, and for always expecting the best from me. Thanks to my little brother, Torry, for making me strive to be the best role model possible. A very special thanks goes to my wife, Anastasia, whose love and companionship helped me through many stressful days and sleepless nights while writing this book.

In addition, I would like to thank all the individuals who directly contributed to bringing this book to fruition. Thanks to everyone at Osborne/McGraw-Hill for putting forth the extra effort to complete this book on schedule, even when I was not. Thanks to Jane Brownlow, Emma Acker, and Julie Smith (my acquisitions editor, acquisitions coordinator, and project editor) for their patience and guidance, and for trusting me to complete this book on time. Thanks to my primary copy editor, Nancy McLaughlin, for her attention to detail, and for making it appear as though I have an English degree of my own. Thanks to the compositors, Kelly Stanton-Scott and John Patrus, and to the illustrators, Lyssa Wald and Michael Mueller, for giving this book its visual appeal. Thanks to Eric Rosenfeld for his sweeping industry insight and expert technical assistance in reviewing this book. Last, but not least, I would like to offer a special thanks to my colleagues Peter Yee, Steve Schmalz, Jim Gray, and Martin Euchner for their help in ensuring the technical accuracy of the material presented in this book.

Preface

Broadband. It's a buzzword on the tip of the tongue of the networking industry. But it's more than just a buzzword - high-performance broadband networks offer many exciting opportunities. They empower service providers to offer bandwidth-hungry multimedia applications that not only stimulate the senses, but that enhance our day-to-day lives. Broadband networks also provide us with near instantaneous access to a multitude of information and services for entertainment, communication, and commerce. This ubiquitous access to information and services does not come without a price, however. The wide variety and constant availability of services means continual threat from thieves, vandals, and eavesdroppers intent on personal gain at *your* expense.

As we come to rely on broadband networks and the services they offer, the security of the underlying network infrastructure becomes paramount. Unauthorized disclosure of sensitive information, denial of availability, and theft of service all pose significant threats to subscribers, corporations, and service providers in terms of financial loss and damaged reputations. Network infrastructures are becoming increasingly complex, and will come to support a wider variety of applications than ever before. As a result, security architects must be well versed in both the fundamentals and subtleties of broadband network security design and implementation.

In addition to countering network threats, real-time multimedia applications throw yet another variable into the mix that service providers, enterprises, and network device manufacturers must account for in their security designs. To ensure the successful operation of real-time applications, security must be designed to account for the individual quality of service (QoS) characteristics of each application. When used improperly, security mechanisms can adversely affect the timely delivery of multimedia content, and even worse, can make network services difficult to use and costly to implement and maintain. Once again, it is the duty of information security professionals to understand the effects of security on QoS, cost, and manageability.

This book addresses all of these concerns and more. It is intended as a no-nonsense guide to designing broadband network security. As such, we limit mathematically theory and other mundane topics in favor of practical and immediately applicable security techniques. Security design is challenging, but with the information and tools provided in this book, you will be well on your way to designing, implementing, and building more secure, higher performing, and less costly broadband network security infrastructures.

Audience

This book is intended for network security architects and hardware design engineers requiring a clear understanding of security in terms of functionality, performance, cost, and manageability. However, these are not the only groups that will benefit from this book. While this book does cover many advanced topics, those readers new to network security will find answers to common questions while learning many important security fundamentals. More advanced readers will benefit from detailed coverage of cryptographic performance, public key infrastructure techniques, network security threats and countermeasures, widely used network security protocols, and detailed case studies of current broadband network security standards and specifications. The book concludes with practical design examples giving readers of all levels the opportunity to apply what they have learned.

What You'll Learn

This book is organized into three main parts with two supplementary appendices. Part 1 of this book introduces readers to the networking and security fundamentals necessary for discussions in ensuing chapters. Chapter 1 provides a brief overview of broadband networks, and discusses their growth and importance. Chapter 2 presents the primary security services, as well as an introduction to cryptography and cryptographic security mechanisms. Chapter 3 discusses the threats that exist on public broadband networks and techniques for countering the threats. It also addresses weaknesses in the TCP/IP protocol suite. Chapter 4 introduces the OSI and TCP/IP network reference models, and discusses the most popular broadband access technologies, including cable, xDSL, fixed wireless, and satellite. The chapter also introduces the QoS concepts of bandwidth, latency, jitter, packet loss, and availability, and how they affect the delivery of multimedia content. Part 1 concludes with an overview of existing broadband security standards and specifications in Chapter 5.

Part 2 supplies the bulk of the discussion of broadband network security design techniques. Chapter 6 discusses a number of popular, general-purpose security protocols - including IP Security, Secure Sockets Layer/Transport Layer Security, and the Kerberos Network Authentication Service - that apply to securing communication in many network environments. Chapter 7 addresses the placement of security services and mechanisms within the various network layers, and discusses how to determine whether host-based security or security gateways are more suited to a particular scenario. The final chapter of Part 2, Chapter 8, delves into the performance, cost, and manageability issues associated with designing security solutions for real-time multimedia applications and other applications with strict QoS requirements. Chapter 8 also discusses the constraints placed on security by embedded devices, and how to balance security and performance within these devices.

Part 3 of this text provides a number of case studies and design scenarios intended to give readers a chance to apply the security fundamentals and advanced design techniques discussed in Parts 1 and 2. The case studies, focusing on current security initiatives in the cable industry, address securing high-speed Internet access (Chapter 9), IP telephony and real-time steaming multimedia applications (Chapter 10), and interactive television (Chapter 11). Chapter 12 presents a step-by-step guide for designing a broadband network security infrastructure from the ground up, and closes with two design scenarios that test the readers' understanding of the topics presented throughout the text.

The two appendices at the end of the book offer supplementary information regarding the TCP/IP protocol suite (Appendix A) and digital certificates and public-key infrastructure concepts (Appendix B).

Part I

Broadband Network Security Fundamentals

Chapter 1

An Overview of Broadband Communication

From the time we are born, communication is a natural part of our daily lives. We learn to speak, read, and write at an early age, and quickly become accustomed to radio, television, and these days the Internet. Cable and satellite TV provide access to hundreds of channels, while modern high-definition televisions offer exceptional picture quality. Not long ago, huge mainframe computers filled entire basements, and were limited to military and academic research. Now, computers have become a medium for entertainment and communication (not to mention being able to fit in the palm of your hand).

Just as important as the television and personal computer are the networks over which they communicate. What good is cable television without the cable? How useful is a computer to the average user if it is not attached to the Internet, or to any network, for that matter? Public networks today are very different from what they were a couple of decades ago. Improvements in physical transmission techniques, bandwidth and service provisioning equipment, and network management and communication protocols allow high-speed networks to support a myriad of real-time multimedia applications. Without these networks, interactive television, videoconferencing, and streaming audio and video might never be possible. Applications such as Internet access, online gaming, and telecommuting would be much less enjoyable without high-speed networks.

The universal availability of information and services enriches our daily lives, but not without risk. This same handy access can be used for unauthorized information disclosure, theft, and espionage. Unsecured computers and networks provide easy targets for malicious intruders. Attacks *will* become more frequent with the increasing number of computers and devices that have dedicated connections to public broadband networks. Broadband access also supplies attackers with faster data transfer, allowing them to capture the entire contents of a computer's hard drive within hours. Poorly implemented network and device security opens the door for service disruption and theft. Not only must the users of broadband services protect their own personal information, but public networks must be designed and implemented with security in mind. The design of high-performance network security infrastructures is the focus of this book.

A Brief History of Telecommunication

Starting in the mid-nineteenth century, communication began shifting toward electronic media. The introduction of the telegraph in 1840 greatly facilitated communication over long distances. Electrical signals traveling near the speed of light transmitted information far more quickly and efficiently than had foot, horseback, rail, or ship. Later, the invention of the telephone allowed transmission of the human voice, thus providing a much more intuitive means of communication than the telegraph. However, the pioneers of electronic communication were not content simply with voice, and in the 1930s, the television emerged as one of the first truly bandwidth-intensive forms of communication. The 1950s and 1960s brought about satellite communication and cable television.

Over the half century that followed, innovation after innovation resulted in technology that simplified and extended our ability to freely exchange information (see Figure 1-1).

The creation of ARPANET (the Advanced Research Projects Agency Network) in 1969 is one of the most influential events in recent telecommunications history. Developed to exchange research information between universities, ARPANET originally connected only four computers, located in California and Utah. It eventually developed into what we now know as the Internet, which connects millions of computers worldwide. During the late 1960s, the closest thing to a public broadband computer network was a 1200-bits-per-second (bps) modem connection over the Public Switched Telephone Network (PSTN). The telephone network was well established relative to state-of-the-art computer networks of the day. This is not to say that broadband networks did not exist—just not as we know them today.

NOTE:

Cable television networks are broadband networks that have been available since the 1960s. However, they were not used for computer communication until recently.

1840 Samuel F. B. Morse patents electromagnetic telegraph

1844 Morse builds first practical telegraph between Baltimore and Washington, D.C.

1876 Alexander Graham Bell granted patent for the telephone

1917 Amplitude Modulation (AM) introduced

1928 Frequency Modulation (FM) introduced

Early 1930s Black-and-white television becomes a reality

1939 National Broadcast Company (NBC) begins broadcasting in the U.S.

1957 Color television and first artificial satellite

Early 1960s Cable TV emerges

1969 Four computers in California and Utah are connected via the Advanced Research Projects Agency Network (ARPANET)

1976 X.25 standard developed for packet-switched services

1980 Introduction of broadband computer network technology

Early 1980s LANs emerge

1984 International Telecommunications Union (ITU) recommends the Integrated Services Digital Network (ISDN) standard, which never takes off due to competing implementations

1985 1000+ hosts on the Internet (Zakon, 2001)

1988 Cable industry discovers fiber optic cabling would increase quality and capacity of cable networks

1989 Digital Subscriber Line (DSL) first developed as means of distributing video content

1992 1,000,000+ hosts on the Internet (Zakon, 2001)

1993 Introduction of the World Wide Web (WWW) and National Center for Supercomputing Applications (NCSA) Mosaic

1994 Digital Broadcast Satellite (DBS) television introduced

1996 Communications Act of 1996 begins widespread deployment of cable modem and DSL services

1998 Digital television networks reach implementation stage

2000 5.1 million households with dedicated broadband Internet access

2001 10+ million hosts attached to the Internet (Zakon, 2001)

2005 191 million broadband devices to be deployed [Kasrel, 2000]

Figure 1-1 Advancements in telecommunications technology over the past century and a half indicate a movement toward broadband services and applications.

That Was Then

Back in the days of text-based user interfaces, a 2400 bps analog modem transmitting data over a telephone line was more than acceptable. At a rate of 2400 bps, a screen (80 characters × 25 lines) of text could be transmitted in just over six seconds. (This is a somewhat simplified calculation that does not account for protocol overhead.) That speed was sufficient for simple applications, such as remote shell access and File Transfer Protocol (FTP), where modest data transfer rates still exceeded the average human's reading and typing speeds.

In the late 1980s and early 1990s, CERN (Conseil Européen pour la Recherche Nucléaire—a European nuclear research organization) developed a navigation system for managing information, which they called *hypertext*. Hypertext allowed links between arbitrary *nodes* (connection points) in documents distributed across many networked hosts. Tim Berners-Lee later coined the phrase *World Wide Web (WWW)* as the name for this program. In 1993, the NCSA (National Center for Supercomputing Applications) released an early copy of the Mosaic web browser for navigating hypertext documents on the Internet. The WWW and the Mosaic browser combined to form a new medium for communicating information.

For years, advancements in analog modem technology seemed to keep pace with Internet applications, which at first were limited to displaying static text and graphics. However, as commerce began spreading to the Internet, the need arose for interactive content that would make users anxious to visit a site—and more importantly, anxious to return. This interactive content has taken the form of more intense graphics, sound, and animation. A 56 Kbps[1] analog modem is capable of transmitting a page of text in well under a second, and a small 10KB image in a couple of seconds. For static websites consisting of text and modest graphics, this rate of transfer is acceptable to most users. However, what happens when you want to download a complex animation or video clip? You wait!

This Is Now

Enter broadband access! Before 1996, major telecommunications providers maintained exclusive control over their networks. The Telecommunications Act of 1996 required these companies to make their networks available for resale by smaller organizations known as *competitive local exchange carriers (CLECs)*. As a result, any organization could offer local telephone service as a CLEC if they had appropriate equipment and funding, and if they satisfied certain government regulations. The growing popularity of the Internet provided an untapped market for CLECs: high-speed Internet access to homes and small businesses. Seeing the need for dedicated, high-speed access, many CLECs began deploying

---

[1] To limit electrical interference, current FCC regulations limit the speed of 56 Kbps modems to 53 Kbps over analog phone lines.

services based on Digital Subscriber Line (DSL) technology. Cable service providers quickly followed suit by offering high-speed access over cable modems. DSL and cable modem deployments both grew considerably during the late 1990s with the cable modem taking an early lead, a position it still holds today. Fixed wireless and two-way satellite technologies, which do not rely upon a wired infrastructure, have also increased in popularity.

Broadband networks have been growing rapidly since 1999. At the end of 2000, over 6 million U.S. households subscribed to a broadband Internet access service (Kennebeck, Marcheck, and Mendelson, 2001). This growth rate is expected to increase quickly as more services and less expensive broadband-enabled devices come to market.

Broadband applications are not limited to Internet access. Cable television, or Community Antenna Television (CATV), has been providing services to subscribers since the 1960s. Recently, cable services have expanded to include high-speed Internet access and digital television. Video- and audio-on-demand will soon be widely available from cable and DSL service providers. Digital telephony is yet another service currently offered by some broadband access providers, allowing them to compete directly with local telephone companies. From interactive television to videoconferencing, telecommuting, and online gaming, if the service requires bandwidth and can be offered over high-speed networks, then it probably is—or will be.

Not only does broadband access provide a dedicated connection to subscription-based services, it enriches the user experience with real-time interaction and high-definition sound and video. The days of waiting 15 minutes for a 2MB video clip to download are all but forgotten. Time to sit back and (almost instantaneously) enjoy the ride!

What is Broadband Access?

While some sources restrict broadband access to Internet connectivity, we will expand the definition to encompass all current and future services offered over broadband networks. These include video, data, voice, and other multimedia services sent over networks capable of transmission in the megabit-per-second range (or nearly so). Some companies, notably cable service providers, offer bundled broadband service packages that include cable television, telephony, and high-speed Internet access over the same connection. Such bundled service packages allow subscribers to receive all major communication services from a single provider.

Existing Broadband Access Technologies

Broadband technologies exist throughout a service provider's network infrastructure, which includes *backbone, distribution,* and *access* networks (see Figure 1-2).

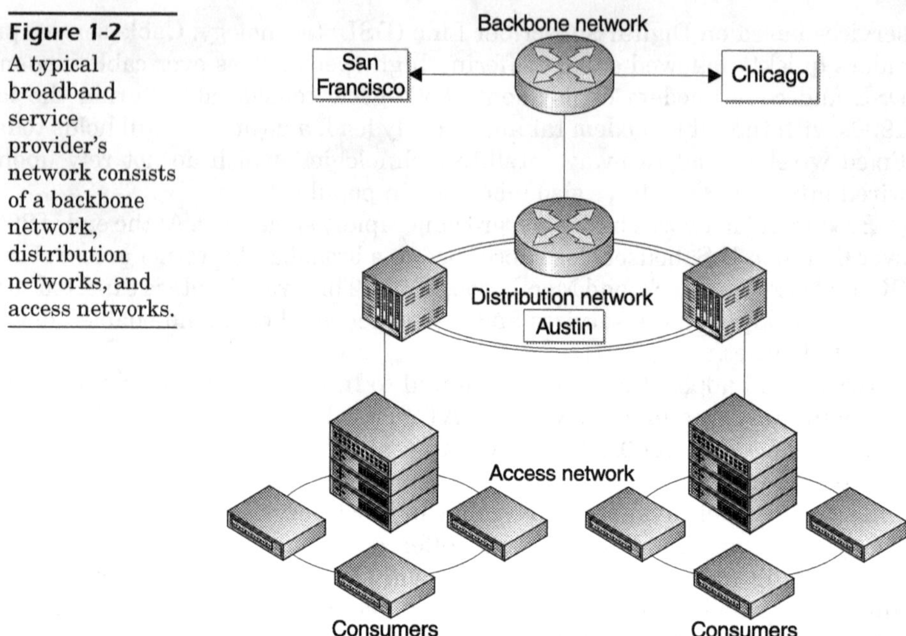

Figure 1-2 A typical broadband service provider's network consists of a backbone network, distribution networks, and access networks.

However, nowhere has the debate over the best broadband technologies been as fierce as in the access network, often called the *local loop* or *last mile*. The access network carries broadband services directly to a subscriber's home or office. The most common broadband access technologies in existence today are cable, DSL, fixed wireless, and satellite. (We will revisit each of these technologies in detail in Chapter 4, during our discussion of public broadband networks.)

Cable

Cable services use the cable television network infrastructure to deliver television, high-speed Internet access, and telephony to subscribers. Until recently, most cable networks provided one-way *downstream* (provider to subscriber) distribution of television signals to homes. However, the advent of high-speed Internet access, as well as the possibility of interactive television, required an upgraded network infrastructure capable of two-way data transmission. To enable two-way transmission, cable operators began installing an additional *upstream* (subscriber to provider) channel, along with the equipment necessary to support that channel. These upgrades are costly, ranging from $200 to $250 per home (Kennebeck, Marcheck, and Mendelson, 2001).

Cable networks are shared media. All users compete for the bandwidth of a single cable, and predetermined bandwidth is not yet guaranteed by the service provider. While cable networks are capable of very high speeds, many subscribers experience poor performance during peak usage hours. Although currently capable of data rates near 30 Mbps, typical data rates for cable services are 1.2–2 Mbps downstream and 128–512 Kbps upstream.

DSL

Digital subscriber line (DSL) uses standard telephone lines to distribute basic telephone services and high-speed Internet access to subscribers. Data transmission occurs at a higher frequency than voice communication, allowing both services to share the same telephone line without interference. Traditionally, DSL service providers have been limited to offering telephone services and Internet access, but they have recently begun offering video-on-demand to compete with the cable market. DSL suffers from a number of deployment hurdles related to transmission distances, which may hinder the future growth of this medium. We will further discuss these difficulties in Chapter 4.

The most common form of DSL, *asymmetric DSL (ADSL),* provides downstream data rates of 1.5–6 Mbps and upstream rates of 128–640 Kbps. The difference in downstream and upstream rates arises from the anticipated usage of residential and small business customers. Typical Internet applications are downstream-bandwidth-intensive, meaning that most of the data transfer occurs from servers to clients. Whether surfing the Web, reading e-mail, or watching a video clip, the average user sends little more than a few commands a minute in the upstream direction. This eliminates the need for a symmetric return path. *Symmetric DSL (SDSL)* is another alternative whose upstream and downstream rates are equivalent. SDSL is better suited to business customers who host websites or share data with their business colleagues. In both these cases, the business may require more upstream bandwidth than that provided by ADSL.

Fixed Wireless

Like cable, fixed wireless microwave transmission provides a single solution for broadband access; it is capable of carrying voice, video, and data simultaneously, at very high data rates. The most common forms of commercial fixed wireless are Local Multipoint Distribution Services (LMDS) and Multichannel Multipoint Distribution Services (MMDS), both of which occupy frequency bands that must be licensed from the FCC (Federal Communications Commission). Unlicensed frequency bands, including the UNII (Unlicensed National Information Infrastructure) and ISM (Industrial, Scientific, and Medical) bands, are available for shared public use; anyone with the appropriate equipment wishing to transmit in these ranges can do so without a license. The licensed bands are restricted to a single provider within a given geographic location. (One might think of such a geographic area as the radius of transmission for an LMDS or MMDS transmitter tower.) This restriction ensures that transmissions from one carrier won't interfere with those of another. Since anyone can transmit in the unlicensed bands, signal quality and available bandwidth tend to degrade as the number of providers increases, and service disruptions may arise.

Fixed wireless is capable of data rates between 2 and 100 Mbps, depending on the frequency used and the distance between the transmitter tower and antenna. Service providers provision the full bandwidth to deliver residential rates of 1.5–5 Mbps, and business rates of 10 Mbps and higher. LMDS is capable of higher data rates than MMDS, but is also more costly. Consequently, it does not provide a viable solution for residential broadband markets. Customers requiring more bandwidth, such as large corporations, migrate toward LMDS; while MMDS and the unlicensed bands remain more suited to residential customers, whose bandwidth needs are not as great.

Despite line-of-sight limitations, fixed wireless technologies offer a number of advantages over traditional wired approaches. Unlike copper wire, which is very susceptible to electrical noise and interference, microwave technology is extremely clear. This greatly improves quality, as fewer bits are dropped during transmission. Wireless solutions are often quicker to deploy and simpler to operate and maintain than their wired counterparts.

Two-Way Satellite

While used for some time to broadcast video signals, Direct Broadcast Satellite (DBS) is also well suited for high-speed Internet access. DBS utilizes a network of *geosynchronous earth orbit (GEO)* satellites for transmitting data directly to subscribers. At a distance of 22,300 miles from the earth's surface, the period of rotation for a GEO satellite is 24 hours, which makes it appear stationary when viewed from the earth's surface. DBS service is available to anyone with a clear view of the sky. This greatly simplifies deployment, and makes high-speed data services available to subscribers who are not serviced by cable or DSL. All that the customer needs is a satellite receiver and a small satellite dish.

First-generation DBS provided one-way downstream communication, with the upstream channel supplied by a dial-up line. Two-way DBS offers communication in the upstream direction via satellite as well. Typically, the return path for two-way satellite runs between 128 and 150 Kbps—more than twice the speed of the fastest dial-up connection—and does not tie up a phone line during use. The downstream data rate for one- and two-way satellite service is much greater, varying from a few hundred Kbps to 1 Mbps and higher. Residential access rates commonly run between 400 and 500 Kbps.

The Future of Broadband

Over 11 million U.S. households are likely to have high-speed Internet access by the end of 2001. This number is expected to quadruple within the next four years (Kasrel, 2000). As the number of broadband subscribers and services grows, our lives will become more information- and service-centric. We will become more and more reliant on broadband access for communication, commercial transactions, entertainment, and research.

Broadband access will not be confined to desktop PCs; portable, handheld devices will provide instant access to broadband services anytime, anywhere, over wireless connections. High-speed access opens the door to distributed computing environments, where handheld devices with little local storage will run applications and store and retrieve data remotely.

Fiber optics, frequently touted as the solution to future bandwidth needs, provides almost unlimited bandwidth for integrated services. Fiber deployment in the local loop will make high-bandwidth residential applications—such as real-time, high-definition videoconferencing over IP networks—a reality.

Fiber Optics

Fiber optic systems use internal reflection to transmit light over long distances with very little *attenuation* (signal loss). The high-end commercial systems that exist today are capable of 400 Gbps and higher data rates over a single optical fiber (Glass et al., 2000). As you can see in Table 1-1, fiber optics provides considerably more bandwidth for broadband applications than other current technologies.

A single fiber optic strand is composed of three concentric layers of material: the core, cladding, and buffered coating. The inner-most layer is the core, which consists of glass. The cladding surrounds the core and has different optical properties than the core. When light enters the strand, it is reflected back into the core once it reaches the boundary between the core and the cladding. *Total internal reflection* results when the boundary reflects all light back into the core. This is what makes fiber optics so efficient for long-distance communication.

Due to the limitations of the switching and multiplexing hardware used in today's broadband networks, we have yet to realize the full potential of fiber optics (Bates, 1999). A single strand of fiber can service many business and residential subscribers, with downstream and upstream data rates of 100–155 Mbps or more. This type of bandwidth allows a single provider to simultaneously offer multiple streaming applications, such as audio- and video-on-demand, to the

Access Technology	Current Maximum Transmission Rate	Typical Upstream Subscriber Data Rate	Typcial Downstream Subscriber Data Rate
ADSL	8 Mbps	1.5–6 Mbps	128–640 Kbps
Satellite	20 Mbps (digital satellite television[2])	400–500 Kbps	128–150 Kbps
Cable	30 Mbps	1.2–2 Mbps	128–512 Kbps
Fixed Wireless	100 Mbps (LMDS) 10 Mbps (MMDS)	10–100 Mbps (LMDS) 1.5–10 Mbps (MMDS)	10–100 Mbps (LMDS) 1.5–10 Mbps (MMDS)
Fiber Optics	400 Gbps	100–155 Mbps	100–155 Mbps

[2] Other satellite technologies (not for residential use) are capable of much higher data rates.

Table 1-1 A Speed Comparison of Current Broadband Networking Technologies

home or office. In a typical household, one family member might be watching their program of choice in the living room, while a second watches a movie on a bedroom television. Still a third family member might be listening to their favorite radio station while playing an online game.

In addition to high bandwidth and low attenuation, fiber optics is more secure than most other transmission media. Even a small impurity in a fiber strand's core can obstruct transmission. While copper wiring, coaxial cable, fixed wireless, and satellite communication might be tapped or monitored without detection, it is very difficult to tap into a fiber optic line without degrading signal quality to unacceptable levels. Because of the extremely small diameter of a fiber, a minor break in the line can completely halt communication.

Today, telecommunication and cable service providers use fiber optics heavily within their backbone and distribution networks. Within public access networks, fiber often coexists in hybrid architectures along with other cabling technologies, such as twisted pair copper or coaxial cable. Access networks consisting entirely of fiber optics are rare, and limited mostly to pilot programs in urban areas that have advanced telecommunication infrastructures. While many providers have plans to upgrade their access networks to fiber optics, the upgrade is prohibitively expensive, and subscribers will not be able to take full advantage of the upgraded networks until fiber deployments run directly to their homes or offices.

The Importance of Security in Broadband Networks

Many broadband users place a great deal of trust in the security of online services and transactions. This trust extends beyond the security of financial transactions, and includes personal information exchanged via e-mail, online chat sessions, and web-based surveys and questionnaires. Users expect a certain level of service availability, as well as assurance that their television, telephone, and Internet access accounts are not being abused by others. No one user, manufacturer, or service provider can achieve these security goals alone; we all share in the responsibility of securing broadband services.

Security and the Average User

When most users think "information security," one of the first things that usually comes to mind is protecting credit card information when purchasing online. Thanks to the publicity of attacks on a number of e-commerce websites, users understand the need for encrypting credit card numbers and other valued information stored in databases and exchanged via public networks. They have heard of Secure Sockets Layer (SSL) technology and its use within popular web browsers to protect their online transactions (see Figure 1-3). From the end

In the HTTPS (secure hypertext transfer protocol) environment, data is encrypted as it travels between the user and the web server.

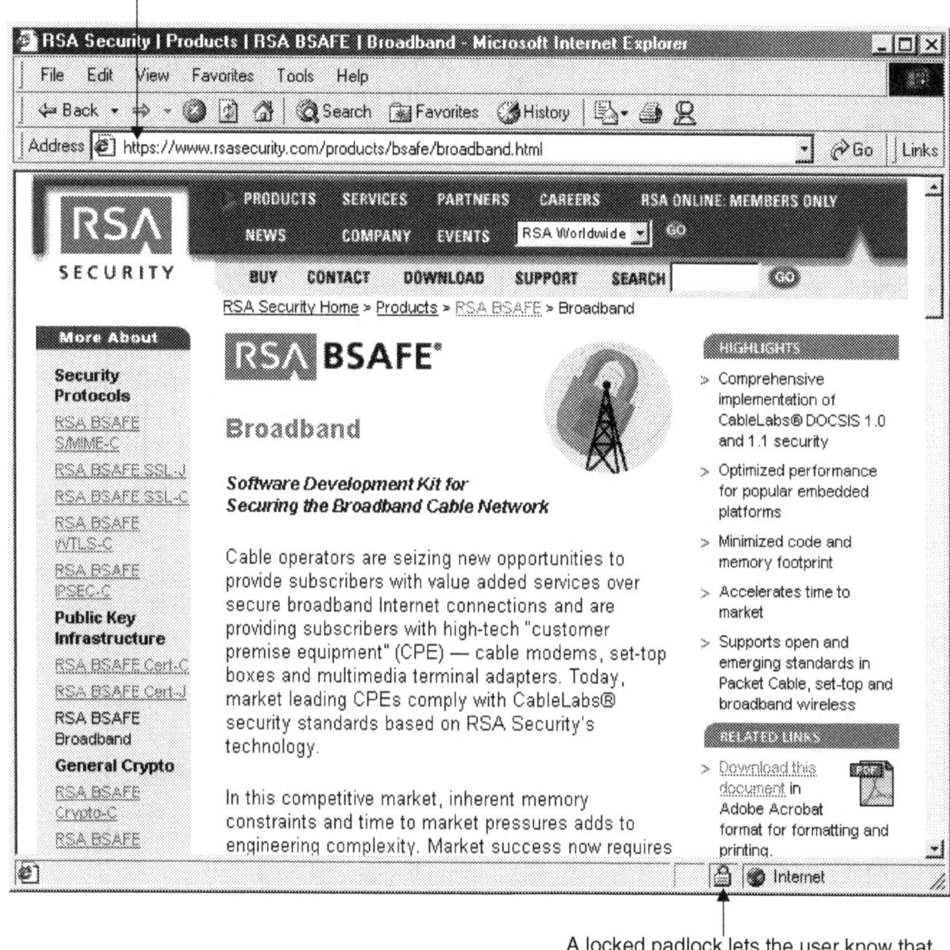

A locked padlock lets the user know that he or she can submit information securely.

Figure 1-3 Most users understand the need for securing information sent over the Internet. SSL-enabled web browsers, such as Microsoft Internet Explorer, indicate a secure connection with the "https://" URL designation and a gold padlock icon.

user's perspective, there is no need to understand the inner details of a security protocol such as SSL. Instead, they should educate themselves regarding how it is used to protect their interests.

High-speed Internet access has made telecommuting practical, and many employees connect to corporate networks from home over *virtual private networks (VPNs)*. VPN software authenticates remote users and encrypts information sent between two computers over a public network such as the Internet. While employees may understand that a VPN protects information during transmission, they may not realize that VPNs are not intended to protect data stored on their personal computers. It is common for an employee's home PC to contain confidential trade secrets, product roadmaps, financial data, and personnel records, all of which may be valuable to an attacker.

Microsoft Windows–based computers running File and Print sharing and other NetBIOS services are frequent targets for attackers. When these services are enabled, they can easily be manipulated to allow remote access to the files and other resources on a PC. Computers that do not run remote access protocols may still succumb to infection by computer viruses capable of destroying files critical to system operation.

These vulnerabilities are most easily addressed by personal firewall and antivirus packages. Personal firewalls lock down a PC by intercepting and blocking unauthorized incoming connections. Most personal firewalls are simple to use, eliminating the need for security expertise. For example, Internet Security Systems' BlackICE Defender (see Figure 1-4) comes configured with four security settings ranging from Trusting to Paranoid. A user simply chooses a level based on his or her security preferences. Most personal firewalls allow the user to disable file sharing and NetBIOS calls, block traffic from particular IP addresses, and log attacks (see Figure 1-5). Antivirus software searches for program code that matches patterns of known viruses, and then either removes or quarantines the code before it can be executed. Firewalls and antivirus solutions are geared at protecting the integrity of a PC by preventing theft and modification of data, and by preventing unauthorized use of the system by attackers.

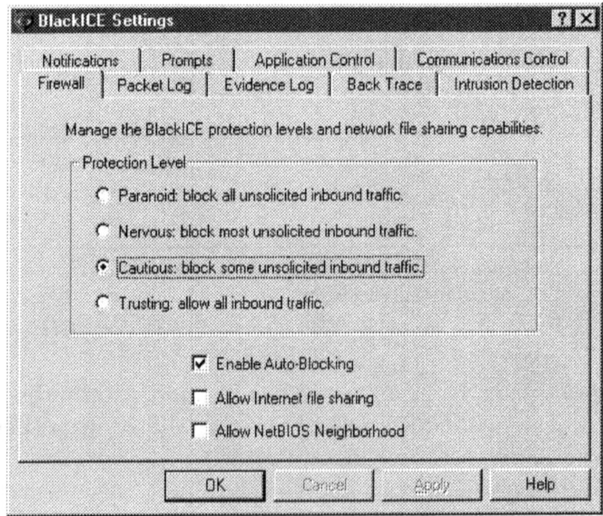

Figure 1-4 Personal firewall software, such as Internet Security Systems' BlackICE PC Protection v3.5, protects personal computers from attackers. Configuration requires little security expertise.

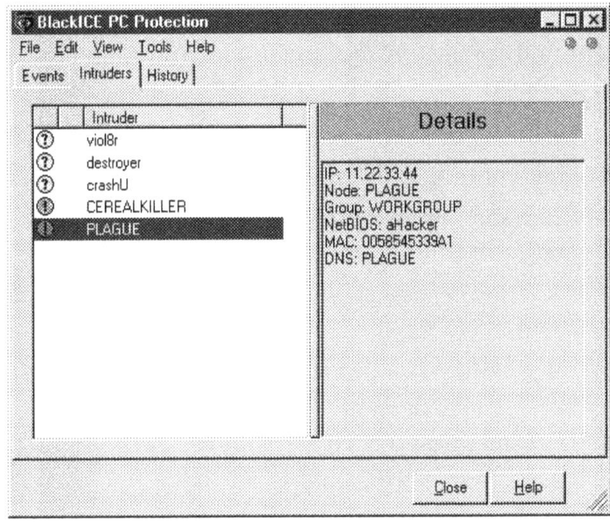

Figure 1-5 BlackICE PC Protection logs each attack and attempts to trace a connection back to the intruder.

Securing the Network Infrastructure

What the average user may not realize is that securing their PCs does not completely isolate them from attack. Users expect the underlying network infrastructure and devices to be secure, often trusting this to be the responsibility of service providers and device manufacturers. Security-savvy users may install antivirus software or personal firewalls to protect their home PCs, but they may never consider the prospect of malicious users stealing service by fraudulently accessing their accounts. They almost certainly cannot prevent an attacker from taking down a network device or server, and denying them service.

Generally, service providers are more concerned with theft of service than end users are, and rightly so—the service provider frequently covers the cost of stolen services. A still greater priority than preventing theft of service is that of keeping attackers or cloned devices from masquerading as network servers and interfering with the overall delivery of service. Such *denial of service (DoS)* attacks result in reduced *quality of service (QoS),* and even in costly service outages. These attacks are a serious threat to any provider because unreliable service negatively affects customer satisfaction and loyalty, and it can permanently damage a company's reputation.

Service theft and DoS attacks affect more than high-speed Internet access. Any network service is a potential target, including television and IP telephony. The growth of broadband services and applications offers many exciting opportunities, but also many challenges. Networks not resistant to eavesdropping, theft of service, and loss of availability are not suited to a society that relies on these services for daily life.

Fortunately, for the average user, it is the job of security experts (like you) to design and implement secure broadband networks. Such experts need more than an understanding of security principles; they must also have an appreciation for how networks operate. Successful security design involves the following:

- Understanding the threats that exist on public networks
- Choosing appropriate security services to protect network applications and devices, and correctly incorporating these services into the network
- Selecting cryptographic algorithms to implement the security services, when necessary
- Employing existing security protocols—such as SSL or IPSec—when appropriate, or developing new protocols when necessary
- Implementing security within the appropriate network layer
- Balancing security and performance for applications with varying QoS requirements (bandwith, latency, jitter, and packet loss) and for embedded devices with limited processing and storage resources

This is by no means a comprehensive list, but it serves to demonstrate the level of consideration and planning that goes into designing security for broadband networks. We will address each of these topics in later chapters.

References

Bates, Regis J. 1999. *Broadband Telecommunications Handbook*. New York: McGraw-Hill.

Glass, Alastair M., David J. DiGiovanni, Thomas A. Strasser, Andrew J. Stentz, Richart E. Slusher, Alice E. White, A. Refik Kortan, and Benjamin J. Eggleton. January–March 2000. "Advances in Fiber Optics." *Bell Labs Technical Journal*.

Kasrel, Bruce. October 2000. "Broadband Content Splits." *The Forrester Report*.

Kennebeck, Keith, Jason Marcheck, James S. Mendelson. 2001. *Residential High-Speed Internet: Cable Modems, DSL, and Fixed Wireless*. Washington D.C.: The Strategis Group.

Zakon, Robert H. 15 April, 2001. *Hobbes' Internet Timeline v5.3*. 10 Aug., 2001 (http://www.zakon.org/robert/internet/timeline/)

Chapter 2

Choosing the Right Tools: Security Services and Cryptography

As information security professionals, we should thoroughly acquaint ourselves with the tools of our trade. Selection of the appropriate tools is critical to the design and implementation of any security architecture, as it greatly affects the performance, usability, and cost of network operation. However, making the right choices is not a simple task given the variety of network applications in existence today. Furthermore, future innovations by hardware vendors and service providers will necessitate creative—and often novel—security solutions. While we should never overlook its importance, security must not interfere with the operation of network services, hinder consumer interaction, or cost too much money.

This chapter begins with an introduction to basic security services and mechanisms. Cryptographic principles, which we will apply throughout this book, are introduced next, followed by a discussion of the many security services that cryptography provides. As you read this chapter, keep in mind the following simple motto: "Too little security is dangerous, too much security can be a hindrance, and the wrong kind of security may be as effective as no security at all."

Security Services and Mechanisms

One of the first steps in designing security is determining the *security services* that must be implemented. The primary services include confidentiality, integrity, authentication, nonrepudiation, authorization, access control, and availability. Each service performs a unique role in securing information systems, and almost any environment can be secured using a combination of these services. Security services accomplish a number of very important objectives:

- Ensuring that confidential information is not viewed by unauthorized entities
- Guaranteeing that a message cannot be altered without detection on its way from sender to recipient
- Authenticating the identity of an individual or the source of a message
- Preventing the sender of a message from being able to deny that he or she sent it
- Determining whether a user has rights to certain resources, and ensuring that only authenticated users gain access
- Making data and network applications and services available to users who request access to them

A *security mechanism* is a tool for implementing a security service. In general, there are multiple mechanisms for implementing each service, and one of the challenges in designing a secure system is selecting the most appropriate mechanisms. In the following sections, we will identify many popular security mechanisms. (We will also revisit many of them later in the chapter, in "The Basics of Cryptography.")

Confidentiality

Confidentiality services offer protection against snooping and eavesdropping by preventing unauthorized disclosure of sensitive information. The standard mechanism for enabling confidentiality is *encryption*. Encryption transforms coherent data into gibberish, so that it makes little sense if stolen or intercepted by an attacker. Encryption protects data at rest (file systems and databases, for example), as well as messages transmitted over communication networks.

Confidentiality services based on encryption also provide limited protection against *traffic flow analysis*—a process by which an attacker tries to deduce valuable information by monitoring the frequency and amount of network traffic flowing between two communicating parties. In addition to masking the contents of a message, some security protocols will conceal its source and destination. Other techniques, such as traffic padding and routing control, can be used to conceal message length and frequency of network traffic.

NOTE:

Traffic padding *is the injection of spurious traffic into a network to hide actual usage patterns.* Routing control *involves directing traffic along a particular path between sender and recipient in order to reduce exposure to eavesdropping.*

Integrity

Regardless of whether information is kept secret or not, it is of little value if it cannot be trusted. Even worse, information that has been tampered with can be used to damage reputations, or to hide malicious or unlawful behavior. Basic *integrity services* ensure that data cannot be easily modified without detection. Many security protocols offer protection against unauthorized insertion, deletion, modification, and replay of transmitted messages. In addition to detecting such intrusions, connection-oriented communication protocols (see Appendix A) guarantee the delivery of messages, both through acknowledgement and through retransmission of lost data. Message digests are the primary mechanism for ensuring data integrity.

Authentication

Authentication services provide a means of verifying the identity of an individual, as well as determining the true source of a transmitted message. (The latter practice is known as *data origin authentication.*) Traditional authentication involves a user presenting a unique set of credentials that distinguishes him or her from other users of a system. Before allowing access to protected resources, the system verifies the identity of the user based on those credentials. In this case, the user could be a human being, device, host, or service.

Many authentication mechanisms exist, and most are based on one or more of the following methodologies: "something you know," "something you have," or "something you are". The most common form of authentication involves memorizing and presenting a user ID and password combination. This is an example of "something you know." Physical hardware devices, such as *time-synchronous tokens* and *smart cards* (see Figure 2-1), are examples of "something you have." *Biometric authentication devices,* which measure physical characteristics or actions (such as fingerprints, voice patterns, hand or facial geometry, and movement), provide examples of the final authentication method, "something you are." *Multi-factor authentication* uses more than one mechanism at a time to identify a user.

Each of the preceding mechanisms authenticates the identity of an individual, not the source of a message. Data origin authentication involves binding a set of credentials to a message in such a way that they cannot be easily removed, and verifying the attached credentials upon receipt of the message.

Popular mechanisms for providing data origin authentication are keyed message digests and digital signatures, both of which also incorporate integrity protection. (We'll discuss these mechanisms later in the chapter; see "HMACs" and "Digital Signatures.")

Nonrepudiation

Nonrepudiation services prevent the sender and recipient of a message from later denying that the message was ever sent or received. For instance, a service provider offering an online service may wish to bill customers based on the number of

Figure 2-1 Smart cards and time-synchronous tokens are examples of "something you have" authentication.

electronic transactions. Nonrepudiation precludes both the customer and the service provider from credibly denying that a transaction occurred at a particular date and time. This facilitates a more binding service and payment agreement between a business and its customers. Digital signatures in conjunction with time stamping are the primary mechanisms for accomplishing nonrepudiation. *Time stamping* associates a transaction with the time and date when the transaction occurred. Without chronological ordering, it is difficult to enforce nonrepudiation.

NOTE:
In practice, nonrepudiation is more a legal term than a technical one, and it implies more than the use of digital signatures and time stamping. Nonrepudiation encompasses an entire set of policies and procedures for establishing and enforcing trusted communication between one or more parties.

Authorization and Access Control

Authorization services determine, after successful authentication, whether a user has certain rights or privileges. Once a user is authorized to access a resource, *access control* mechanisms, such as static *access control lists (ACLs)* and *dynamic policy managers*, determine the level of access granted to that user. Users who do not successfully authenticate are generally denied access altogether. Access is typically restricted to some combination of read, write, erase, and execute privileges. (Read, write, and erase privileges apply to stored data, and execute privileges apply to applications.)

There are two basic types of access control: *discretionary access control (DAC)* and *mandatory access control (MAC)*. DAC allows users to define who can access information that they themselves have created. MAC, on the other hand, controls access to *all* information, based on a set of rules established by the administrator of a system. Environments (universities, for instance) where users regularly share unclassified information may benefit from the flexibility of DAC. However, MAC provides more secure and centralized management of access control policies.

The level of access associated with a resource varies, depending on the sensitivity of the information and on each user's responsibilities or position within an organization. In general, users should be given no more privileges than are necessary for fulfilling their responsibilities. This concept is known as the *principle of least privilege*. For example, administrators generally require more access to hosts and network services than line workers do. Conversely, most administrators perform only a subset of the administrative activities within an organization, and do not require complete, unrestricted access to all resources—just to those required for doing their jobs.

Availability

Malevolent attacks and poor planning often result in destruction of data and unacceptable interruptions in network service. Associated monetary losses may range from a few hundred dollars, for recovering data from a corrupted hard drive, to tens or hundreds of thousands of dollars in sales revenue for a disabled web server. *Availability services* guarantee network resources and data are present when and where an authorized user requests access to them.

An important class of attacks, known as denial-of-service (DoS) attacks, can completely disable a network host by exploiting weaknesses caused by poor security or software development practices. Many security protocols attempt to subvert DoS attacks by combining the benefits of confidentiality, integrity, authentication, and access control. Sometimes, however, these basic security services are not enough to ensure that resources remain available to legitimate users. Hardware and software that is prone to crashing, or that cannot efficiently and reliably cope with large volumes of network traffic, leave a system vulnerable to attack. For this reason, it is important to follow proper software design methodologies.

Not all lack of availability stems from malicious attacks. Loss of power and power surges can cause data corruption, service outages, and damage to computer equipment. It is always advisable to employ uninterruptible power supplies (UPSs) with surge protection, and to perform regular backups of crucial data.

Redundancy is another key factor in maintaining availability. In the event of a hardware or systems failure, redundant systems provide a fail-over mechanism. Companies with bright IT professionals replicate the operation of mission-critical services and applications to more than one server at a time, so that if a server crashes, another one can take its place.

The Basics of Cryptography

Cryptography underlies many aspects of information security, and is a fundamental building block within any security infrastructure. Most security protocols use cryptography, in one form or another, to provide at least one of the major security services. Such services are crucial in combating common network-based attacks (as described in Chapter 3). Communication protocols employing cryptographic principles help reduce or eliminate the likelihood of unauthorized information disclosure, DoS attacks, and theft of service.

Before we begin our discussion of cryptography, we should acquaint ourselves with some crypto lingo. *Encryption* is the process of masking coherent text or data by converting it into an apparently random sequence of characters or bits. An *encryption algorithm*, also called a *cipher,* is a computational procedure that transforms an input value, or collection of values, into an output value, or collection of values. A *cryptographic service provider (CSP)* is a set of software libraries containing one vendor's implementation of various cryptographic algorithms. The "secret" input to an encryption algorithm is known as *plaintext,* or *cleartext*. Plaintext can be human-readable text or binary data (0s and 1s). *Ciphertext* is the output that results when an encryption algorithm is applied to plaintext. We use the following notation for representing encryption:

$$\mathrm{E}(\mathrm{P}) = C$$

Here, E represents the encryption process, P represents plaintext, and C represents ciphertext.

Reversing the process to produce plaintext from ciphertext is called *decryption.* We represent the decryption process as follows:

$$\mathrm{D}(\mathrm{C}) = P$$

A *key* is a numeric value or set of values that, when combined with plaintext, produces ciphertext. Different keys result in different ciphertext. When we include keys, our notations for encryption and decryption become $\mathrm{E}_k(P) = C$ and $\mathrm{D}_k(C) = P$, where E_k and D_k are the encryption and decryption operations, respectively, using key k.

Figure 2-2 illustrates the behavior of a basic encryption algorithm. Notice that the encryption operation accepts plaintext and a key as input, and then outputs ciphertext. The corresponding decryption operation produces plaintext from ciphertext and a key.

CAUTION:

In most cases, the security of strong encryption algorithms rests in keeping the key, and only the key, secret. Trying to improve the security of an algorithm by hiding its implementation details or increasing its complexity (in what is known as security through obscurity) rarely produces the desired result. Without public scrutiny, weaknesses in the algorithm may not be discovered until they are exploited by an attacker. Furthermore, if an algorithm is very complex, then testing its effectiveness becomes much more difficult.

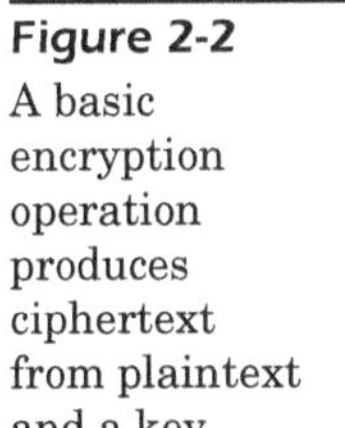

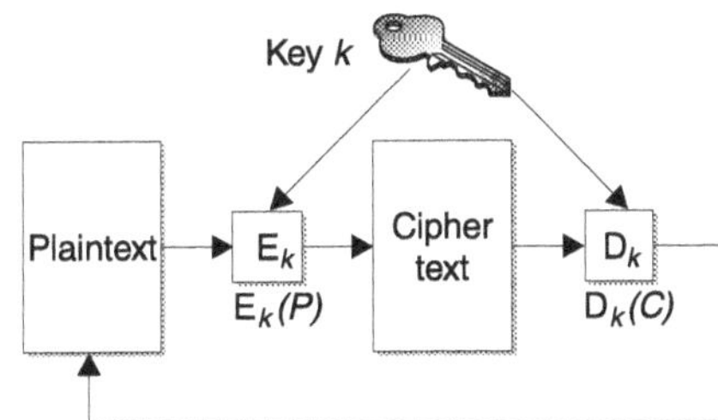

Figure 2-2 A basic encryption operation produces ciphertext from plaintext and a key.

One popular method for attacking cryptographic algorithms is called an *exhaustive key search*, in which an attacker attempts to decrypt a message by systematically trying every key in the keyspace. The *keyspace* of an algorithm is the range of all possible key values. For example, a key length of 56 bits results in a keyspace of 2^{56} possible keys. The analysis of cryptographic algorithms to identify weaknesses is known as *cryptanalysis*. The goal of cryptanalysis is to find methods of attacking cryptographic algorithms that are more efficient than exhaustive key searches. (We'll discuss cryptanalysis further in Chapter 3.)

There are two general types of encryption algorithms: symmetric and asymmetric. When the encryption and decryption operations use the same key, the algorithm is termed *symmetric*. An *asymmetric algorithm,* also known as a *public-key algorithm,* uses different keys for encryption and decryption. The key used for encryption is the *public key*, and the key used for decryption is the *private key*.

Random Number Generation

Much of the functionality and security of cryptographic operations depends on the generation of random data. *Random number generators (RNGs)* supply the basis for key generation, which is arguably the single most important operation in all of cryptography. Disclosure of keying information results in a single point of failure in the security of most cryptographic algorithms. If an attacker obtains a key, or can easily guess what it is, then he can decrypt any data that has been encrypted with that key.

An ideal RNG will generate output that is unpredictable and unbiased. *Unpredictable* output provides no way to anticipate future output by analyzing past outputs. When output values are evenly distributed throughout the range of possible outputs, the output is termed *unbiased* because no one output is any more likely to occur than another. In addition, the output of a RNG will vary from case to case, even under the same initial conditions. This last characteristic indicates true randomness, and is difficult to accomplish unless a hardware-based generator is used.

Pseudorandom number generators (PRNGs) provide the same basic service as RNGs, except that they are based on deterministic algorithms. The output of a PRNG should be unpredictable and unbiased, and it will appear truly random at first glance. However, repeated use of the same input value, or *seed*, will result in identical output every time. This is the primary difference between random and pseudorandom number generators. Since truly random behavior is difficult to model on a computer, the repeatable nature of PRNGs makes them well suited for software-based applications.

The real security of PRNGs relies on the selection of appropriate seed values. For a PRNG to be effective, the seed used to initialize it must be extremely difficult to guess. Even though the operation of a PRNG is deterministic, we can still produce a random output by supplying a random seed. For cryptographic purposes, it is not sufficient to seed a PRNG using something as simple as the current system time. Many functions for obtaining system time have accuracy limits of 1/1000 or 1/10,000 of a second; some are even less accurate. An attacker who can estimate the approximate time at which the seed was obtained, and who knows how to map the system time to a seed value, will have little difficulty decrypting an intercepted message. He or she can simply try different seed values until one of them generates a key that successfully translates the message. If the attacker can estimate the system time to within ±10 seconds, for instance, then successful decryption will require, at most, 200,000 such guesses—so even on a meager desktop computer, cracking the message will take very little time.

Typically, you will combine information from many sources to generate a single seed value. The term *entropy* describes the level of randomness or disorder in a system. The higher the entropy, the more random is the system. In the world of encryption, we seed PRNGs by gathering entropy. Entropy-gathering techniques commonly include a combination of the following activities:

- **Taking system snapshots** Application window positions, memory and system process statistics, or system date and time
- **Sampling over time** Processor utilization or hard drive timing
- **Measuring external inputs** Mouse movement, keystrokes, or audio levels from a microphone

---

CAUTION:

When supplying a seed value to initialize a PRNG, always gather entropy from a combination of sources. The increased randomness will make it very difficult (if not impossible) for an attacker to guess the seed.

An alternative to relying on system parameters for producing secure seed values is the use of a hardware-based entropy generator. Hardware-based solutions provide an extremely high level of entropy for seeding PRNGs. For example, some encryption products, such as RSA Security's BSAFE® Crypto-C, include an interface to the random number generator built into every Intel Pentium® III (and higher) processor. Intel's RNG captures thermal noise, converts it to random data, and places that data in a hardware buffer accessible by Crypto-C.

Commercially available white noise generators provide a suitable level of entropy easily captured by a sound card input device, such as a microphone. Some generators can even be attached to serial ports. However, for high-security applications, you should avoid white noise products that use digital sources of randomness because their output is periodic. Analog generators, which capture the electrical noise emitted by internal components, create a much less predictable source of entropy.

Symmetric-Key Cryptography

Symmetric-key cryptography uses the same key for encryption and decryption; it is often referred to as *secret-key* or *conventional* cryptography. When using symmetric encryption, the sender and recipient must agree on an encryption algorithm and a shared secret key prior to communication. There are two types of symmetric algorithms: stream ciphers and block ciphers.

Stream Ciphers

A *stream cipher* transforms plaintext into ciphertext one bit or byte at a time by combining the plaintext with a *keystream*. A keystream generally consists of a pseudorandom bit pattern, such as the output of a PRNG. The combination often takes the form of an *exclusive OR (XOR)* operation. An XOR statement has four possible forms:

$$0 \oplus 0 = 0$$

$$0 \oplus 1 = 1$$

$$1 \oplus 0 = 1$$

$$1 \oplus 1 = 0$$

NOTE:
The output of an XOR operation has an equally likely possibility of being zero or one; the output depends only on the plaintext input and the keystream. This makes for an unbiased distribution of values in the ciphertext.

The recipient of the encrypted data calculates the same keystream that the sender used, and XORs the ciphertext with the keystream to decrypt the data, as shown in Figure 2-3.

Stream Cipher Keystream Generation Stream ciphers offer a more practical approach to encrypting large messages than one-time pads (see sidebar entitled One-Time Pads for more information). A stream cipher uses a pseudorandom keystream, which makes key management much less of a hassle. A keystream is generated by supplying a relatively short key to a keystream-generation algorithm.

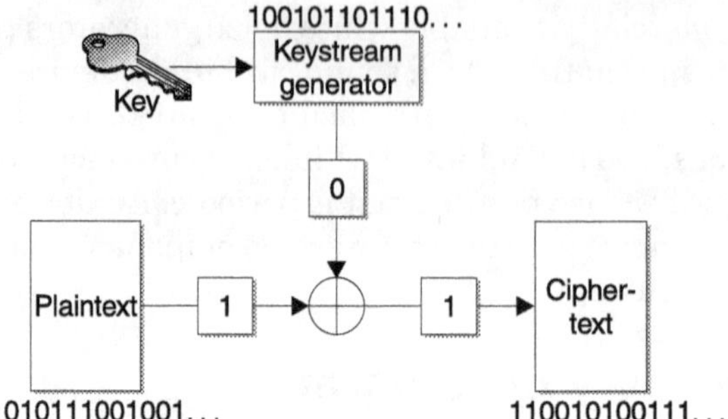

Figure 2-3 A typical stream cipher encrypts one bit or byte of data at a time by XORing plaintext input with a keystream.

(128-, 192-, and 256-bit keys are common today.) Using the key as a seed, the keystream generator expands the key into a much, much longer keystream. Although not as secure as a one-time pad, the keystream that results from an appropriately chosen key size may not repeat for days or weeks on a high-speed communication channel. The larger the key, the less frequently the keystream repeats itself. As for key management, it is much easier to transmit a small seeding key than an entire keystream.

CAUTION:
When using stream ciphers, be sure that the keystream never repeats during a single encryption session. An attacker with knowledge of plaintext contained within one section of a message might be able to obtain a piece of the keystream by XORing with the corresponding ciphertext. The recovered keystream fragment could then be used to decrypt future ciphertext once the keystream begins to repeat. Also, never encrypt multiple messages with the same encryption key. Otherwise, the key will generate identical keystreams for each message, and can be exploited in the same fashion as a repeating keystream.

One-Time Pads

One-time pads use a completely random keystream, or *pad*, to generate a perfectly secure encrypted message. The keystream must be the length of the *entire* plaintext message, and *must not* repeat. Use of a completely random, nonrepeating keystream produces ciphertext that is immune both to cryptanalytic attack and to exhaustive key searches. Regardless of the computational resources at his or her disposal, an attacker not possessing the keystream can only guess what the original plaintext might have been. A keystream of this type can be practical for encrypting short messages, but key management quickly becomes difficult as the size of a plaintext message grows. Since the keystream cannot be generated by a deterministic algorithm, the sender and recipient must exchange the entire keystream prior to communication.

The generation of the keystream is very important, since the security of the algorithm depends on the randomness of the keystream. The more random the keystream, the harder it will be for an attacker to decrypt the ciphertext using cryptanalytic techniques. For example, a cipher using a keystream consisting entirely of zeros produces output identical to its input. Using a keystream of all ones simply flips each corresponding bit in the output. If the keystream is truly random, then the algorithm will be completely secure (as with a one-time pad). Properly implemented stream ciphers lie somewhere between these two extremes.

Synchronous vs. Self-Synchronizing Stream Ciphers A *synchronous stream cipher* generates a keystream independent of the plaintext and ciphertext. Both the encryption and decryption ends generate the same keystream using a shared secret key and a keystream generator. With no relationship between the keystream and data, the encryption and decryption keystreams must be synchronized prior to decryption; otherwise, the resulting plaintext will not match the original plaintext. Successful transmission requires a synchronization mechanism external to the basic XOR operation of the stream cipher. Insertion or deletion of ciphertext bits during transmission will cause loss of synchronization. Most stream ciphers are synchronous.

A *self-synchronizing stream cipher* uses a keystream that depends on previously generated ciphertext bits. This allows the decryption and encryption keystreams to synchronize automatically after receiving a certain number of ciphertext bits. Self-synchronization, however, results in the propagation of bit errors during decryption. (One wrong ciphertext bit affects many plaintext bits.) Synchronous stream ciphers do not have this problem. Due to their ability to recover from loss of synchronization, self-synchronizing stream ciphers are more susceptible to data modification attacks (insertion, deletion, and replay) than are synchronous stream ciphers, so they may require additional use of origin authentication and data integrity mechanisms (Menezes, van Oorschot, and Vanstone, 1996).

RC4 The most commonly used stream cipher is RC4. RC4 was created by Ron Rivest in 1987, and is a trade secret of RSA Security, Inc. ("RC" stands for "Rivest Cipher.") The internal implementation details of the algorithm were kept secret until 1994, when anonymous hackers posted them to the Internet. RC4 is a synchronous stream cipher that uses a variable-length key, commonly 40–256 bits long, and a keystream independent of plaintext input. As with most stream ciphers, RC4 runs efficiently in software, and is considerably faster than even the fastest block cipher. (We will discuss the performance characteristics of RC4 in Chapter 8. For a discussion on implementing the RC4 algorithm in software, refer to Schneier, 1996.)

Block Ciphers

A *block cipher* operates on a fixed-length block of plaintext input to produce a ciphertext block of the same length. Figures 2-4 and 2-5 illustrate the basic

Figure 2-4
Typical block ciphers encrypt one full block of data at a time to produce a block of ciphertext. Common block sizes are 8 and 16 bytes.

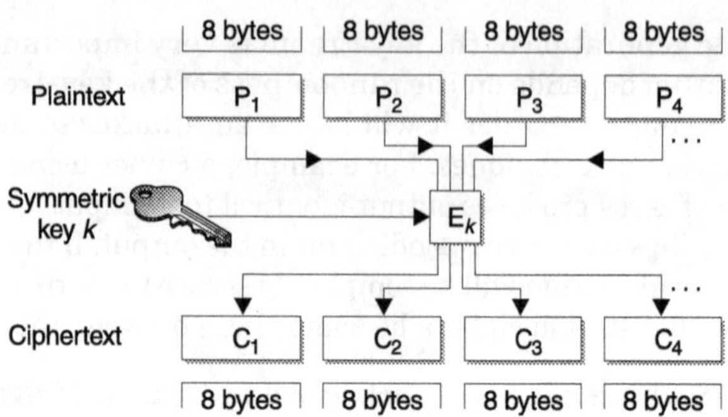

operation of a block cipher. A *block* is a collection of bits, and the length of the block is the *block size*. A typical block size for many block ciphers is 8 or 16 bytes (64 or 128 bits). For a basic block cipher, a given plaintext block will encrypt to the same ciphertext block when using the same key. However, there are operational modes that add an element of randomness to the ciphertext output even when the same key is used. (We'll discuss this shortly, in "Modes of Operation.")

In their simplest form, block ciphers break contiguous plaintext input into multiple blocks, encrypting each block independently. For example, suppose you want to encrypt 240 bytes of plaintext using a block size of 8 bytes. A block cipher would grab the first 8 bytes of plaintext, and encrypt them by combining them with a key, or with values stored in a *key table*. The result would be an 8-byte block of ciphertext. The cipher would continue grabbing 8-byte plaintext blocks and encrypting them until no plaintext remained. The encryption process would produce 30 blocks (or 240 bytes) of ciphertext in all—exactly the same number as in the plaintext input. Decryption would follow the exact same steps, with ciphertext as input and plaintext as output.

NOTE:
Most symmetric ciphers generate a key table from a single key during algorithm initialization—a process known as key setup. *The use of key tables provides additional resistance against cryptanalytic attacks on the algorithm (Burnett and Paine, 2001).*

Figure 2-5

For basic block ciphers, decryption is the exact opposite of encryption. Decryption produces one block of plaintext from a block of ciphertext.

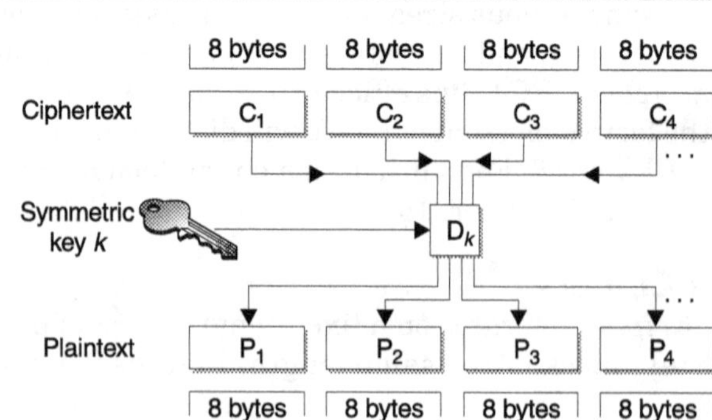

Block ciphers work well as long as the amount of plaintext input is an exact multiple of the block size. However, special handling is required if the input length is not a multiple of the block size. Using our earlier scenario, 243 bytes of plaintext would result in 30 complete ciphertext blocks. How would you encrypt the remaining 3 bytes? Since most block ciphers operate on entire blocks only, you would need to add 5 bytes of padding to the final block. The purpose of *padding* is to fill in the remaining bytes with known data before the block is encrypted. The padding can be any value, as long as both sides understand how to interpret and remove the padding during decryption.

Modes of Operation

Block ciphers can operate in a variety of different modes. Each mode has its own unique characteristics, and some modes allow block ciphers to simulate the behavior of stream ciphers. We will concern ourselves only with the four main modes of operation: electronic codebook (ECB), cipher block-chaining (CBC), cipher feedback (CFB), and output feedback (OFB). Each of these modes of operation is as secure as the underlying block cipher, and some modes provide additional resistance to cryptanalysis.

Electronic Codebook Mode The simplest operational mode for block ciphers is *electronic codebook (ECB)*. In ECB mode, each block of plaintext is encrypted independently using the underlying block cipher. A given plaintext block encrypts to the same ciphertext when using the same key, and vice versa. Figure 2-6 demonstrates ECB mode operation.

NOTE:

Key reuse is not quite as dangerous with block ciphers as with stream ciphers. Although care should be taken to minimize key reusage, more than one data set can be encrypted using the same key when block ciphers are used. This will simplify key management considerably in the case of encrypted file systems or databases that contain many files or fields.

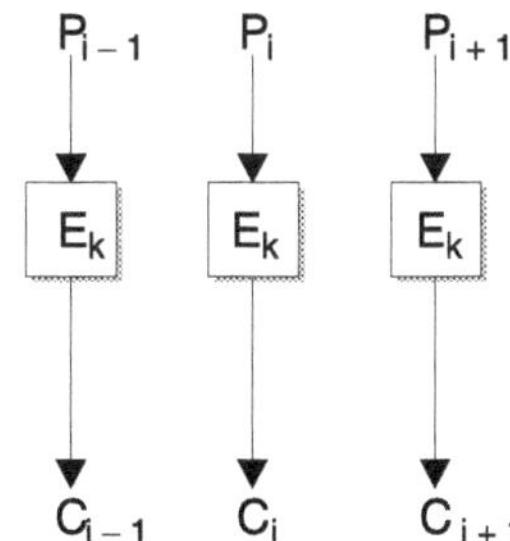

Figure 2-6 ECB mode is the simplest of all block cipher modes. One block of plaintext encrypts directly to a block of ciphertext.

Since ECB mode encrypts all plaintext blocks independently, blocks can be encrypted out of order or at the same time. Encrypting more than one block at a time is known as *parallelization*. Both encryption and decryption operations can be easily parallelized in ECB mode.

NOTE:
Parallelization offers much higher performance than block-by-block encryption, but requires a multiprocessor environment or a single processor capable of executing multiple instructions simultaneously.

While it is the simplest block cipher mode, ECB is also the least secure. Because ECB mode does not conceal plaintext patterns, it makes redundant plaintext data susceptible to cryptanalysis. Blocks can also be removed, repeated, or swapped without knowledge of the key. For example, an attacker who knows the messaging format used by two communicating parties can alter the name of the sender in a message by replacing the appropriate ciphertext blocks in the original message with ciphertext blocks containing someone else's name. If the attacker obtains the necessary ciphertext by capturing network traffic, there is no need for him to produce any ciphertext. Consequently, he never needs to know the encryption key. This technique, known as *block replay*, is difficult to detect when using ECB mode, but requires that the encryption key change infrequently.

CAUTION:
ECB mode does not conceal plaintext patterns, so it is inappropriate for encrypting long messages or large datasets.

Cipher Block-Chaining Mode In *cipher block-chaining (CBC)* mode, each plaintext block is XORed with the previously calculated ciphertext block prior to encryption; this technique is known as *chaining*. (See Figure 2-7.) Chaining prevents block-replay attacks by introducing a feedback mechanism into the block cipher. In effect, each ciphertext block depends on all previous plaintext blocks. Although blocks can be removed from the beginning or end of the ciphertext, CBC mode prevents manipulation of plaintext blocks. Chaining serves to conceal any plaintext patterns that may have existed before the plaintext was XORed with the ciphertext. Like ECB mode, CBC mode processes entire blocks of data, and cannot operate on partial blocks.

Since no ciphertext exists for encrypting the first plaintext block, an *initialization vector (IV)* must be supplied as input. An IV is nothing more than random data equal in length to the block size. More than one message can be encrypted with the same key; however, each message should use a different IV. When the same key and IV are used for multiple encryptions, a given plaintext input will always encrypt to the same ciphertext. However, when the same key

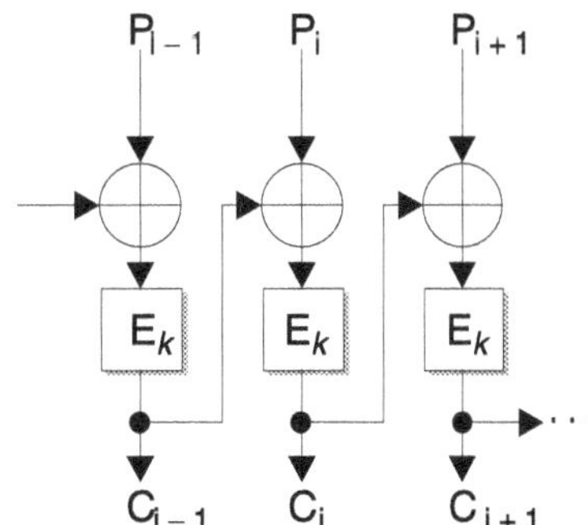

Figure 2-7 CBC mode conceals plaintext patterns—and thus prevents block-replay attacks—by incorporating a cryptographic feedback mechanism (an XOR operation).

is used for multiple messages, distinct IVs help to mask plaintext patterns common to more than one message. If not masked, these patterns can leave the messages vulnerable to cryptanalytic attack. An IV need not be kept secret, but its integrity should be verified.

CBC mode is fast, requiring only the additional XOR operations, which most computers perform very efficiently. Chaining makes parallelization during encryption very difficult. The decryption process, on the other hand, can be easily parallelized.

During encryption, a single bit error in the plaintext (for example, a toggled bit) will result in numerous errors in the ciphertext, due to the chaining property. This is not of significant concern because these errors reverse themselves during decryption, leaving only the original error. Conversely, a single bit error in the ciphertext propagates into multiple bit errors in the plaintext. The number of plaintext bits affected depends on the block size. The plaintext block that corresponds to the ciphertext block containing the error will be completely corrupted. The next plaintext block will then contain a single bit error in the same position as the original ciphertext error. This behavior, known as *error propagation*, can make CBC mode (as well as ECB and CFB modes) unsuitable for applications not tolerant to bit errors during transmission. Except for the two corrupted blocks, the cipher will recover and will eventually flush out the error. CBC mode decryption will also recover following the insertion or deletion of entire blocks of ciphertext, as long as the block boundaries remain intact.

CAUTION:
Adding or removing individual ciphertext bits causes block boundaries to shift. During decryption, this shifting will result in complete corruption of the data in all subsequent blocks. With the exception of 1-bit CFB, none of the block cipher modes discussed here can recover from this type of error.

Cipher Feedback Mode *Cipher feedback (CFB)* mode allows a block cipher to act as a self-synchronizing stream cipher. This mode does not need a complete plaintext block for processing, so it is valuable for applications that require immediate data transmission. CFB mode algorithms are denoted by the number of bits they encrypt at a time: n-bit CFB encrypts n bits at a time. CFB mode is not efficient for encrypting a few bits at a time, since each encryption of n bits requires a complete encryption of the block cipher.

NOTE:
The throughput of CFB and OFB mode ciphers (unlike that of CBC mode ciphers) is reduced by a factor of m / n, *where* m *is the block size of the cipher, and* n *is the number of bits encrypted (Menezes, 1996).*

A common value for n is 8 (the length of an ASCII character). Figure 2-8 illustrates CFB mode operation.

As in CBC mode, XORing conceals plaintext patterns, and feedback prevents direct manipulation of plaintext blocks. A unique IV should be used for each encrypted message. CFB mode exhibits error propagation when decrypting ciphertext, so a ciphertext error during transmission will result in a block containing a single bit error followed by an entire block of corrupted data. CFB mode will recover from the insertion or deletion of entire blocks of ciphertext as long as the block boundaries are not modified. A word of caution when using block ciphers in CFB mode: unlike CBC mode, which prevents block-replay attacks, CFB mode allows an intruder to add or delete ciphertext blocks without destroying the integrity of the entire message. A malicious user may use this trait to conceal or destroy critical portions of a message (name, account number, dollar amount, and so on) to slow or invalidate an online transaction.

Output Feedback Mode *Output feedback (OFB)* mode enables a block cipher to run as a synchronous stream cipher. As with synchronous stream ciphers, OFB mode relies on an external mechanism for synchronizing the encryption and decryption keystreams. An IV is required, and it should be unique (although it does not need to be secret). Block ciphers run in n-bit OFB mode, which means that n bits of each output block are placed in a feedback register for the next encryption (as illustrated in Figure 2-9).

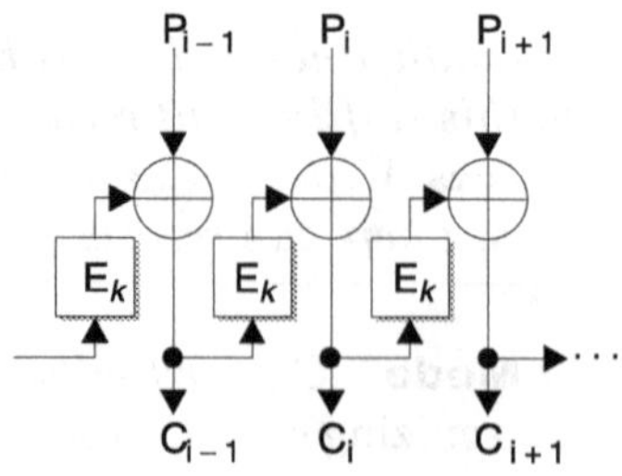

Figure 2-8
CFB mode allows a block cipher to operate as a self-synchronizing stream cipher.

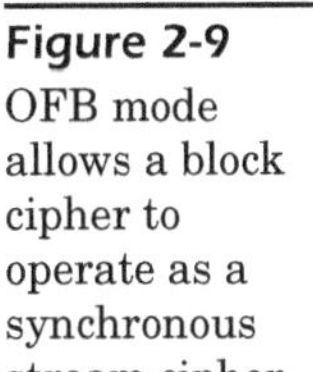

Figure 2-9 OFB mode allows a block cipher to operate as a synchronous stream cipher.

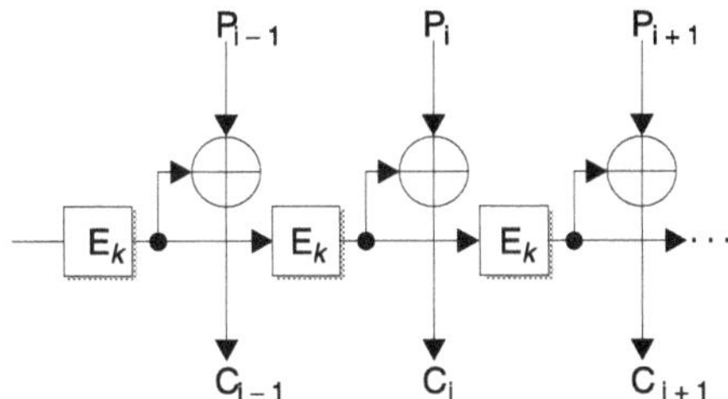

CAUTION:
For security reasons, n should not be less than the block size (in bits) when n-bit OFB mode is used. Values other than n shorten the cycle of the keystream generator, resulting in a less-than-optimal keystream length; in other words, the keystream will begin to repeat more quickly (Davies and Parkin, 1983).

OFB mode has *no* error propagation, so a single bit error in ciphertext results in a single bit error in the decrypted plaintext. Loss of synchronization is fatal when using OFB mode.

CAUTION:
Similar to a synchronous stream cipher, OFB mode generates a keystream. Care should be taken to ensure that the keystream does not repeat itself when long messages are encrypted; otherwise, OFB mode is susceptible to the same weaknesses inherent in stream ciphers.

Table 2-1 compares the four modes of block cipher operation that we've focused on in this section.

Mode	Plaintext Patterns Concealed?	Self-Synchronizing?	Error-Recovering?	Error Propagation	Parallelized Execution
ECB	No	N/A	N/A	One block	Encrypt and decrypt
CBC	Yes	No	Yes	One bit + one block	Decrypt only
CFB	Yes	Yes	Yes	One block + one bit	Decrypt only
OFB	Yes	No	Yes	None	Neither

Table 2-1 A Comparison of Block Cipher Modes of Operation

A Comparison of Common Block Ciphers

Table 2-2 provides a comparison of various block ciphers in terms of relative strength and performance. (In Chapter 8, we'll consider their raw encryption and decryption rates, key setup times, and code size.)

Table 2-3 lists characteristics of common block ciphers. Notice the wide variety and range in key sizes and numbers of rounds. As mentioned in the earlier "Block Ciphers" section, most block ciphers use block sizes of 64 or 128 bits.

NOTE:

In general, the larger the key size, number of rounds, and block size, the more secure the algorithm. While larger values do improve security, they tend to reduce performance.

It's worth mentioning here that RC6 is based on RC5, and supports similar customization in terms of key size, number of rounds, and word size. RC6 was a finalist for the Advance Encryption Standard (AES), and the values listed for RC6 in the table reflect AES requirements. While many implementations of RC6 use these parameters, it is conceivable that other parameter values (and in particular, larger key sizes) could be used as well.

Algorithm	Relative Strength[1]	Relative Performance
DES	Low	Moderate
3DES	Moderate–very high[2]	Very slow
RC2	Moderate–high	Fast
RC5	High–very high[3]	Moderate–very fast
RC6	High	Very fast
AES (Rijndael)	High	Fast
IDEA	High	Slow
Blowfish	High–very high	Very fast

[1] Relative strength is based on block size, key size, number of rounds per encryption, and known cryptanalytic weaknesses of the algorithms.

[2] The strength of 3DES depends on whether it is being used with two or three keys.

[3] For both RC5 and RC6, security depends on the parameters chosen (key length and number of rounds).

Table 2-2 A Comparison of Block Ciphers in Terms of Relative Security and Performance

Algorithm	Effective Key Length	Block Size	Number of Rounds	Additional Customization	Patented?
DES	56 bits	64 bits	16	None	No
3DES	112 bits (2 keys) or 168 bits (3 keys)	64 bits	48	Number of keys	No
RC2	8–1024 bits	64 bits	18	None	Trade secret
RC5	0–255 bytes	64 or 128 bits	Variable; typically 12 or 16	Word size of 16, 32, or 64 bits	Yes
RC6	128, 192, or 256 bits	128 bits	Variable; typically 20	Word size; typically 32 bits	Yes
AES (Rijndael)	128, 192, or 256 bits	128, 192, or 256 bits	10–14[4]	None	No
IDEA	128 bits	64 bits	8	None	Yes
Blowfish	32–448 bits	64 bits	16	None	No

[4] AES uses 10 rounds for 128-bit key/block size, 12 rounds for 192-bit key/block size, and 14 rounds for 256-bit key/block size.

Table 2-3 Common Characteristics of Block Ciphers

Symmetric-Key Distribution and Management

Suppose two employees, Alice and Bob, need to exchange a confidential e-mail concerning an up-and-coming company merger. Their employers both use symmetric encryption to secure e-mail. Alice encrypts the e-mail using a random key that she generates, and then sends the e-mail to Bob. Before Bob can decrypt the e-mail, he must obtain the encryption key generated by Alice. Alice could place the key on a floppy disk, walk to Bob's desk, and hand the disk to Bob. This method might be acceptable if Alice and Bob are in the same building, but what if one is in London and the other in New York?

Key-encryption keys offer an alternative to exchanging keying information by hand. A *key-encryption key (KEK)* is a key used specifically for encrypting other symmetric keys. For security, KEKs are often longer than the keys they encrypt, and need not be used for the same algorithm. It's common for a slow, but very secure symmetric cipher, such as 3DES, to encrypt the key for a faster, weaker cipher, such a DES. The faster algorithm improves performance when encrypting large amounts of data, while the more secure algorithm provides long-term key protection. Performance of the slower algorithm is rarely an issue, since the cipher is used infrequently, and only for encrypting small bits of data.

CAUTION:
When the KEK technique employs the use of weaker algorithms for bulk encryption, it should be limited to encrypting data that must remain encrypted only for short periods of time.

Returning to our example, Alice and Bob could exchange a KEK once, and then use it for encrypting any number of symmetric keys. Alice could then e-mail the encrypted key along with the message. Not a bad idea, but what if Alice and Bob have never communicated before? This scenario demonstrates the difficulties associated with distributing a shared secret key for use in symmetric encryption. As we will soon see, public-key cryptography provides a solution to this problem. (See the upcoming section, "Key Distribution, Exchange, and Agreement.)

NOTE:
When using symmetric-key cryptography, the communicating parties must agree on a shared key before they begin encrypted communication. Exchanging the key requires a prior relationship between the sender and recipient.

Another difficulty in using symmetric encryption is key management. As users are added to a system, the number of keys grows very quickly. For example, in a small company with 25 employees, each employee would have to manage 24 keys (one for every other employee). The company as a whole would need 600 keys. While these figures are not overwhelming, imagine a larger company with 2000 employees. The total number of keys required would skyrocket to 3,998,000! Once again, public-key cryptography solves this problem.

NOTE:
Symmetric-key cryptography requires that each user obtain a key from every other user with which he wants to communicate. For n *users, this results in* $n - 1$ *keys per user. However, the total number of symmetric keys needed for the entire system is* $n^2 - n$. *As we will soon see, public-key cryptography reduces the total number of keys in a system to* n *key pairs.*

Message Digests

The primary mechanisms for achieving message integrity are *message digests*. Message digests are based on *hashing functions,* which convert variable-length input data into a fixed-length output. An *ideal* message digest, *M,* has the following characteristics given input x:

- Given any length input, *M* will produce a fixed-length output.
- *M* is a one-way function.

- Small changes in input to M result in drastically different output.
- The output of M is collision free.
- Calculation of $M(x)$ should be relatively easy.

A message digest generates a fixed-length output from an arbitrary-length input. Some message digests will accept an infinite length input, while others place theoretical limitations on the input length, which, for most intents and purposes, can still be considered infinite. For example, SHA-1 will accept an input of up to $2^{64} - 1$ bits in length (more than 2.3 billion MB of data!). The output, $M(x)$, is a fixed length that depends only on the algorithm used. Two common message digests, MD5 and SHA-1, produce 128-bit and 160-bit digests, respectively.

The second characteristic of a message digest is that it is a *one-way* function. This means that given $M(x)$, it is mathematically infeasible to calculate x. Someone who obtains the output of a digest operation cannot reverse the operation to obtain the input. The only way to determine x is to digest input values until the output matches.

To add a little perspective, imagine that a friend has challenged you to a contest. Your friend has generated a digest, and now she wants you to guess the input. Since the output is a fixed length regardless of the size of the input, you have no way of knowing whether the digest represents a single word, a novel, or an entire encyclopedia. Assuming that the message is relatively short, you supply a few guesses to a program that calculates a digest value. After analyzing the output for several minutes, you conclude that random guessing will get you nowhere. There does not seem to be any correlation between the input and output. Changing a single letter in a sentence or paragraph seems to completely change the output. (In fact, an ideal message digest should flip 50 percent of the output bits when a single input bit is flipped. If this were not the case, the algorithm would be statistically biased aiding a cryptanalytic attack.) After many hours of trying, you conclude that you will never guess the input, and decide to call it quits. This is a wise decision on your part, as you could easily spend the rest of your life trying to win this challenge.

The output of an ideal message digest is unique, or *collision free*, meaning that two different inputs will not result in the same output. In other words, there are no two inputs x and y with $x \neq y$, such that $M(x) = M(y)$. For a digest function to be truly collision free, the output length would have to be infinite, which would make the use of digests very impractical (read "worthless"—picture yourself growing old waiting for a digest operation to complete).

Popular digest algorithms take a more practical approach. While their output is finite, they are capable of producing such a multitude of distinct outputs that locating two that match (not using the same input) is extremely difficult, and guessing two inputs that produce the same output is nearly impossible. A message digest with an n-bit output can produce 2^n different digests. For SHA-1, the number of possible outputs is 2^{160}, which provides an extremely high level of assurance that the output will be unique. In the next section, we'll discuss how this advantage, along with one-way operation, can be used to ensure message integrity.

NOTE:
In practice, the number of messages needed to find a collision is much less than the number of possible outputs. The well-known "birthday problem" of probability states that you have greater than a 50 percent chance of finding two matching outputs after trying 1.2 times the square root of the total number of possible outputs. For a message digest that produces an n-bit output, the total number of distinct digests is 2^n, which means that you will likely find a collision after trying $1.2 \times 2^{n/2}$ messages. For 160-bit SHA-1, this equates to approximately 2^{80} (or 1.2×10^{24}) messages.

Message digests provide a unique representation of the input data. This feature is quite useful when a representation of the original data, and not the data itself, is acceptable. Take, for instance, user passwords stored in a database. Authentication systems generally store a hash of a user's password because knowing the hash does not provide any information useful for deriving the password. When a user logs into a system, the authentication program hashes the user's password and compares the result to what is stored in the database of hashed passwords. If the hashes match, then the user is authenticated. If the hashes do not match, then the user has supplied the incorrect password. As we will see in our discussion of digital signatures, message digests can be used as a condensed representation of larger data sets.

Table 2-4 compares the relative strength and performance of many popular message digests.

Algorithm	Output Digest Length	Relative Strength[5]	Relative Performance
MD2	128 bits	Moderate	Very slow
MD4	128 bits	Low	Very fast
MD5	128 bits	Moderate	Fast
SHA-1	Variable; 160, 256, 384, and 512 bits are common	High–very high	Very slow–moderate[6]
RIPEMD	128 and 160 bits	High	Moderate

[5] Relative strength is based on output digest length and known weaknesses in the algorithm. In general, longer digests are more secure.

[6] Performance depends on the length of the output digest.

Table 2-4 A Comparison of Common Message Digests

CAUTION:

Many weaknesses have been identified in the MD4 algorithm, so you should avoid using it in high-security environments.

HMACs

So, how do you guarantee the integrity of a message? For each message, you can generate a message digest, and transmit the original message and digest together. To verify that the message has not changed in transit, you would create a digest from the message and compare it to the one accompanying the message. Although an attacker could not change the original message and still produce the same digest, what would prevent her from altering the original message, generating a new digest, and swapping these with the originals? The answer is this: nothing. An attacker could swap both without detection.

A digest itself provides little guarantee that a message has not been altered in transit. One possible solution is to use a *hashed-based message authentication code (HMAC)*, often called a *keyed message digest*. An HMAC combines a message with a shared secret key prior to performing the digest operation, as shown in Figure 2-10. Inclusion of the secret key prevents an attacker from intercepting and modifying the original message undetected. Without knowing the secret key, an attacker must combine the original message with possible keys until he calculates a matching digest. At this point, he has either found a weakness in the collision-free operation of the message digest or, more likely, determined the correct key.

You can, of course, achieve the same level of assurance by using a symmetric- or public-key encryption algorithm to encrypt the digest. Although an attacker could then alter the original message and generate a new digest, she could not re-encrypt the digest without the key. Logically, this precaution is the same as using an HMAC. However, encryption operations are usually more computationally intensive than digest operations, and therefore require more processing time (or resources, depending on how you look at it).

NOTE:

A message digest ensures only the integrity of a message; it will not provide confidentiality. If confidentiality is required, the original message should be encrypted as well.

If the attacker eventually discovers the key, he might launch an attack well after the original message has been received and verified. For example, two companies transmitting financial transactions throughout the course of a day might not notice a bogus transaction as long as the attached MAC is verified. An attacker with knowledge of the shared key might inject such transactions. However, the sender and recipient could counter this attack by frequently changing

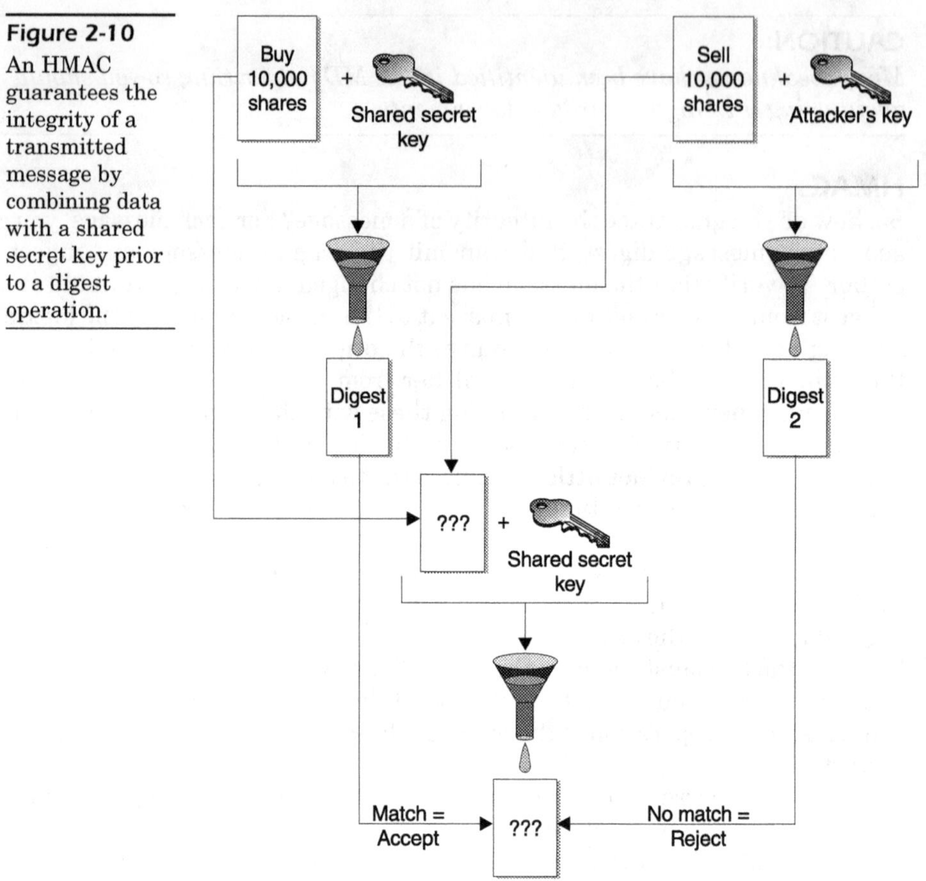

Figure 2-10 An HMAC guarantees the integrity of a transmitted message by combining data with a shared secret key prior to a digest operation.

the shared secret key. As with encryption algorithms, the longer the key, the longer it will take an attacker to uncover the key. Since digest operations are very efficient, the addition of long keys adds very little to the time required to calculate the digest. RFC 2104 further describes HMAC operation (Krawczyk, Bellare, and Canetti, 1997).

Public-Key Cryptography

Public-key cryptography, also known as *asymmetric encryption,* uses the concept of public and private keys. As discussed in the earlier section "The Basics of Cryptography," the public key is often used for encryption, and the private key for decryption. Unlike symmetric-key cryptography, a complex mathematical relationship exists between the public and private keys, making calculation of one from the other extremely difficult. In crypto lingo, the mathematical relationship is referred to as a *hard problem*. For some algorithms, the hard problem involves factoring very large numbers into prime factors; for other algorithms, it involves the calculation of discrete logarithms or other difficult mathematical operations. All hard problems share a common characteristic: they are problems for which no known shortcuts exist. Conceptually, they may not be too difficult; however, solving them is computationally expensive and prohibitive.

What are the benefits of using different keys for encryption and decryption? The private key usually belongs to an individual, and only that individual has knowledge of the key. (Devices such as cable modems, cellular phones, and VPN-enabled routers might also contain private keys.) The individual's public key, on the other hand, can be exchanged freely. Anyone wishing to send a secure message to another user simply encrypts the message using the recipient's public key. The only person who can decrypt the message is the owner of the corresponding private key, as illustrated in Figure 2-11.

NOTE:
A public-key infrastructure (PKI) *provides a framework for requesting, distributing, managing, and revoking digital certificates.* A digital certificate *binds an individual's public key to his or her identity. Most PKIs publish digital certificates to a centralized directory, so that users can quickly locate the public key of another individual based on their physical identity. (Refer to Appendix B for a detailed discussion of PKI.)*

The use of distinct key pairs results in security unmatched by conventional encryption techniques. In addition to encryption, public-key cryptography is used for key exchange, key agreement, and generation of digital signatures.

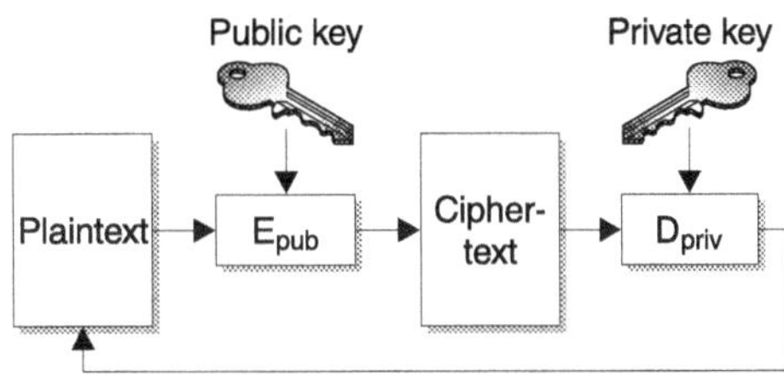

Figure 2-11 Public-key cryptosystems use different keys for encryption and decryption.

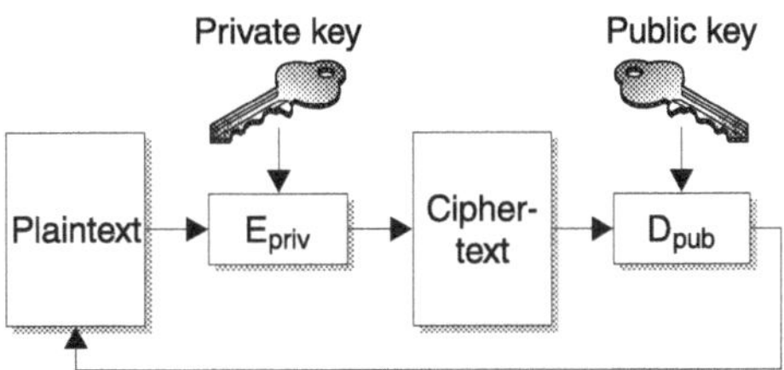

Key Distribution, Exchange, and Agreement

One of the difficulties in using symmetric-key cryptography is secure distribution or exchange of keying information. Both the sender and recipient must exchange a common key, which is randomly generated by either the sender or the recipient. If the sender generates the key, how does he transmit it to the recipient, or vice versa? Public-key cryptography solves this problem through its use of key pairs.

NOTE:

Kerberos is a network authentication protocol used to securely distribute keys using only symmetric algorithms, such as DES. (We'll discuss Kerberos in Chapter 6.)

Say that Alice and Bob want to exchange a key for use with a symmetric cipher. First, Alice randomly generates a shared secret key. She then encrypts the shared secret key, using Bob's public key. Upon receipt, Bob extracts the shared secret key by performing a private-key decryption. Voìla! The symmetric key has been exchanged securely, and Alice and Bob can begin encrypted communication, using the symmetric key. In fact, this is a simplified version of the process by which many security protocols, such as SSL (discussed in Chapter 6), exchange traffic encryption keys. As we discussed in the preceding section, using public-key algorithms for key distribution, which involves a small amount of data, and symmetric-key algorithms for encrypting the bulk of the traffic makes sense, given their relative performance.

NOTE:

Public-key cryptography requires no prior relationship between the sender and recipient. However, as we will see in Chapter 3, it does rely on the use of trusted third parties to avoid man-in-the-middle attacks.

Key agreement varies slightly from key distribution and exchange, in that keys are not physically transmitted from one party to another. Instead, both parties share public values that, when combined with private values, generate a shared secret key. Although both parties must still exchange public keys, they generate the shared secret key independently. As a result, there is no way for an attacker to intercept the shared secret key. Diffie-Hellman is such an algorithm.

NOTE:

The Diffie-Hellman (DH) algorithm was originally proposed by Diffie, Hellman, and Merkle, and is credited as the first public-key algorithm. The relationship between the public and private values used by DH is based on the calculation of discrete logarithms. For a detailed discussion of the Diffie-Hellman algorithm, refer to (RSA Laboratories, 1993a).

Digital Envelopes

The speed of asymmetric algorithms makes them impractical for encrypting bulk data. A more common approach is to use an asymmetric algorithm in conjunction with a symmetric algorithm by creating a *digital envelope*.

NOTE:

Public-key operations are very slow relative to symmetric-key operations, and especially to those using a private key. Encryption using a public key can take upward of 100 times longer than a symmetric-key encryption. A private-key operation can take more than 1000 times longer.

A digital envelope addresses slow performance by allowing each cryptographic algorithm to do what it does best: symmetric ciphers perform bulk encryption, and asymmetric ciphers enable key distribution. The process of creating a digital envelope, which is illustrated in Figure 2-12, follows these steps:

1. The sender, Alice, first encrypts the message, using a symmetric cipher.
2. Alice then encrypts the symmetric cipher key, using the public key of the intended recipient—in this case, Bob.
3. The combination of the encrypted message and the symmetric key is a digital envelope, which Alice sends to Bob.

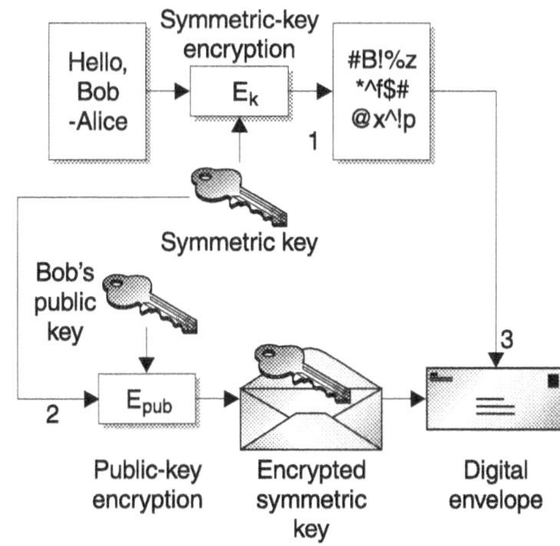

Figure 2-12 Digital envelopes combine the speed of symmetric cryptography with asymmetric key distribution.

Bob performs the opposite steps (as shown in Figure 2-13) to obtain the contents of the envelope:

1. First, Bob separates the encrypted message and symmetric key.
2. He then decrypts the symmetric session key, using his private key.
3. Finally, Bob decrypts the message, using the recovered symmetric key.

Anyone with the recipient's public key can create an envelope, but only the individual with the corresponding private key will be able to decrypt the contents. This is somewhat analogous to sending a letter in the mail. A friend wishing to send you a letter first handwrites or types the letter (generation of plaintext), and then folds the letter so that it can be inserted into an envelope. At this point, the letter cannot be read unless someone unfolds it (decrypts it). Your friend inserts the letter into an envelope and seals the envelope (public-key encryption) using his tongue (a convenient public key for you, since most people have access to one). A mail truck then delivers the mail to your house (transmission). Once you receive the letter, you use your trusty letter opener (private key) to open the envelope (private-key decryption). Once you have access to the folded letter, you must first unfold it to read it (symmetric-key decryption). Unfolding the letter is exactly the reverse of folding, hence a symmetric operation. This requires knowledge of how to unfold the letter (the symmetric key).

Digital Signatures

Earlier in this chapter, we learned that public keys are generally used for encryption, and that private keys are used for decryption. This makes sense for securing transmitted messages, since only the owner of a private key can decrypt a

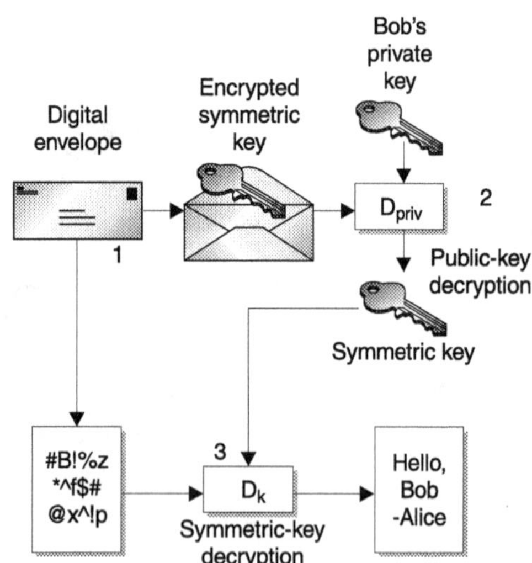

Figure 2-13
Recovering the contents of a digital envelope involves decrypting the session key, and then decrypting the message.

message that has been encrypted with his or her public key. A private key can just as easily be used for encryption, and a public key for decryption. I bet I can guess what you're thinking now—"Why would anyone want to encrypt using their private key? Since the public key *is* publicly available, couldn't anyone decrypt the message." As we'll soon see, the intended purpose of encryption with a private key is not to provide confidentiality but, rather, authentication and nonrepudiation.

Let's look at an example. Bob decides he wants to send a message to Alice. He wants Alice to know that he sent the message, so he encrypts it using his private key. After receiving the message, Alice decrypts it, using Bob's public key. If the result is legible (and not gibberish), then Alice can be somewhat certain that no one but Bob could have sent the message. It will also be difficult for Bob to later deny having sent the message to Alice in the first place. This certainty, however, relies on the level of trust placed in Bob to properly protect his private key. What we've just described is the basic process for generating a *digital signature*. Digital signatures are the electronic analog to handwritten signatures. Since only Bob possesses his private key, he is the only one capable of generating the corresponding digital signature.

CAUTION:
Just as with handwritten signatures, if a digital signature can easily be reproduced by many people, it will not be of much benefit. For this reason, private keys must be safely stored, and the owner of a key should never disclose the key to another user.

Recall that public-key operations are slow when compared with symmetric-key operations and message digests. In this example, Bob does not gain much from encrypting the entire contents of a message. (We have already established that doing so will not provide confidentiality, and will be computationally prohibitive.) A better solution is to encrypt a condensed representation of the original message. In effect, this will provide the same level of security as encrypting the entire message, since Bob is really only concerned with the use of the private key, not with the amount or type of data encrypted. Luckily, we have just the right tool for the job: a message digest. By encrypting the digest, Bob can conserve precious computational resources, and still ensure authentication and nonrepudiation. The most popular signature-generation algorithm is RSA. (Ron Rivest, Adi Shamir, and Leonard Adleman invented the RSA algorithm.) The process of creating and verifying an RSA digital signature is shown in Figure 2-14. Here are the steps involved:

1. Bob generates a condensed representation (a digest) of the original data by supplying the message as input to a message digest algorithm, such as MD5 or SHA-1.

Figure 2-14 Generating a digital signature involves passing the message through a message digest, and then encrypting the resulting digest with a private key.

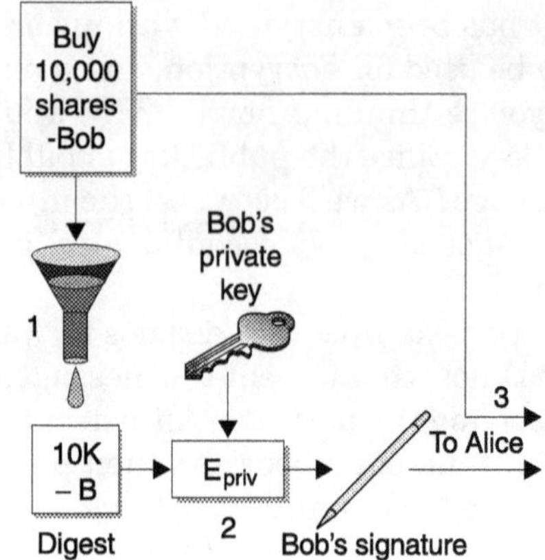

2. He then encrypts the digest, using his private key to produce a digital signature.
3. Bob sends both the message and the signature to Alice.

When Alice receives Bob's message, she verifies the RSA digital signature (as shown in Figure 2-15):

1. Alice creates a message digest (Digest 1) of Bob's original message, using the same algorithm that Bob has used.
2. She then decrypts the digital signature using Bob's public key to recover Bob's digest (Digest 2).
3. Alice compares Digest 1 and Digest 2; if they match, the signature verifies correctly.

NOTE:
The RSA Laboratories Cryptography FAQ v4.1 (RSA Laboratories, 2000) offers the following description of the RSA algorithm:

> *The RSA algorithm works as follows: take two large primes, p and q, and compute their product n = pq; n is called the modulus. Choose a number, e, less than n and relatively prime to (p–1)(q–1), which means e and (p–1)(q–1) have no common factors except 1. Find another number d such that (ed – 1) is divisible by (p–1)(q–1). The values e and d are called the public and private exponents, respectively. The public key is the pair (n, e); the private key is (n, d). The factors p and q may be destroyed or kept with the private key.*

The security of the RSA algorithm rests in the inability of an attacker to factor the very large modulus, n, into its prime factors p and q.

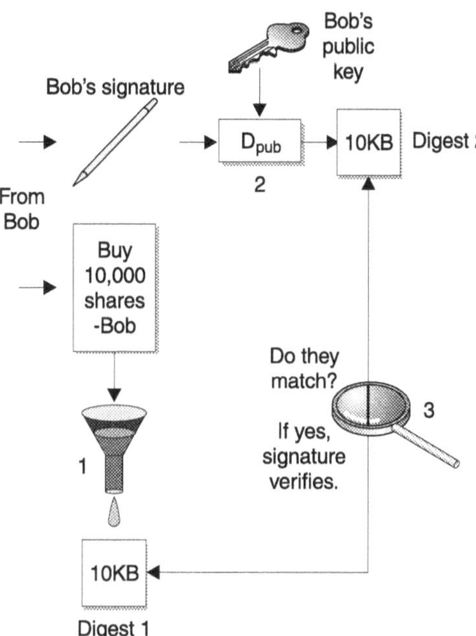

Figure 2-15 Alice verifies Bob's digital signature by comparing digests.

In addition to authentication, use of a message digest has the added benefit of ensuring the integrity of the message. The process of encrypting a digest mimicks the performance of an HMAC. If the signature verifies correctly, then it is extremely unlikely that the message has been altered in transit.

In recent years, the popularity of digital signatures has grown in conjunction with code signing. Microsoft's Authenticode allows developers to sign ActiveX controls and Java applets. The signature attached to each ActiveX control or Java applet is verified before the code is allowed to run within Microsoft web browsers. Netscape's Object Signing feature offers similar assurance for Java applets running within Netscape browsers. Cable modem vendors have begun placing signature verification software on their modems to verify signed firmware upgrades. A signature must verify correctly; otherwise, the cable modem will not install the upgrade. (We'll discuss secure software upgrades in Chapter 9).

A Comparison of Common Public-Key Algorithms

Table 2-5 provides a comparison of RSA, Diffie-Hellman (DH), Elliptic Curve Cryptography (ECC), and the Digital Signature Algorithm (DSA). "Hard Problem" refers to the category of mathematical problem on which the security of the algorithm is based. The values for relative performance refer to operations using the private and public keys, respectively. (We will discuss the performance of each of these algorithms in further detail in Chapter 8.) Private-key operations are generally slower than public-key operations (with the exception of DSA and other public-key algorithms based on DSA).

Algorithm	Key Sizes	Hard Problem	Security Procedures	Relative Performance (Private/Public)
RSA	Variable. Common values are 768, 1024, and 2048 bits.	Factoring large number into prime factors	Encryption, key exchange, and digital signatures	Slow/very fast
DH	Variable. Common values are 768, 1024, and 2048 bits.	Calculating discrete logarithms	Key agreement	Slow/very slow
ECC	Variable. Common values include 160–256 bits and higher.	Calculating elliptic-curve discrete logarithms	Encryption, key exchange, and digital signatures[7]	Very fast/Slow
DSA	512–1024 bits modulus. (Must be a multiple of 64.)	Calculating discrete logarithms	Digital signatures	Moderate/very slow

[7] By itself, ECC does not provide encryption, key exchange, or digital signature capabilities. ECC simply describes the underlying mathematical engine. Common algorithms utilizing ECC are Elliptic Curve AES (ECAES) for encryption, Elliptic Curve Diffie-Hellman (ECDH) for key agreement, and Elliptic Curve DSA (ECDSA) for digital signature generation and verification.

Table 2-5 A Comparison of Various Public-Key Algorithms

RSA provides the most versatility of any public-key algorithm. It can be used for encryption and decryption, key exchange, and signature generation and verification. RSA is the most commonly used of all public-key algorithms. It is supported and in use by every major web server and client browser to enable SSL authentication and key exchange.

ECC has gained recent attention in conjunction with its use in embedded devices, and particularly in wireless device. The speed of other public-key algorithms in generating digital signatures may prohibit their use on slow processors, such as those contained within mobile phones. While ECC signature verification is slower than that of RSA, it is much faster when generating signatures. Methods for computing elliptic-curve discrete logarithms are less efficient than those for factoring or computing conventional discrete logarithms, which allows the use of smaller key sizes (RSA Laboratories, 2000a). 163-bit ECC is generally considered as secure as 1024-bit RSA in terms of the computational time required to determine a key by means of an exhaustive key search.

Public-Key Cryptography Standards

The Public-Key Cryptography Standards (PKCS) are specifications created by RSA Laboratories for promoting public-key cryptography. They define and outline the usage of various public-key algorithms, cryptographic message formats, certificate and certificate request syntaxes, and cryptographic token interfaces. PKCS currently consists of 11 de facto standards, with several more in development.

PKCS #1: RSA Cryptography Standard (RSA Laboratories, 2001) defines public-key cryptography based on the RSA and Multi-Prime RSA algorithms. *PKCS #1* specifies two key types, RSA public key and RSA private key, as well as *cryptographic primitives* for implementing various RSA operations. Cryptographic primitives are the mathematical operations underlying the basic *schemes* of encryption and signing. Encryption and decryption operations compose the encryption scheme, and signature generation and verification operations compose the signing scheme. PKCS #1 also defines the syntax in Abstract Syntax Notation One (ASN.1) for encoding keys and identifying schemes within a message (refer to Kaliski, 1993 for a discussion of ASN.1).

PKCS #3: Diffie-Hellman Key Agreement Standard (RSA Laboratories, 1993a) describes the use of the Diffie-Hellman algorithm for key agreement in terms of parameter generation, the two phases of Diffie-Hellman key agreement, and object identifiers for ASN.1 encoding.

NOTE:

You may be wondering about PKCS #2 and PKCS #4—both describe the RSA public-key algorithm, and have since been incorporated into PKCS #1.

PKCS #5: Password-Based Cryptography Standard (RSA Laboratories, 1999a) discusses the implementation of password-based encryption to protect sensitive information, such as private keys. PKCS #5 covers key derivation, encryption and message authentication schemes, and ASN.1 encoding for keys and schemes. *Key derivation* involves producing a *derived key* from a *base key* (a password, for instance), a salt value, and an iteration count. The encryption scheme employed by PKCS #5 is symmetric.

PKCS #6: Extended-Certificate Syntax Standard (RSA Laboratories, 1993b) enhances the power of X.509 public-key certificates by supplying additional information as a set of attributes. The X.509 certificate and set of attributes together form an extended certificate. The X.509 version 3 (v3) standard similarly augments digital certificate activity by specifying the use of extensions. The Public-Key Infrastructure (X.509) Working Group within the IETF, also known as the PKIX Working Group, profiled the use of X.509 v3 certificates for use on the Internet (Housley et al., 1999).

PKCS #7: Cryptographic Message Syntax Standard (RSA Laboratories, 1993c) provides the syntax for describing data with cryptography applied to it. Six content types currently exist: data, signed data, enveloped data, signed-and-enveloped data, digested data, and encrypted data. A *content type* simply describes the contents of a cryptographic message, and one content type can be nested within another. For instance, the signed data content type indicates that a message contains data and an encrypted digest of the data. The encrypted message digest acts as a digital signature identifying the signer(s). PKCS #7 specifies object identifiers for ASN.1 encoding of each content type, and has been adopted and extended by the IETF as an Internet standard (Housley, 1999).

PKCS #8: Private-Key Information Syntax Standard (RSA Laboratories, 1993d) applies ASN.1 encoding to private-key and encrypted private-key information. A password-based encryption algorithm, as described in PKCS #5, can be used to encrypt sensitive private keys.

PKCS #9: Selected Object Class and Attribute Types (RSA Laboratories, 2000b) defines the **pkcsEntity** and **naturalPerson** object classes, as well as their associated attribute types and their ASN.1 encoding. It also introduces a number of attribute types useful for describing PKCS-related data in conjunction with PKCS #7, PKCS #10, PKCS #12, and PKCS #15.

PKCS #10: Certification Request Syntax Standard (RSA Laboratories, 2000c) describes the syntax for certificate requests sent to a *certification authority (CA)*. A CA is responsible for issuing X.509 public-key certificates. A certificate request consists of the subject's name and public key, certain attributes describing the subject, a signature algorithm identifier, and a digital signature on the request information. (The *subject* is the entity requesting a certificate from a CA.) The CA is responsible for authenticating the subject and verifying the signature attached to the request before issuing a certificate. Signature verification provides *proof of possession* (assurance that the requestor has access to the corresponding private key). PKCS #10 also provides object identifiers for ASN.1-encoding certificate requests. (See Appendix B for a discussion of digital certificates and public-key infrastructure [PKI] concepts.)

PKCS #11: Cryptographic Token Interface Standard (RSA Laboratories, 1999b) presents an *application programming interface (API)* to devices capable of storing cryptographic information and performing cryptographic operations. An API provides an abstract representation of an algorithm's internal details in terms of specific input requirements and the resulting output. This interface, called Cryptoki (short for *cryptographic token interface*), abstracts the details of cryptographic hardware devices, such that an application does not need intimate knowledge of a device in order to interoperate with it.

PKCS #12: Personal Information Exchange Syntax Standard (RSA Laboratories, 1999c) describes the syntax for safely transporting personal information, such as private keys and certificates. The standard includes a number of privacy and integrity modes for securing and authenticating personal information using (a) public-key cryptography and digital signatures or (b) passwords and keyed MACs. Password-based encryption builds on PKCS #8. Microsoft Internet Explorer is an example of an application that supports PKCS #12 for importing and exporting credentials within a web browser.

PKCS #13: Elliptic Curve Cryptography Standard discusses elliptic-curve cryptographic operations, including key generation, public-key encryption, key agreement, digital signatures, and ASN.1 encoding. (As of the writing of this book, this specification is still under development.)

PKCS #14: Pseudorandom Number Generation Standard addresses the proper implementation of robust pseudorandom number generators. This standard will not outline entropy gathering techniques. (As of the writing of this book, this specification is still under development.)

PKCS #15: Cryptographic Token Information Format Standard (RSA Laboratories, 2000d) provides a common file and directory structure for storing information on cryptographic tokens. Such information can take the form of public/private and symmetric keys, certificates, data, PINs, and *biometric templates* (which are used for comparing physical characteristics during biometric authentication). This allows a single application to locate information on cryptographic tokens from many different vendors.

Federal Information Processing Standards and Certification

The National Institute for Standards and Technology (NIST) has developed the Federal Information Processing Standards (FIPS) as guidelines for representing, communicating, and securing data in information systems. Many of these standards are based on ANSI standards. Three main categories of FIPS publications exist:

- **Hardware and software standards/guidelines** Syntax and semantics for database manipulation, exchange of electronic data, and graphical system modeling
- **Data standards/guidelines** Codes and representations for calendar data, geographic regions under possession of the United States, metropolitan areas, congressional districts, and so on.
- **Computer security standards/guidelines** Guidelines for access control, cryptography, general computer security, risk analysis, and contingency planning for open systems

---

NOTE:
An open system *is one based on interoperable standards and readily available, nonproprietary components. This is the opposite of a* closed system. *Unix-based systems, for example, are generally open, while Windows-based systems are closed.*

A number of FIPS publications relate directly to cryptographic algorithms, software, and hardware. NIST's Cryptographic Module Validation (CVM) Program provides testing and certification in the following areas:

- *FIPS 140-1 and FIPS 140-2: Security Requirements for Cryptographic Modules* provides security requirements for maintaining confidentiality and integrity of unclassified data when designing and implementing hardware- and software-based cryptographic modules; FIPS 140-2 supercedes FIPS 140-1.

- *FIPS 46-3: Data Encryption Standard (DES)* describes the operation of the DES and Triple DES (also called the Triple Data Encryption Algorithm, or 3DES) algorithms. It also discusses the migration of cryptographic systems from DES to Triple DES.
- *FIPS 81: DES Modes of Operation* defines the use of DES in ECB, CBC, CFB, and OFB modes.
- *FIPS 186-2: Digital Signature Standard (DSS)* discusses generation and verification of digital signatures using the Digital Signature Algorithm (DSA), RSA, and Elliptic Curve DSA (ECDSA) algorithms.
- *FIPS180-1: Secure Hash Standard (SHS)* lists enhancements to the SHA-1 message digest algorithm.
- *FIPS 185: Escrowed Encryption Standard (EES)* describes the use of key escrow and the Skipjack algorithm in decrypting intercepted telecommunications by law enforcement agencies.

We will refer to these standards throughout this book, since many government and financial institutions employ only FIPS-compliant cryptographic implementations. Many security-related specifications also reference FIPS compliance.

Store-and-Forward vs. Session-Based Encryption

Store-and-forward describes messaging systems, such as e-mail, where messages are frequently created on one host, forwarded to a second host for storage, and retrieved by a third host. In the case of e-mail, an e-mail client encrypts a message (or a symmetric key when using digital envelopes), using the public key of the recipient, before sending it to a mail server. After retrieving the message, an e-mail client decrypts the message, using the recipient's private key. This type of encryption requires management of keying information to determine which key has been used to encrypt a given message. Secure Multipurpose Internet Mail Extensions (S/MIME) (which we will discuss in Chapter 7) is a security protocol designed for this exact purpose. In most cases, store-and-forward operations do not affect the performance of network communication, since they can be performed offline. Store-and-forward encryption techniques are also appropriate for encrypting file systems, databases, and other forms of persistent data.

NOTE:

PKCS #7 defines messaging formats for signed, enveloped, encrypted, and digested data. These message formats are appropriate for store-and-forward applications.

The goal of *session-based encryption* is to protect data while it is in transit. Security protocols, such as SSL and IPSec, generate a random session key during session establishment. This session key is often referred to as a *traffic encryption key (TEK)*. The key is used to encrypt data just before it leaves the sender, and to immediately decrypt the data when it reaches the recipient. Once the communication session ends, the key is destroyed and can no longer be used to encrypt or decrypt data. Session-based encryption provides a one-time, secure tunnel between the sender and recipient. Each session uses a different key to limit the exposure of data in the event of key compromise. VPNs use session-based encryption to transparently secure communication between remote hosts over public networks (a topic we will discuss further in Chapter 6). Key management is very simple for session-based encryption. Since there is no need to store the key, it is generated, used, and immediately destroyed.

Choosing the Appropriate Cryptographic Tools

Now that we have introduced the tools of the trade, let's discuss when each tool is most appropriate. Table 2-6 summarizes the security services addressed by cryptography.

Using Stream Ciphers

Due to their relative simplicity, stream ciphers are easily implemented in software. They require much less code than typical block ciphers, resulting in a smaller code footprint, as well as a reduction of the time spent on development and quality assurance testing. Another effect of their simplicity is enhanced performance. Stream ciphers are ideal for encrypting session-based network traffic, as they are much faster than block ciphers.

Table 2-6 Popular Security Services and Their Corresponding Cryptographic Mechanisms

Security Service	Security Mechanisms
Confidentiality	Symmetric-key and public-key encryption
Integrity	Message digests (when combined with a shared secret or encryption)
Identity authentication	Digital signatures—proof of private key possession ("something you have")
Data origin authentication	HMACs and digital signatures
Nonrepudiation	Digital signatures

NOTE:
For most modern stream ciphers, key size does not affect raw encryption and decryption rates, so security can be increased without long-term performance penalties. However, key setup time increases with key size resulting in longer initialization times.

Synchronous stream ciphers work well in situations where error propagation is unacceptable, and where external keystream synchronization mechanisms are available. Self-synchronizing stream ciphers are ideal in the absence of external synchronization, but should be restricted to communication channels with very low error rates, or to applications that are not sensitive to error propagation.

REMEMBER:
When using a stream cipher for session-based encryption, take care to avoid keystreams that are shorter than the length of the data to be encrypted.

Data-modification attacks are easily detected when synchronous stream ciphers are used, because such attacks cause loss of synchronization. Self-synchronous stream ciphers, on the other hand, automatically recover from synchronization errors after receiving a certain number of ciphertext bits. As a result, they are susceptible to block-replay and other data-modification attacks. In the absence of additional integrity checks, it is best to use synchronous stream ciphers.

One-time pads provide an extremely high level of security, but are only appropriate for short messages, due to the difficulty involved in managing very long keystreams. One-time pads are more of academic interest than they are practical.

Using Block Ciphers

Block ciphers work well for store-and-forward applications and for encrypting persistent data. They are much less sensitive to key reusage than stream ciphers are, so a single key can be used to encrypt multiple data sets. This reduces some of the headaches associated with key management. Certain block cipher modes emulate synchronous and self-synchronizing stream cipher behavior, but are rarely as fast as stream ciphers of equivalent functionality.

Choosing a Block Cipher Mode

ECB mode is well suited for high-performance applications that encrypt short data sets, such as encryption keys, or data not exhibiting plaintext patterns. ECB operation is easily parallelizable for both encryption and decryption, thus allowing simultaneous processing of plaintext and ciphertext blocks.

TIP:
As we will discuss in Chapter 8, compression algorithms remove redundancy, and may make ECB mode viable for encrypting larger data sets.

CBC, CFB, and OFB modes conceal plaintext patterns by incorporating a feedback mechanism. Each mode requires an IV that should be unique for every message. Feedback limits parallelization, and it is difficult to perform simultaneous encryption operations when using CBC or CFB modes. Decryption operations, on the other hand, can be easily parallelized. Neither encryption nor decryption operations can be parallelized in OFB mode, so no benefit is gained from OFB in a multiprocessor environment.

REMEMBER:
ECB mode does not conceal plaintext patterns.

CBC is probably the best general-purpose mode, but it must encrypt and decrypt entire blocks at a time. While this poses no problem for encrypting files, databases, or e-mail messages, CBC does not work well with applications that transmit data in chunks smaller than the block length. CFB mode is better suited to this type of environment. 8-bit CFB works well for terminal applications that transmit single characters (i.e., keystrokes) at a time.

TIP:
CBC is a good general-purpose block cipher mode. CFB mode works well for applications that frequently transmit data units smaller than the block size, such as single characters.

ECB and CBC modes require padding, so their ciphertext output is sometimes longer than their plaintext input. This may pose a problem when input and output must be written to the same memory location. For example, suppose you are using a dedicated hardware register with a fixed length. Whenever the plaintext does not end on a block boundary, padding is necessary. If the plaintext input already fills the buffer, the addition of padding will result in an overflow when you attempt to write ciphertext back to the register. CFB and OFB modes cause no such difficulty; they simulate stream ciphers, so their input and output are of equal length.

OFB mode has no error propagation. This makes it suitable for applications such as streaming voice and video, where error propagation leads to noticeable degradation in signal quality. ECB, CBC, and CFB modes can still be used with these applications, but should be coupled with a reliable network transport layer, such as the Transmission Control Protocol (TCP). Reliable delivery protocols detect errors and retransmit corrupted data. Keep in mind, however, that data retransmission introduces delays that might also result in signal degradation for streaming applications. (See Appendix A for a discussion of TCP and UDP. See Chapter 6 for a discussion of reliable and unreliable transport protocols.)

TIP:
Lack of error propagation makes OFB mode suitable for streaming multimedia applications, even over unreliable network transport layers, such as the User Datagram Protocol.

Using Message Digests

When combined with encryption or shared secret keys, message digests provide message integrity and data-origin authentication. A digest is a unique, condensed representation of a larger data set, which may be appropriate when only a representation is required (for example, when you are storing and comparing hashed password values, or generating digital signatures).

Using Public-Key Algorithms

Computationally intensive public-key algorithms are not appropriate for encryption of large data sets. Their usage should be limited to key exchange, digital signature generation and verification, and the creation of digital envelopes. Operations using private keys are frequently the slowest of all.

CAUTION:
Digital signature generation requires encryption with a private key, so you should limit their use on devices with slow processors and in environments with stringent timing requirements.

Keep in mind the following when choosing between digital signatures and HMACs for origin authentication:

- Due to their use of shared secret keys, HMACs allow only the original recipient to verify the authenticity and integrity of a message. Persistent digital signatures can be verified by anyone with access to the signer's public key.
- Digital signatures require a slow public-key operation, while HMACs use a very efficient digest operation.

NOTE:
PKCS #8 and PKCS #12 address secure storage of private keys.

Interoperability Notes

When developing open systems, choose standardized algorithms, such as DES, 3DES, AES, and RSA. These algorithms are widely available from many vendors and, with the exception of AES, have been subjected to years of public scrutiny. You may wish to include support for proprietary or patented algorithms (such as RC5 and RC6) for added security or performance, but make them optional. RC4 may be an exception to this rule of thumb. While RC4 is still a trade secret of RSA Security, Inc., it is the most commonly used stream cipher, and is widely supported by many applications and security products.

How Secure Is Too Secure?

There will be times when you must balance the level of security in your organization against implementation costs and system performance. Some general guidelines exist (Schneier, 1996) for choosing the appropriate level of cryptographic security:

- In general, the cost of securing your data should not exceed the perceived value of that data. (Keep in mind that the value of most data depreciates over time.)
- The amount of time required to decrypt your data (via cryptanalysis or an exhaustive key search) should exceed the amount of time for which the data must remain secure. In most cases, this guideline pertains directly to the time required for an exhaustive key search, so choose keys of sufficient length. Avoid excessively long keys; they may result in slow performance.
- The amount of data encrypted with a single key should be less than would be required to decrypt the data through a cryptanalytic attack. This is especially true for stream ciphers whose keystreams may repeat, and for block ciphers running in modes (such as ECB) that do not conceal plaintext patterns.

Here are two additional points worth considering:

- Key lengths for symmetric-key and public-key algorithms cannot be directly compared. When choosing a key for a symmetric cipher (such as RC4, RC5, RC6, AES, and Blowfish), keep in mind that key lengths of 56 bits and lower are considered very weak, and that lengths of 128 bits and higher are considered strong. For public-key algorithms (such as DH and RSA), a 512-bit key length is considered weak, a 768-bit key may be acceptable for low-security applications on slower processors, and 1024-bit keys and higher are recommended for secure applications. Remember that ECC key sizes are much smaller than those for RSA and DH. Key sizes that are too large can result in unacceptably slow performance for some algorithms.

- Loosely speaking, algorithms using larger block sizes and more encryption rounds are more resistant to cryptanalysis. For algorithms that allow you to adjust the round count, increasing the number of rounds improves security but reduces performance. For RC5, Ron Rivest suggests using no less than 12 rounds. This may be adequate for applications requiring higher performance; however, a round count of 16 is probably a better choice for most other applications.

CAUTION:
Algorithms using 56-bit keys are no longer considered secure. Distributed.Net and the Electronic Frontier Foundation (EFF) have successfully cracked DES in 22 hours and 15 minutes. Using a worldwide network of nearly 100,000 PCs and a specially designed machine for cracking DES, known as "Deep Crack," they exhaustively searched the keyspace at a rate of 245 billion DES keys per second (RSA Security, 1999).

References

Burnett, Steve, and Stephen Paine. 2001. *RSA Security's Official Guide to Cryptography*. New York: McGraw-Hill.

Davies, D. W., and G. I. Parkin. 1983. "The Average Cycle Size of the Key Stream in Output Feedback Encipherment." *Advances in Cryptology—Crypto '82*. New York: Plenum Press.

Housley, R. June, 1999. *RFC 2630: Cryptographic Message Syntax*. Internet Engineering Task Force.

Housley, R., W. Ford, W. Polk, and D. Solo. January, 1999. *RFC 2459: Internet X.509 Public Key Infrastructure Certificate and CRL Profile*. Internet Engineering Task Force.

Kaliski, Burton S., Jr. November, 1993. *A Layman's Guide to a Subset of ASN.1, BER, and DER*. RSA Laboratories.

Krawczyk, H., M. Bellare, and R. Canetti. February, 1997. *RFC 2104: HMAC: Keyed-Hashing for Message Authentication*. Internet Engineering Task Force.

Menezes, A., P. van Oorschot, and S. Vanstone. 1996. *Handbook of Applied Cryptography*. Boca Raton: CRC Press.

RSA Laboratories. May, 2000. *Frequently Asked Questions About Today's Cryptography, Version 4.1*.

—5 January, 2001. *PKCS #1 v2.1: RSA Cryptography Standard.*

—1 November, 1993. *PKCS #3 v1.4: Diffie-Hellman Key Agreement Standard.*

—25 March, 1999. *PKCS #5 v2.0: Password-Based Cryptography Standard.*

—1 November, 1993. *PKCS #6 v1.5: Extended-Certificate Syntax Standard.*

—1 November, 1993. *PKCS #7 v1.5: Cryptographic Message Syntax Standard.*

—1 November, 1993. *PKCS #8 v1.2: Private-Key Information Syntax Standard.*

—25 February, 2000. *PKCS #9 v2.0: Selected Object Classes and Attribute Types.*

—26 May, 2000. *PKCS #10 v1.7: Certification Request Syntax Standard.*

—December, 1999. *PKCS #11 v2.10: Cryptographic Token Interface Standard.*

—24 June, 1999. *PKCS #12 v1.0: Personal Information Exchange Syntax.*

—6 June, 2000. *PKCS #15 v1.1: Cryptographic Token Information Syntax Standard.*

RSA Security, Inc. 19 January, 1999. "RSA Code-Breaking Contest Again Won by Distributed.Net and Electronic Frontier Foundation (EFF)." *RSA Security News.* 11 September, 2001 (http://www.rsasecurity.com/news/pr/990119-1.html).

Schneier, Bruce. 1996. *Applied Cryptography*. New York: John Wiley and Sons, Inc.

—6 January, 2001. PKCS #1 v2.1: RSA Cryptography Standard.

—1 November, 1993. PKCS #3 v1.4: Diffie-Hellman Key-Agreement Standard.

—25 March, 1999. PKCS #5 v2.0: Password-Based Cryptography Standard.

—1 November, 1993. PKCS #6 v1.5: Extended-Certificate Syntax Standard.

—1 November, 1993. PKCS #7 v1.5: Cryptographic Message Syntax Standard.

—1 November, 1993. PKCS #8 v1.2: Private-Key Information Syntax Standard.

[illegible]

[illegible]

[illegible]

Chapter 3

The Need for Security: Network Threats and Countermeasures

The growing dependency of commerce, communication, and entertainment on computer networks emphasizes the need for reliable and robust security. This is especially true for companies whose livelihood relies on secure electronic services and transactions. Network downtime, stolen data (financial records, proprietary and customer-related information, and so on), and defacement of websites can result in financial loss and undermine the public's confidence in a company's ability to conduct business. In the case of service providers, end users may be hesitant to use a service that they view as insecure. To maintain customer loyalty, network design engineers, security architects, and administrators face many challenges in protecting consumer data and network resources. These tasks will only become more difficult as network environments grow more diverse, and more necessary as we come to rely on public networks for almost every facet of daily life. The first step to overcoming these challenges is to understand your enemies and their tactics.

In this chapter, we'll discuss the threats and vulnerabilities that exist on public networks. What we cover here is by no means a comprehensive list, but it describes many of the avenues used by attackers to gain unauthorized access to resources and to disrupt network service. We begin with an overview of the "Five W's"—who, what, why, when, and where—addressing common motivations for attack, when attackers are likely to strike, and from where. Next, we introduce the primary categories of attack and their corresponding countermeasures.

TCP/IP, the most commonly used network protocol suite on public broadband networks, is inherently insecure, so we conclude the chapter by tackling a number of vulnerabilities specific to the TCP/IP suite.

As you read this chapter, keep in mind that a security infrastructure is only as strong as its weakest link. Skilled attackers can quickly locate the weak link in a security infrastructure, and they will concentrate their efforts on that link because as it gives them the highest possibility of success. For instance, in Chapter 2, we discussed the importance of strong pseudorandom number generation; the security of many cryptographic operations relies on robust PRNGs. Using a predictable or biased PRNG to create encryption keys or keystreams weakens even the most secure of encryption algorithms.

Security design has many such subtleties and interdependencies, and care must be taken to ensure that all components of a security solution meet or exceed the minimum requirements defined for the solution as a whole. Employing multiple security mechanisms to protect crucial resources can help alleviate unforeseen weaknesses. However, use discretion—redundancy often introduces undue processing strain, and may slow network performance.

Who, What, and Why? Attackers and Their Motivations

What is a hacker? Many people perceive hackers as withdrawn and rebellious teenagers bent on stealing private information and defacing websites as practical jokes or to demonstrate their abilities to friends. Quite to the contrary, many hackers have no intention of disturbing computer systems or destroying data (nor must they be under the age of 18). The term *hacker* originally described an individual who enjoyed learning about and playing with all things related to computers. The earliest hackers were often skilled with electronics and understood the operation of simple computers, inside and out. Their intentions were not malicious, and hackers circumvented security measures only when those measures stood in the way of exploring a computer system.

Today, however, rationales vary greatly from one hacker to the next. Most still view hacking as an intense game of exploration. They find the thrill of the chase irresistible, and in lieu of the excitement of discovery, they often overlook the ramifications of being caught. A hacker with a conscience may seek redemption by notifying a computer system's owner of vulnerabilities once those vulnerabilities have been exploited. Even after gaining unauthorized access, the hacker may feel as though he or she has provided a beneficial service to the victim. Other so-called ethical hackers won't break into a system without first obtaining permission and offering to immediately share their findings with the owner. Some "hacker" consultants, in fact, make a living charging for this type of service.

While most hackers are simply in search of an adrenaline rush, others claim to live and act by a code of ethics. Steven Levy originally described the Hacker Ethic in his book *Hackers: Heroes of the Computer Revolution* (Levy, 1994). The philosophy consists of six idealistic tenets, many of which still apply to current-day hackers:

1. Access to computers—and anything which might teach you something about the way the world works—should be unlimited and total.
2. All information should be free.
3. Mistrust Authority—Promote Decentralization.
4. Hackers should be judged by their hacking, not bogus criteria such as degrees, age, race, or position.
5. You can create art and beauty on a computer.
6. Computers can change your life for the better.

The objectives of hackers range from harmless to destructive, and their skill levels are equally diverse. Although the intentions of most hackers may be relatively benign, there are many vandals, thieves, and spies with far more devious goals. These individuals, called *crackers*, might not think twice about destroying data or halting a network if it helps them achieve their final objective or conceal their whereabouts. As noted by Richard Power in his book *Tangled Web: Tales of Digital Crime from the Shadows of Cyberspace* (Power, 2000), many people insist on the distinction between "noble" hackers and their less-honorable cousins, crackers:

> The term *hackers*, of course, has become somewhat hackneyed. Some in cyberculture distinguish between *hackers* and *crackers*. The politically correct use refers to those who break in simply to explore as *hackers* and to those who break into systems to steal or destroy information as *crackers*. But even those hackers who break in just to explore are guilty of at least breaking and entering.

Despite the noble intentions of early hackers, the motives of current-day hackers are often questionable at best. The already fuzzy line separating hackers and crackers blurs further with each publicized attack. The means employed by the two groups are often indistinguishable, and even the most humane of hackers can cause accidental destruction. Table 3-1 lists some of the common motivations for attacking computer systems and networks. Table 3-2 categorizes various groups of attackers according to skill level and associated threat.

Reason	Description
Joy riding and bragging rights	The attacker views hacking as an exciting game of discovery and exploration, and does not intentionally wish to damage the property of others; the attacker wishes to demonstrate technical prowess to his or her peers. This group includes casual hackers and many novices.
Revenge	The attacker feels as though he or she has been treated unfairly, and is seeking retribution or financial compensation. The most common members of this group are disgruntled employees.
Theft	The attacker is in search of personal gain through unauthorized financial transactions or theft of intellectual property, proprietary information, user data, or network services. This group includes career criminals, hired professionals, and organized crime rings.
Politics	The attacker defaces a website, destroys data or computer systems, disrupts service, or reveals confidential information (for example, sensitive financial data, personnel records, or medical histories) to further a political agenda. This group includes political activists and members of terrorist organizations.
Corporate espionage	The attacker is hired to gather information on a rival company (for instance, information regarding upcoming mergers, earnings reports, or unfavorable transactions) that, when leaked to financial markets, may affect stock prices or sabotage the company's public image. The attacker may also be hired to steal intellectual property or trade secrets. This group includes hired professionals.
Information warfare	The attacker is a member of a military organization focusing on intelligence gathering, dissemination of misinformation, military or corporate sabotage, and/or destruction of commercial and communication infrastructures. This group consists primarily of highly trained and experienced government spies, intelligence officers, and members of terrorist organizations.

Table 3-1 Common Motivations for Computer Attacks

CAUTION:

Although technically not an attack, legitimate users, such as employees and paying subscribers, may unintentionally destroy data or disrupt service. This is especially true when systems administrators give inquisitive users more access than is necessary for doing their work.

Group	Details	Skill Level	Threat
Casual hackers and crackers	Varying levels of skill, ranging from beginners to seasoned veterans. Often rely on widely available automated tools to locate and exploit weaknesses.	Low–high	Low–moderate
Employees and insiders	Direct access to internal resources. May have detailed knowledge of a company's computer systems and security mechanisms.	Low–high	Moderate–high
Thieves and career criminals	May be highly skilled at evading discovery and capture. Detailed understanding of financial and accounting systems.	Moderate–high	High
Corporate spies and other hired professionals	Proven level of skill. Often insiders with direct access to confidential information.	Moderate–very high	High–very high
Foreign governments and terrorist organizations	Highly trained with proven level of skill. Focused on intelligence gathering and effective information warfare tactics.	Very high	Very high

Table 3-2 Categories of Computer Attackers

The employees/insiders category of attackers deserves special consideration. Companies are often too trusting of their employees, believing that they would never stoop so low as to attack their employer. This "Not Us" attitude lulls some companies into a false sense of security. As a result, they neglect internal threats altogether until those threats become reality. By that time, severe and irreversible damage may already have been done.

Internal attacks are rarely publicized to the extent of their Internet-based counterparts, and fear of wide-open Internet connections causes many organizations to invest heavily in perimeter defenses. They may shelve efforts at promoting stronger authentication and encryption within the company in lieu of firewalls and VPN equipment, neither of which prevent an employee from altering, stealing, or disclosing private information stored on a company PC. Perimeter defenses are highly effective at deflecting external threats, yet they offer little protection against disgruntled employees, corporate spies, or rogue administrators with trusted access to internal resources. Insider attacks are dangerous for several reasons:

- Insiders do not have to overcome perimeter defenses.
- Insiders have less restricted access to internal resources than outsiders do. In many instances, insider access is complete and unhindered.
- Insiders often have detailed knowledge of internal networks and the locations of valuable resources.
- Insiders can easily determine when administrators and other security personnel are out of the office.

When? "The Network Administrator Went Home Hours Ago..."

Traditionally, experienced attackers launch attacks when their activities are least likely to be noticed. Having exploited a vulnerability to gain access to a system, a cracker will introduce new weaknesses or backdoor utilities, which he or she will later use to gain access after the original hole has been discovered and patched. Intruders cover their tracks by editing system log files and concealing hacking tools; and to prevent others from gaining access, they may even patch the holes that they've originally exploited. All of this takes time, and skilled security personnel can easily spot such activity when monitoring user activity and system processes in real time. However, many organizations cannot afford—or see no need—to staff security administrators around the clock for monitoring network operations. Knowing this, intruders tend to initiate attacks late at night or in the early morning hours. Holidays provide additional opportunities for attack. (We'll discuss an example of this later in the chapter, in the section entitled "The Mitnik Attack.") In many cases, intruder activity is logged and discovered the following morning, but the damage may already have been done.

Although the attack itself may last only seconds, an intruder will often spend many weeks mapping out the network and searching for vulnerabilities. These activities may include scanning hosts and routers for exploitable services and known vulnerabilities, determining the topology of the network, uncovering trust relationships between networked hosts, and searching the Web for subtle clues that may assist in the attack. Such activities seldom affect system operation, and may occur during peak operational times when it is difficult to isolate scanning activity amidst dozens of valid user sessions. Although these activities are harmless by themselves, they usually indicate an active attack in the making.

For the majority of broadband consumers and service providers, the cold reality is that an attacker might strike at any time. The number of home users and small businesses with always-on broadband connections provides many opportunities for attackers. Intruders know that the average consumer lacks security expertise and may never detect an attack regardless of when it occurs. Many service providers are ill equipped to detect or prevent service theft on their access networks; such theft may go unnoticed for months or years. Service disruption poses another significant threat to providers. Attackers wishing to disrupt network service will strike during times of peak network usage, when their attacks will be most effective.

Where? The Internet's a Big Place!

Historically, the attacks experienced by most organizations originate from within the confines of their own walls. Insiders have greater access to internal resources than strangers do, and they are more likely to know where those resources are located and what security measures are in place to protect them. Even so, external attacks are becoming more and more frequent, thanks to the popularity and ubiquity of the Internet.

In the past, remote network access via the Internet was not always available. Before an attacker could strike, he or she had to first discover the telephone number of a modem attached to the internal network. Without a modem, launching an attack required physical access to a computer. Today, most companies and gov ernment agencies have direct connections to the Internet, and many of these organizations lack proper perimeter security for keeping out intruders. As a result, Internet-based attacks have become very popular and will continue to become more common as our dependence on network services grows.

Sophisticated attackers will launder their connections through many different computer systems, making their whereabouts extremely difficult to trace. Attacks that seem to originate from the local university may, in fact, stem from a foreign country. Over the Internet, an attacker can strike from next door, across the country, or the other side of the world with equal ease. Signals traveling at the speed of light traverse the globe in a matter of milliseconds, and Internet users rarely incur long distance charges. Even when intrusions are detected, it is difficult for a single organization to trace an attack back to its source. A successful trace requires coordination of each group involved. The relative ease of Internet-based attacks highlights the need for preventative security measures and limits the effectiveness of reactive techniques.

Broadband Access vs. Dial-up Access

Although this book focuses primarily on infrastructure security, the effects of broadband and dial-up access on end-user security warrant a brief discussion. Broadband access provides much higher transfer rates than dial-up connections do. An attacker who has compromised a host attached to a broadband connection can quickly download data. The effects of faster downloads are twofold:

- In the same amount of time required by a dial-up connection, an intruder can download much more data.

- When looking for a particular piece of sensitive material, an intruder requires much less time to extract data and, consequently, may be less easily discovered.

Always-on connections simplify an attacker's task by making it easier to locate a compromised system. Not only is the connection active around the clock, the IP address of an always-on connection seldom changes. Users who leave their computers powered on and attached to a broadband connection risk attack at any time. Since dial-up users rarely remain connected for longer than necessary, they are less susceptible to attack. In addition, the IP address assigned by an *Internet service provider (ISP)* changes from one dial-up connection to the next. Table 3-3 compares the characteristics of dial-up and broadband connections.

Categorizing Common Attacks

Now that we have established the Five W's, let's turn our attention toward *how* attackers compromise computer systems. There are many tools and techniques available to attackers who wish to obtain unauthorized access to network resources. These tools include utilities for monitoring traffic and system operation, and automated hacking tools for gaining access to a host, taking control of communication sessions, and disabling a host or an entire network. A small percentage of attackers demonstrate great skill, creativity, and agility, allowing them to surmount security obstacles on-the-fly. However, a far greater percentage of novice attackers rely on automated tools to discover and exploit security vulnerabilities. Most network-based attacks fall into one or more of the following categories:

- Eavesdropping
- Impersonation
- Denial of service (DoS)
- Data modification

Characteristic	Dial-Up	Broadband
Connection type	Dial-on-demand	Always-on
IP address	Changes on each call.	Static or infrequently changing
Relative connection speed	Low	High
Remote control potential	Computer must be dialed in to control the system remotely.	Computer is always connected, so remote control can occur anytime.
ISP-provided security	Little or none	Little or none

Table 3-3 Comparison of Dial-Up and Broadband Access Services (CERT Coordination Center, 2001)

- Packet replay
- Routing

Passive Attacks vs. Active Attacks

Before we begin our discussion of specific threats and countermeasures, we should introduce the concepts of passive and active attacks as originally proposed by Stephen Kent (Kent, 1977). A *passive* attack involves capturing or monitoring transmitted data as it flows from source to destination over a network. The goal of this type of attack is to obtain information from the transmitted data without altering its contents or affecting its delivery. Such data may include proprietary or otherwise confidential information, username and password combinations, and network topology information. Most passive attacks go completely unnoticed, and are extremely difficult to detect—especially on wireless networks. For this reason, preventing passive attacks usually involves protecting data while in transit rather than detecting intruders.

An *active* attack involves the insertion, deletion, modification, or interruption of transmissions between communicating parties. Various types of active attacks include service disruption, alteration of message content while in transit, retransmission of data to gain unauthorized access or defeat security measures, and transmission of data from a host masquerading as a trusted system (to name only a few). Active attacks pinpoint weakness in services and programs, and are much more invasive than their passive counterparts. Many of these weaknesses are preventable, but many have yet to be discovered. For this reason, avoiding all active attacks on even a small network is a daunting, if not impossible, task.

NOTE:

In general, passive attacks are hard to detect, yet easy to prevent. Active attacks, on the other hand, are easier to detect, but are more difficult to prevent.

Eavesdropping

Eavesdropping is the primary form of passive attack, in which communication sessions are monitored in real time or captured for offline analysis. On shared media networks, traffic directed at a specific host is visible to all hosts within the same network *segment,* or *broadcast domain.*

NOTE:

For IP, broadcast addresses are usually addresses with the host portion set to all ones. For a Class C network address of 192.1.1.0, the broadcast address is 192.1.1.255. All hosts within the network segment will respond to messages sent to this address. (Refer to (Stevens, 1994) for a description of address classes.)

With a network monitoring utility known as a packet sniffer, it is relatively simple to view data traversing a network segment. A *packet sniffer* intercepts traffic flowing across a network adapter interface and converts it to a human-readable form (see Figure 3-1). Some monitoring utilities come complete with graphical user interfaces that greatly simplify analysis. This is somewhat analogous to a telephone wiretap, where electrical signals are converted to audio. Network administrators use packet sniffers to troubleshoot network problems and monitor traffic patterns. In general, an operating system will configure network adapters to read the contents of traffic addressed only to its own MAC address. In order for a packet-sniffing utility to capture all traffic, it must place the network adapter into promiscuous mode. Once in *promiscuous mode,* a network

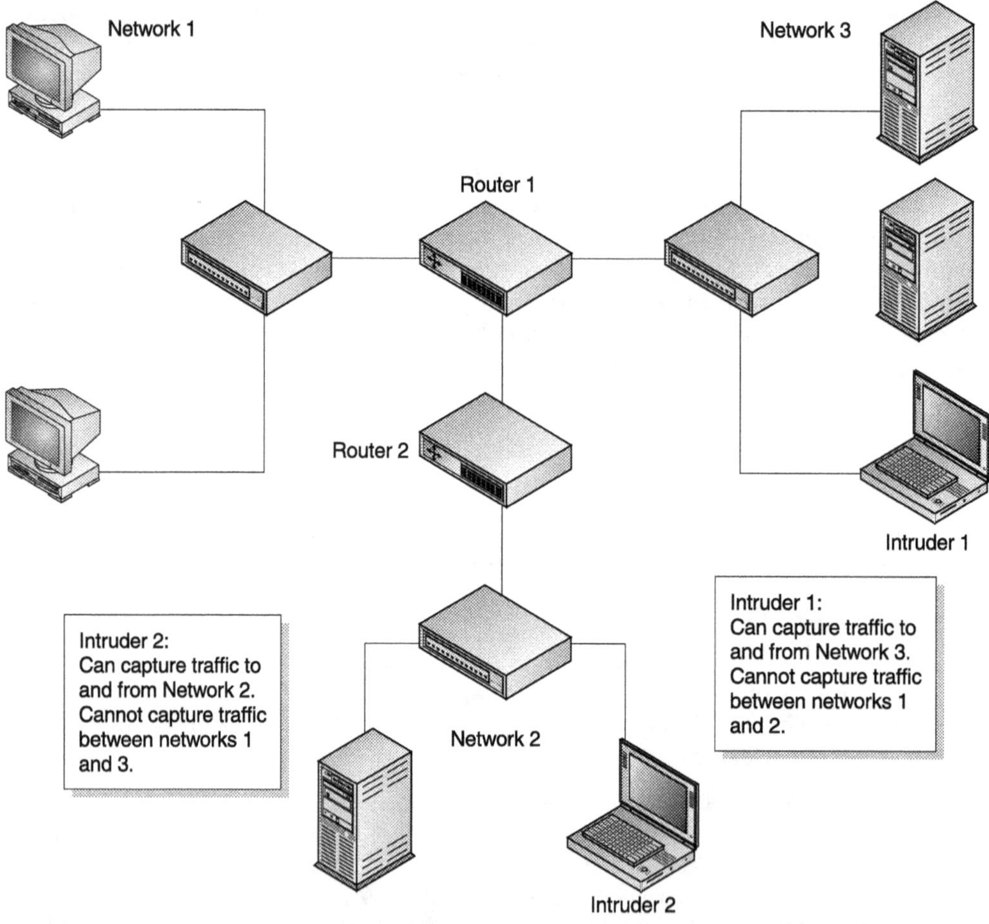
Network 1
Network 3
Router 1
Router 2
Intruder 1
Intruder 1:
Can capture traffic to
and from Network 3.
Cannot capture traffic
between networks 1
and 2.
Intruder 2:
Can capture traffic to
and from Network 2.
Cannot capture traffic
between networks 1
and 3.
Network 2
Intruder 2

Figure 3-1 It is simple for an intruder to attach a packet-sniffing host to a hub or concentrator and capture traffic on a network segment.

adapter intercepts and reads all data packets flowing across the network segment to which the host is attached. These packets need not originate or terminate on the segment; they just need to pass through the segment en route to their final destination.

Most packet sniffers allow the user to filter traffic based on the source and destination address, port number, protocol, and upper-layer application (such as FTP, Telnet, HTTP or SMTP). Captured data can be stored and analyzed for sensitive information. For example, many applications, such as FTP and Telnet, transmit user credentials as plaintext. Using a packet sniffer, an intruder can capture an FTP session and reconstruct the session later to obtain the username and password combination that was used to authenticate to the FTP server. It is not necessary to capture an entire session because user credentials are generally exchanged early on. Some of the more advanced packet-sniffing utilities reconstruct entire HTTP sessions, allowing an attacker to view complete web pages visited by another user. Figure 3-2 illustrates the output of a popular Windows-based packet-sniffing utility after it has captured a Telnet authentication sequence. Notice the unencrypted username and password.

Countermeasures for Eavesdropping Attacks

Eavesdropping attacks are very difficult to detect. They do not affect the integrity of transmitted data, nor do they interfere with network operation. Without viewing the processes running on a host, the only telltale sign of a packet-sniffing utility is a network adapter running in promiscuous mode. During normal operation, network adapters rarely run in promiscuous mode; it usually indicates the use of a packet sniffer and should trigger an immediate alarm. Other eavesdropping techniques, such as wiretapping and capturing electromagnetic (EM) emissions, do not make use of a promiscuous mode network adapter, so are not detectable by this type of alarm.

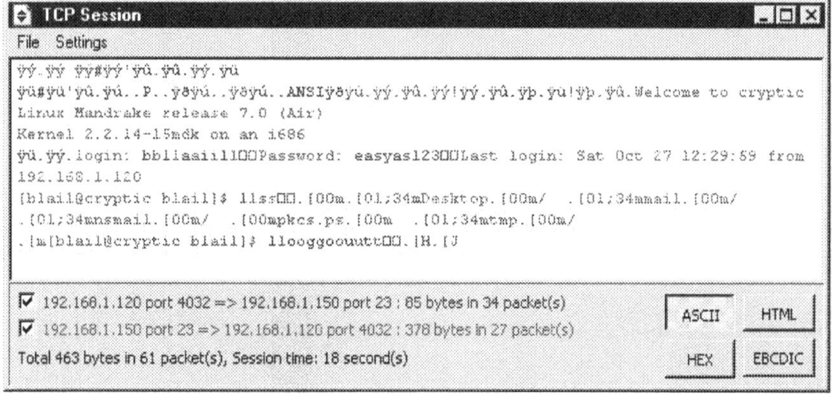

Figure 3-2 Output from a popular packet-sniffing utility, CommView from Tamsoft, Inc. demonstrates how easy it is for an attacker to capture unencrypted usernames and passwords during an authentication session.

Although difficult to detect, eavesdropping is simple to prevent. Confidentiality services provide protection against eavesdropping attacks. Even when an attacker succeeds in capturing data, it is very difficult for an attacker to glean valuable information if that data is encrypted (see Figure 3-3). A determined individual might perform offline cryptanalysis of the data and, depending on the strength of the encryption scheme used, might derive the key originally used to encrypt the data. However, with properly chosen algorithms and key lengths, cryptanalysis becomes extremely costly and time consuming. (In this context, "costly" refers to the amount of processing power required to derive the encryption key or partially decrypt the data.)

REMEMBER:

When choosing cryptographic security for communication networks, balancing security and performance is important. Just as choosing too weak an encryption scheme or too short a key makes data susceptible to cryptanalysis, choosing too slow an algorithm or too long a key may adversely affect network performance. This is especially true for real-time voice and video communication.

In addition to protecting transmitted data, confidentiality services (in conjunction with traffic padding and routing control) help prevent the analysis of traffic patterns over a network. Traffic patterns provide subtle clues as to the

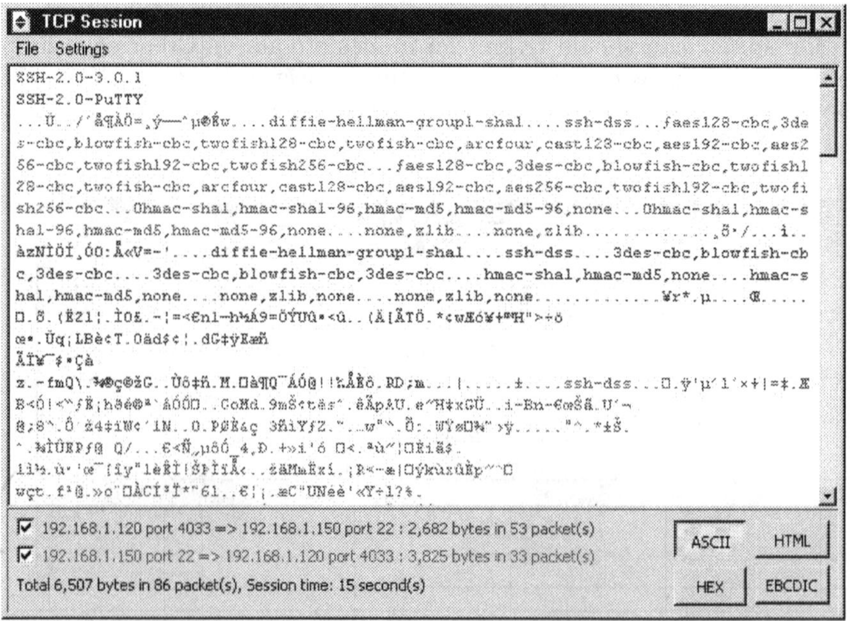

Figure 3-3 The same authentication session as shown in Figure 3-2, with encryption applied

nature of communication between two entities. For example, if an intruder notices large amounts of traffic between two companies at regular intervals, he or she may infer that the companies are secretly conducting business. The leakage of such information may result in fluctuating stock prices due to speculation. Some security protocols, such as IP Security (IPSec), provide tunneling modes that mask the true source and destination addresses of data packets. (We will discuss IPSec in more detail in Chapter 6.)

To capture traffic on a wired network, an attacker must remotely compromise a networked host, or physically attach a computer to a network. Firewalls and host hardening may prevent an attacker from remotely installing a packet-sniffing utility, but neither will stop an attacker from walking into an office and attaching a notebook computer to an unused network jack. Many businesses implement physical security mechanisms ensuring that only employees have access to company facilities. Nonetheless, attackers posing as maintenance or janitorial staff may be granted access to restricted areas. In addition to encrypting network traffic, limiting physical access to network jacks and disabling unused jacks makes it much more difficult for passersby to snoop around the network.

The use of Layer 2 switching devices reduces the exposure of sensitive information to eavesdropping. Unlike the hubs and concentrators employed by shared media networks, switches do not replicate traffic to all ports unless the traffic is addressed to the broadcast address. Instead, a switch directs traffic from one port directly to the port to which the destination host or router is attached. An attacker attempting to capture data by running a packet sniffer on one of the switch ports will see only those packets directed to or from his computer. Keep in mind that traffic may still be captured if it traverses a shared media network on the way to its destination. Although switches reduce the effectiveness of eavesdropping attacks, they are not dedicated security devices. Many switches include traffic monitoring modes for capturing all traffic that passes through the switch. This functionality is intended for network performance monitoring and troubleshooting, but it can also be exploited by an attacker.

The use of fiber optic cabling limits the capturing of EM emissions by eliminating EM radiation outside the fiber strand. In addition, it is extremely difficult to tap into a fiber strand without introducing significant signal degradation. Although costly, fiber optics provides a viable solution for securing high-bandwidth intercampus or building-to-building network segments.

Impersonation

The goal of *impersonation* attacks is to bypass authentication systems and exploit trust relationships. Impersonation involves supplying false credentials to gain access to a protected system or resource, and may include any of the following:

- Guessing or obtaining a transmitted username and password combination.
- Stealing a physical authentication mechanism, such as a token or smartcard.

- Determining the secret key shared by two communicating parties. (For information on secret keys, review the "HMACs" discussion in Chapter 2.)
- Stealing the private key of a user or device, and using the key to decrypt data or generate digital signatures.
- Substituting credentials in a public-key cryptosystem. This is known as a *man-in-the-middle* attack (and we'll discuss this type of attack later in the chapter, in the section entitled "Man-in-the-Middle Attacks").
- Capturing and retransmitting authentication information. (We'll address these activities in the section entitled "Packet Replay.")
- Falsifying the source address of a data packet. This technique is known as *address spoofing* (and we'll discuss it in the section entitled "Address Spoofing").

Simple password-based authentication systems provide little protection against a determined attacker. As mentioned in the preceding section, it is very simple for an attacker to obtain username and password combinations by using a packet sniffer, and to later supply those credentials to authenticate to a valid user's account. Password guessing is the simplest form of impersonation attack. An unsophisticated attacker may gain access simply by guessing passwords at random until finding a match. While this technique is not very efficient, the use of poorly selected passwords greatly improves the chances of random guessing. Users often select passwords based on the names of family members or pets, interests, hobbies, or words found in a dictionary.

CAUTION:

There is no substitute for a well-chosen password. Automated tools exist for systematically trying every password combination in sequence until a valid password is found. For example, the program might try the following sequence: a, b, ...z, aa, ab, ..., az, ba, bb, ...bz, and so on. This type of attack, known as a brute force attack, will eventually succeed, given enough time and computational resources. This same type of attack can be used to determine encryption keys. Dictionary attacks are similar to brute force attacks, except that the password-guessing utility derives passwords from a precomputed word list (for example, the dictionary file included with most Unix operating systems). These types of attacks have a surprisingly high success rate.

Many network devices contain private keys, which they use to authenticate users to other network elements via public-key cryptography. In some cases, network service providers associate these private keys with the accounts of paying subscribers. An attacker who gains access to one of these private keys, and has some familiarity with programming electronic devices, may create a *clone*. To the service provider, cloned devices are indistinguishable from the original.

Since they can be used to steal service via the accounts of valid subscribers, cloned devices pose a significant financial threat to service providers.

Countermeasures for Impersonation Attacks

There are a number of basic precautions that help combat impersonation attacks. Most of these measures center on protecting user and device credentials:

- *Enforce proper password policy.* This is your first line of defense. Requiring that passwords have minimum lengths of six to eight characters and incorporate nonalphabetic characters reduces the effectiveness of brute force password-cracking techniques. Frequent password updates and the use of account *lockouts* (disabling an account after a certain number of invalid login attempts) also improve security. Bear in mind, however, that too many restrictions hinder usability, and as a result, end users may attempt to circumvent security measures. At the very least, they'll gripe about them.
- *Encrypt traffic flowing across the network.* It is especially important to protect authentication sessions, thus preventing intruders from capturing plaintext username and password combinations.
- *Closely protect keying information, such as shared secrets and private keys.* Cryptographic mechanisms, such as HMACs and digital signatures, provide suitable authentication, but only when keys are closely guarded. As we discussed in Chapter 2, PKCS #5 password-based encryption can be used to secure keying information. PKCS #8 and PKCS #12 deal specifically with safely storing and transporting private key information. *Hardware security modules (HSMs)* provide an extremely high level of key protection. HSMs are dedicated devices for generating keys and performing cryptographic operations in hardware. An attacker cannot physically extract encryption or signing keys that have been generated within an HSM.
- *Choose appropriate key lengths.* In addition to decrypting confidential data, an attacker who determines a shared secret value or private key by using a brute force attack can masquerade as another user by generating valid message authentication codes or digital signatures. Longer keys are less susceptible to brute force attacks; but, as we have mentioned, they may slow the performance of some cryptographic algorithms.

We will discuss countermeasures for specific impersonation attacks (for example, address-spoofing and man-in-the-middle attacks) later in this chapter.

Denial of Service

Denial-of-service (DoS) attacks slow system performance and service response times, and may temporarily render a system or service completely unusable.

Typical DoS attacks work by flooding a host or device with a large number of incoming requests. As the target host attempts to respond to the queuing requests, its processor and memory utilization increase, and its response time decreases. In extreme cases, all available resources on the target host are consumed, and the system crashes. Other DoS attacks involve clogging networks by generating excessive network traffic, consuming valuable disk storage space, or physically damaging a computer's hardware or network connection. Some of the more sophisticated attacks generate very little traffic by efficiently exploiting weaknesses in specific implementations of the TCP/IP stack. Some common DoS techniques and tools are TCP SYN flooding, Ping of Death, Teardrop.c, Bonk, Snork, Land.c, Winnuke, Winfreez, and Smurf; we will discuss many of these in detail in the following sections.

DoS attacks launched from a single host can be very effective. However, their effects are multiplied when many hosts participate. A *distributed denial-of-service (DDoS)* attack is a coordinated attack, using many compromised hosts, that an intruder launches against a single target. To accomplish this task, the intruder must gain access to each host independently, and install a piece of remotely controllable software known as an *agent*. Each agent responds to a request from a special host known as a *handler*. To launch an attack, the intruder connects to the handler via a client application and issues an attack request. It is the handler's responsibility to propagate the request to each agent. Figure 3-4 demonstrates this process. In February 2000, many major e-commerce companies, including Yahoo!, eBay, CNN, Amazon, and E*Trade, reported successful DDoS attacks against their websites. Those attacks, some lasting many hours, resulted in sluggish server performance and inability to respond to customer requests. Some commonly used DDoS tools are Tribe FloodNet (TFN), Tribe FloodNet 2K (TFN2K), Trinoo, Trinity, Stacheldraht, Omega, Shaft, and Mstream.

DoS and DDoS attacks are rarely used alone. They tend to cripple the target host, making exploration of the system difficult to impossible. The typical joy rider or thief finds little gratification in completely disabling a host or service. Instead, an attacker will make use of a DoS attack in conjunction with another type of attack, such as session hijacking (which is described in the section entitled "Session Hijacking," later in this chapter).

One notable example of a DoS attack is the Internet worm created by Robert T. Morris, Jr. The worm took advantage of the Unix sendmail and finger services to gain unauthorized access to computer systems. Once on a system, the worm executed processes on the local host and attempted to gain further access to remote hosts. Each of the processes executed on the local host would, in turn, execute processes of their own. In a relatively short period, the worm utilized enough processing time to slow the host to a crawl. (For a detailed account of the Morris worm, refer to Power, 2000.)

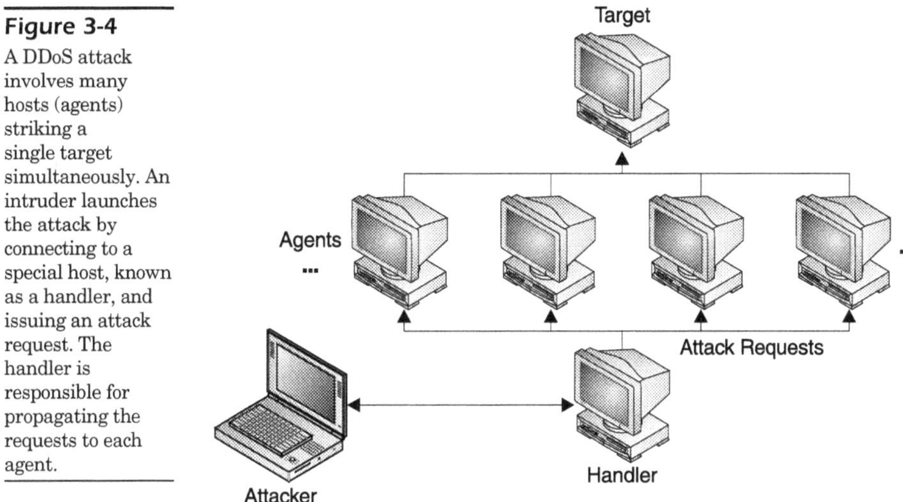

Figure 3-4
A DDoS attack involves many hosts (agents) striking a single target simultaneously. An intruder launches the attack by connecting to a special host, known as a handler, and issuing an attack request. The handler is responsible for propagating the requests to each agent.

NOTE:
One account from the University of Utah describes the effects of the Morris worm on one of their hosts. Ninety minutes following infection by the worm, the system was unusable (Power, 2000).

Although DoS attacks are infrequent, the service outages and downtime that they cause can result in tens or hundreds of thousands of dollars in lost revenue for large e-commerce companies and network service providers. In addition, the negative publicity associated with such disruption can permanently damage a company's reputation. DoS attacks are very difficult to prevent, and, in many cases, the best that can be done is to contain their effects. Throughout the remaining sections of this chapter, we will address countermeasures for many common DoS attacks.

Data Modification

Data modification involves the insertion, deletion, or alteration of stored or transmitted data. Using data modification, an attacker might change the value of an online transaction, alter the sender name or the contents of an e-mail message, modify a billing records database, or completely garble a message. By changing the contents of a command sent to a router or a switch, an attacker might reconfigure the device to reroute traffic, steal network service, or modify the quality-of-service characteristics of an existing account. An attacker may also conceal malicious activity by modifying unprotected log files.

Some data modification inadvertently results from noisy communication channels. Electrical interference or excessive traffic may introduce dropped packets or flipped bits (that is, causing ones to become zeros, or vice versa). Moreover, certain encryption modes produce a great deal of error propagation. Modification to a single bit of transmitted ciphertext may affect multiple bits of plaintext during decryption. When there are no appropriate integrity mechanisms in place to retransmit corrupted data, frequent error propagation may completely garble messages. Even a single instance of propagation can destroy critical data, such as an account number in a financial transaction.

Countermeasures for Data-Modification Attacks

Integrity mechanisms prevent the modification of stored and transmitted data from happening without detection. It is often very difficult to stop an attacker from physically modifying data, and most integrity mechanisms do not preclude modification. Instead, they make changes to the original data easy to detect. As we discussed in Chapter 2, message digests are the primary cryptographic mechanisms for providing message integrity. Recall that two messages varying by even a single bit will produce completely different message digests. Digital signatures and HMACs combine message digests with encryption or shared secret keys to prevent attackers from modifying messages without detection.

REMEMBER:

Digital signatures require a relatively slow private key encryption operation, which may make them impractical for high-performance real-time applications. HMACs are faster, but still reduce the throughput of a connection.

HMACs provide integrity only during a communication session. They are not appropriate for stored data due to their use of a shared secret key, which is usually destroyed by the sender and recipient once the data has been transmitted and received. Digital signatures base integrity on a public/private key pair, which remains after the termination of a connection. As a result, digital signatures are well suited for ensuring the integrity of stored data. We will discuss the raw performance characteristics of each of these mechanisms in Chapter 8.

When communication channels are noisy, reliable Transport layer communication protocols, such as TCP, retransmit lost or dropped packets. TCP also provides a basic level of message integrity through its use of simple checksums. For more robust integrity protection, IPSec employs cryptographically strong

HMACs for ensuring the integrity of encapsulated data and IP header information. When using unreliable transport protocols, you should choose cryptographic modes that introduce little or no error propagation. Stream ciphers, as well as block ciphers running in OFB mode, are appropriate for noisy channels.

Packet Replay

In a *packet-replay* attack, legitimate data packets are captured and retransmitted in an attempt to gain access to a network resource, or to duplicate a previous transaction. Replay attacks are most often used to exploit flaws in authentication mechanisms by mimicking captured authentication sessions. For example, some authentication schemes provide no method of distinguishing one connection attempt from another if they originate from the same source address. Knowing this, an intruder can gain access by replaying an authorized user's login sequence. Network protocols that don't employ sequence numbers (UDP, for instance) are highly susceptible to replay attacks. Figure 3-5 illustrates a typical replay attack.

Countermeasures for Packet-Replay Attacks

There are many ways to prevent packet-replay attacks. Two of the most common approaches are timestamping and sequence numbering. *Timestamping* associates

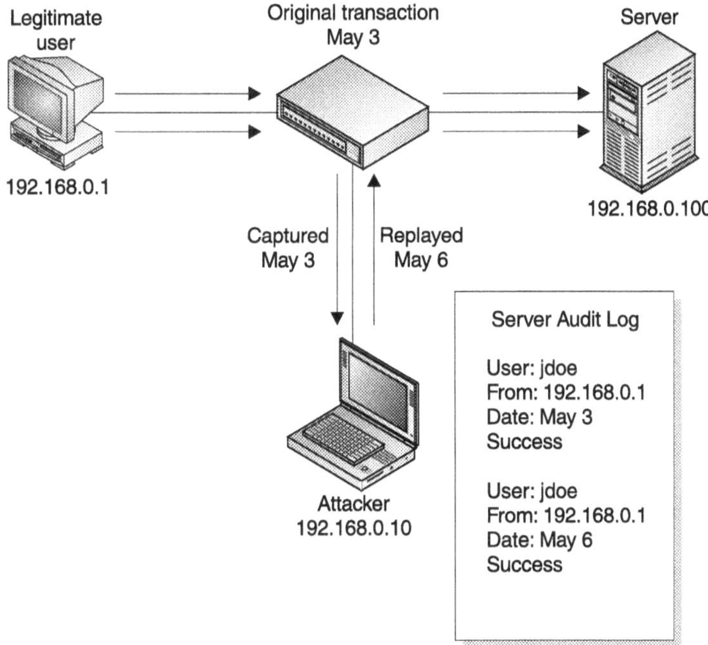

Figure 3-5 Using a packet-replay attack, an attacker can capture a transaction and later replay the captured data to duplicate the transaction.

a transaction with the exact time and date that the transaction occurred. This requires more than just attaching the current system time to a message. In order for timestamping to be effective, it must incorporate a message integrity mechanism, such as an HMAC or digital signature, to ensure that an attacker cannot modify a timestamp once it has been issued. The timestamp must also come from a common trusted source.

Reliable transmission protocols use *sequence numbering* to assist in detecting lost packets and reconstructing messages from packets that arrive out of order. This feature exists in reliable transport layer protocols, such as TCP. However, IP does not use sequence numbering, and protocols running on top of IP that do not utilize sequence numbers are prone to replay attacks. IPSec includes an anti-replay mechanism that assigns a sequence number to each transmitted packet. As each packet arrives, the receiving host tracks the sequence number and compares it with a list of valid sequence numbers. A packet containing a sequence number that has already been received, or that is outside a window of acceptable sequence numbers, is flagged and rejected.

The use of encryption often hinders replay attacks by concealing the true nature of a transaction. When encrypted, a login session may be indistinguishable from an e-mail message or online stock trade. An intruder who cannot determine the purpose of a packet might not know how to use the packet in a replay attack. However, encryption alone will not prevent replays. If the encryption key does not change from one connection to the next, an attacker may retransmit an encrypted login sequence and achieve the desired result without ever decrypting the data. An effective solution is the use of challenge/response phrases or time-synchronous values (such as RSA Security's SecurID) that change for each connection.

Routing Attacks

Routing attacks take many forms, most of which focus on rerouting network traffic or denying service. Intruders who modify routing information may configure routers to pass traffic through arbitrary network segments for monitoring. They may also delay or prevent delivery by directing traffic along slow network segments or into routing loops. A packet trapped in a *routing loop* is bounced between routers until its time-to-live expires and it is discarded. An attacker can alter or override routing information by any of the following methods:

- Logging into a router and manually reconfiguring its settings
- Injecting bogus routing advertisements to modify routing tables
- Using IP source routing to specifying arbitrary paths through a network

Routers use *routing protocols* to exchange network topology information with neighboring routers. This information is used to calculate the best path for directing traffic over a network. Routing protocols send route updates at regular intervals, or when the topology of a network changes. This process allows a router to dynamically advertise new or unresponsive network segments without interaction from network administrators. Unfortunately, some routing protocols, such as Router Information Protocol version 1 (RIPv1), have no method of authenticating route information. An attacker who injects bogus RIP updates into a network can modify the routing tables of all routers attached to a network segment. Unless configured otherwise, each of the infected routers will then include this bogus information in updates sent to adjacent routers, thus causing the infection to spread. In Figure 3-6, the attacker uses an RIP update to change the default gateway of Router 1 to point to Host 1. This allows the attacker to divert and monitor all traffic destined for the true default gateway (Router 2) through her computer. This technique can be used to reroute traffic through arbitrary network segments, to poison routes by marking them as unreachable, or to create routing loops as DoS attacks.

Every Internet Protocol header includes an options field for specifying special handling requirements for each packet. One of the available options is source routing. Using *source routing,* the sender of a packet can specify the route that the packet should take to its final destination, instead of letting each router decide the best path according to its routing tables. The intended purpose of source routing is to allow the sender to override incorrect entries in routing tables or to debug network routing problems, such as routing loops. However, source routing is much more commonly used by attackers to direct traffic through a particular network segment. Regardless of how efficient or logical the path taken by the packet, the destination host must respond using the inverse of the route specified by the source (Braden, 1989). An attacker spoofing the address of another host or hijacking a TCP session can route traffic directly through a network segment

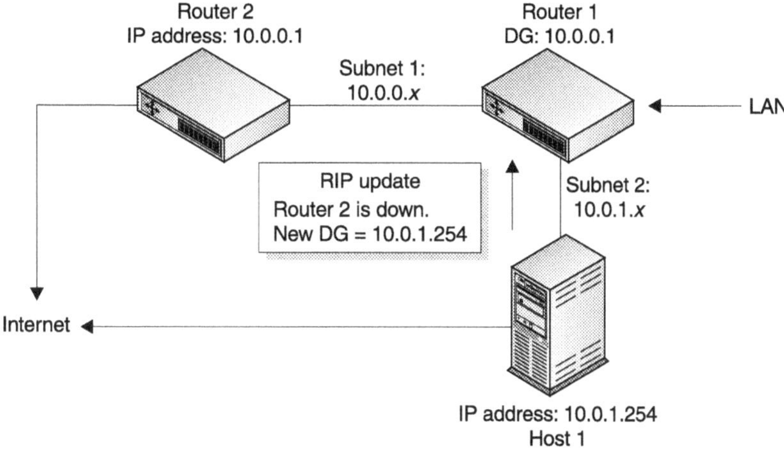

Figure 3-6 By injecting bogus RIP advertisements into a network, an intruder can alter the contents of the routing tables and redirect network traffic.

containing a packet sniffer. This will allow the attacker to view each response from the destination host and will help ensure that the attacker can respond before the spoofed host does (see Figure 3-7).

Countermeasures for Routing Attacks

RIP version 2 (RIPv2), Open Shortest Path First (OSPF), Enhanced Interior Gateway Routing Protocol (Enhanced IGRP), Border Gateway Protocol (BGP), and Intermediate System-to-Intermediate System (IS-IS) all include options for route authentication. Most of these routing protocols support simple passwords and message-digest–based authentication. Both methods require that a route advertisement contains some unique identifying information, or the packet will be discarded. Simple passwords offer limited protection, since they are transmitted in plaintext along with the route updates. Anyone with a packet sniffer attached to the network can obtain a password.

The use of message digests, such as MD5 and SHA-1, provide a more robust solution. Given the routing information, a key, and a key ID, a routing protocol supporting this form of authentication will create a keyed message digest and then append the digest to the route advertisement. The key is a password shared between neighboring routers and is never transmitted over the network. Given the one-way behavior of message digests, there is no way for an attacker to obtain the shared key from a digest. The receiving router verifies the authenticity

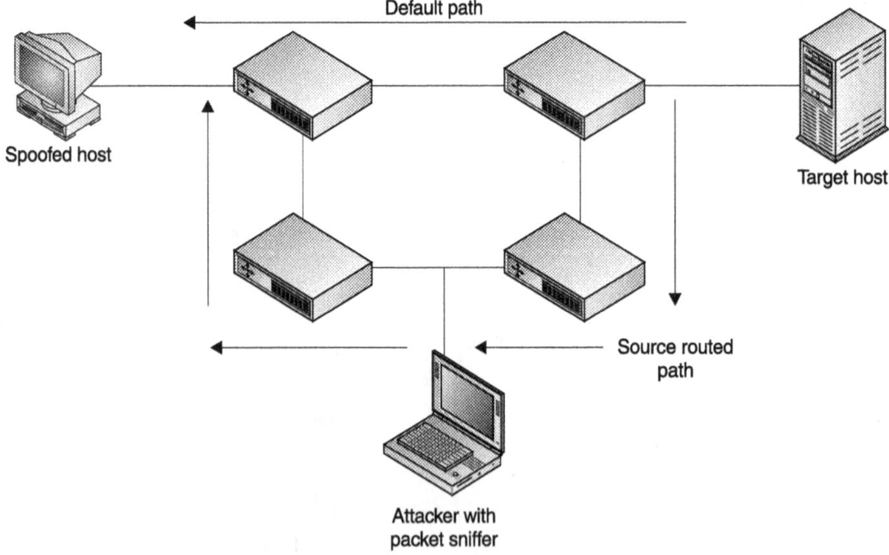

Figure 3-7 The IP source routing option allows an attacker to explicitly specify the path that a packet takes to and from a destination host. Source routing is used in address-spoofing and session-hijacking attacks.

of the packet by using the shared key to calculate a similar digest, and then comparing the calculated digest with the one received in the update. This procedure is identical to that used by hashed-based message authentication codes (which were discussed in Chapter 2).

In addition to route authentication, the following recommendations are useful (Furlan, 2001):

- Create a specific network subnet for exchanging routing information. Only routers should be attached to this subnet.
- Create an exclusive list of all routers that are allowed to advertise route updates, and post the list to each router.
- To prevent a router from inadvertently responding to bogus updates, disable routing protocols on interfaces that are attached to subnets with no other routers.

Finally, be aware that IP source routing is rarely used for legitimate purposes; it should be disabled in most environments.

NOTE:
Most current TCP/IP implementations automatically ignore source routing information, so source routing attacks may not pose much of a threat. However, some older implementations may still be vulnerable to this type of attack.

TCP/IP-Specific Attacks

Starting in the late 1960s, government-funded research organizations began investing considerable time and effort in developing packet-switched networks. The TCP/IP protocol suite evolved out of the need for a common set of packet-based communication protocols. Security was not a major concern in the development of TCP/IP, since public networks in the early 1970s exchanged mostly unclassified information that was valuable only to academic communities. As a result, the suite is susceptible to a number of threats:

- Address spoofing
- Sequence number prediction
- Various TCP/IP DoS attacks
- Session hijacking

Address Spoofing

Many network services, such as the Berkeley Unix “r” utilities, authenticate clients based on source IP address. (To distinguish the Berkeley utilities from standard Unix commands, the Berkeley command names are prefaced with the letter "r.") This method of authentication provides little protection against sophisticated attackers because IP addresses are easily falsified, or *spoofed*. *Address-spoofing* attacks are commonly used to exploit known trust relationships between two hosts (see Figure 3-8). For instance, if an attacker issues a command to a remote computer, making it look as though the message has come from a trusted host, the remote computer may execute the command without question. Address spoofing also conceals the identity of an intruder, making it difficult to pinpoint the source of an attack.

CAUTION:
Address-based authentication is a bad idea. By spoofing an IP address, an attacker can make a packet look as though it came from any host on the Internet, including one you trust. Routers and other packet-filtering devices should always be configured to drop packets from an external network that contain source addresses on an internal network.

Spoofing attacks can be launched from any location as long as the attacker does not need to view responses from the target host. If the attacker knows how the victim will respond, he can generate the appropriate acknowledgments and counter responses to maintain the communication session. The output of many network-based programs can be anticipated in advance. An attacker may be able to achieve the desired result by issuing only a single command on a remote host, so there is no need for the attacker to respond.

For this type of attack to be successful and to go undetected, the spoofed host must not respond to replies from the target host. Upon receiving an unsolicited packet, a spoofed host running TCP will issue a connection reset. If the target host receives such a reset request, it will immediately end the attack by dropping the connection. To avoid this problem, spoofing attacks are usually launched while the spoofed host is unavailable—for instance, when the host has been powered down for maintenance. Often, a spoofing attack is combined with a DoS or routing attack to either temporarily disable the spoofed host or divert traffic away from it. Common DoS techniques used in conjunction with address spoofing include the following:

- Sending packets containing the spoofed source address to multiple hosts. Each host will respond to the source address in the packet, flooding the spoofed host with traffic. (We'll discuss this kind of attack—called a Smurf attack—later in the chapter.)
- Sending a packet with the broadcast address as the source to the spoofed host. Some hosts will respond by sending a reply to the broadcast address, to which all hosts on the network segment will reply in turn, flooding the spoofed host with traffic

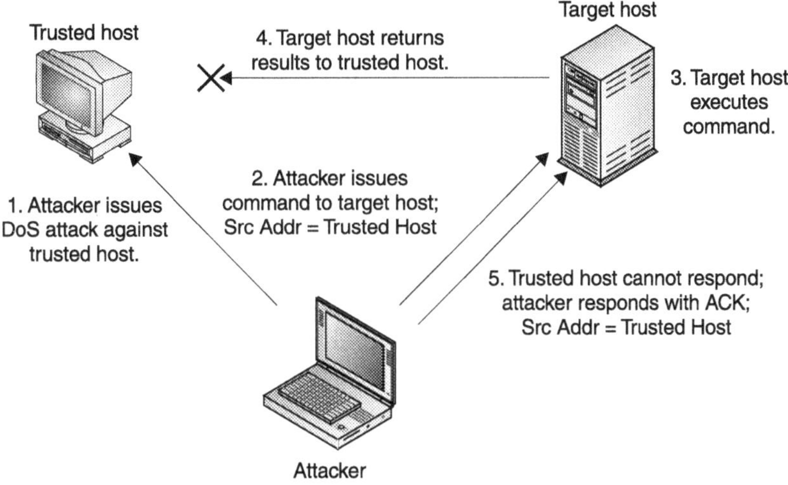

Figure 3-8 Address-spoofing attacks allow intruders to exploit trust relationships based on IP addresses.

In addition to flooding the victim, both of these attacks will generate excessive network traffic that may clog a slow network connection.

NOTE:
If you've been paying close attention, you might have noticed that the DoS attacks discussed earlier in this chapter use address spoofing. Although DoS attacks are often used to facilitate address-spoofing attacks, they themselves may employ spoofed source addresses.

TCP Sequence Number Prediction

The TCP/IP protocol provides basic protection against source address spoofing. Every TCP packet contains a sequence number, which helps identify the session to which the packet belongs. Sequence numbers are also used in reconstructing the original message from packets that arrive out of order. During connection establishment, communicating hosts exchange *initial sequence numbers (ISNs),* as illustrated in Figure 3-9. Host A sends a TCP SYN request containing an ISN to Host B. Host B replies by acknowledging receipt of Host A's ISN, and provides its own ISN. Notice that the acknowledgement contains the sequence number of the next packet that Host B expects to receive. Host A then acknowledges receipt of Host B's ISN, and the handshake is completed. Each subsequent packet from Host A to Host B, and vice versa, contains a unique sequence number. If either host receives an acknowledgement containing an invalid ISN, it will issue a connection reset. An intruder will have a difficult time maintaining a spoofed connection if he or she cannot predict the ISN.

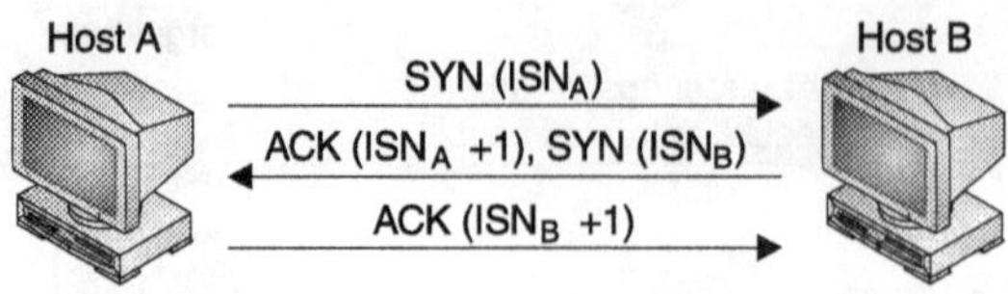

Figure 3-9 During a TCP three-way handshake, initial sequence numbers are exchanged.

Unfortunately, sequence numbers alone do not provide adequate protection against spoofing attacks. Many TCP/IP implementations use very predictable methods for incrementing ISNs, the most common approach being to add a constant value to the preceding ISN. A sophisticated intruder might easily determine the increment value by sending a few packets to the target host and analyzing the ISNs contained in the responses. If the method used to establish the ISN is predictable, Host A could easily supply a forged source address, wait an appropriate time for Host B to reply, and then respond by acknowledging the anticipated ISN. If Host A is correct in its prediction of Host B's ISN, the handshake will be completed successfully, and Host B will continue to accept packets from Host A. Using this technique in combination with address spoofing, an intruder can connect to a host and execute commands without ever receiving responses from the remote host.

CAUTION:

The User Datagram Protocol (UDP) is easily spoofed due to the absence of sequence numbers. UDP also lacks acknowledgements, so there is no need for an attacker to positively acknowledge every reply from the target host.

Session Hijacking

Session hijacking is a variation of IP address spoofing in which an attacker disconnects two communicating hosts or takes control of one half of a communication session. This form of attack allows an intruder to bypass authentication mechanisms by hijacking a session following successful authentication by the victim. Although numerous scenarios exist, two common session-hijacking techniques are presented here.

Scenario 1: Disconnecting a Host (Dave and Moussa, 2001)

1. The attacker monitors communication between a client and server to obtain their IP addresses and port numbers.
2. The attacker listens for an ACK packet from the client containing the next sequence number that the client expects to receive.

3. The attacker then sends an RST packet to the client, using the sequence number obtained in step 2. The packet contains the source address of the server and the destination address of the client.
4. When the client receives this packet, it immediately closes the connection, thinking that the server has requested the reset. All future packets from the server will be ignored.

Scenario 2: Taking Control of a Session (Joncheray, 1995 and Chambers, Dolskey, and Iyer, 2001):

1. The attacker monitors communication between a client and server to obtain their IP addresses and port numbers.
2. The attacker alters sequence number counters of the client and the server by inserting packets containing null data into the communication stream. Null data packets do nothing more than increment the sequence number. As a result, the client and server enter a desynchronized state, where the expected and received sequence numbers do not match. If the sequence numbers fall outside of the current TCP window, each host will begin ignoring packets from the other.

REMEMBER:

For flow control, TCP maintains a variable-length window of valid sequence numbers. The size of the window equals the maximum amount of data that a host can send before waiting for an acknowledgement. Packets containing sequence numbers within the window are accepted, while those outside the window are ignored.

---

3. The attacker can now "reinstate" the connection with the server by inserting spoofed packets that contain the sequence numbers expected by the server. These packets may also contain commands to be executed on the server.
4. Since no RST packet has been sent, the legitimate client waits for the connection with the server to time out. In the meantime, the attacker may decide to resynchronize the session and return control back to the client (Dittrich, 1999).

NOTE:

The desynchronization technique used in Scenario 2 may result in a storm of ACK packets between the client and the server as each side notifies the other of the sequence number it expects to see in the next packet. These packets are not retransmitted if lost, so the exchange of ACK packets halts as soon as one of the packets is dropped.

Session hijacking requires that the attacker be on the path between the spoofed and target hosts. For routable protocols, such as IP, different packets within the same communication session may follow different paths across the Internet, depending on how they are directed by intermediate routers. For this reason, a session-hijacking attack is most effective when launched near one of the endpoints of the communication path, where an attacker has a high likelihood of intercepting all traffic (see Figure 3-10). As an alternative, IP source routing can be used to direct traffic through a particular network segment. Hunt and Juggernaut are popular automated session-hijacking utilities.

The Mitnik Attack

One of the most famous attacks in the history of computer security was launched by Kevin Mitnik against a computer belonging to Tsutomu Shimomura, a well-known

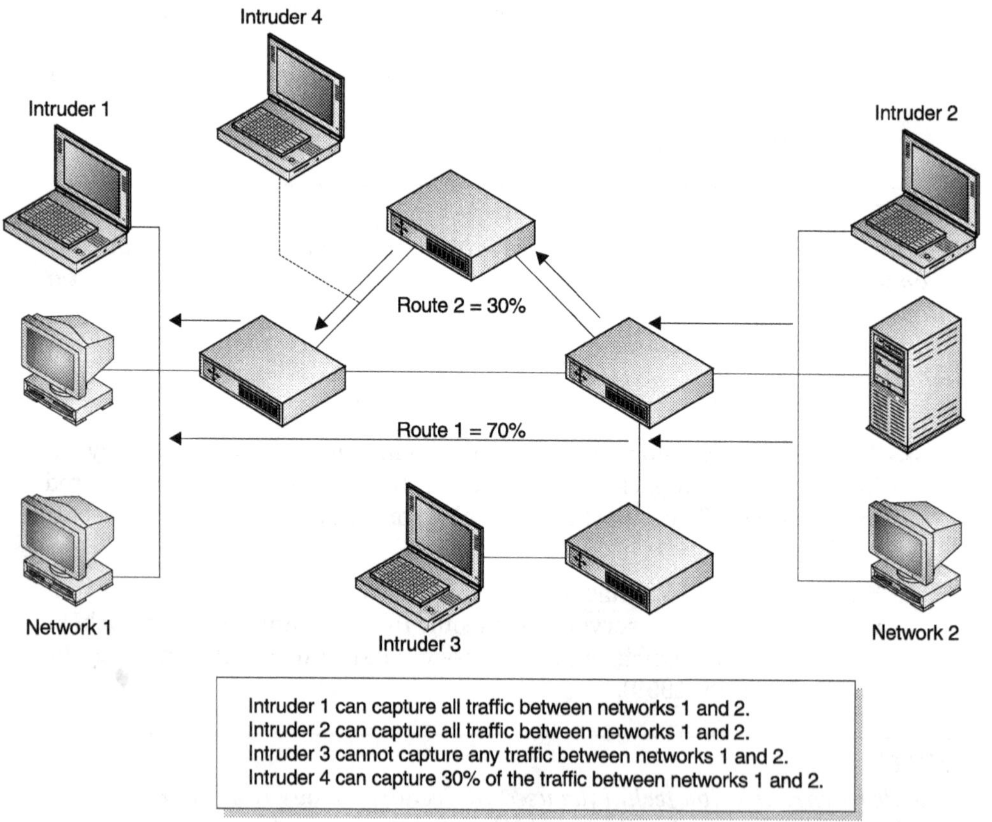

Figure 3-10 Session-hijacking attacks are most effective near the endpoints of a communication session. The closer an attacker is to an endpoint, the higher the possibility of capturing all traffic between Networks 1 and 2.

computer security expert. The attack combined DoS and TCP address spoofing to exploit a trust relationship between Shimomura's computer and another host on the Internet. By examining various logs files, Shimomura managed to piece together the steps taken by Mitnik:

1. Mitnik breaks into a host named toad.com.
2. He issues commands to discover trust relationships between Shimomura's computer and another host on the Internet.
3. Mitnik disables the trusted host by using a TCP SYN flood (as described in the upcoming section, "TCP/IP Denial of Service").
4. To determine the initial sequence number, Mitnik sends 20 packets to Shimomura's computer, looking for a pattern. He determines that the sequence number grows by 128,000 for each successive packet.
5. Mitnik sends spoofed packets to Shimomura's computer, making them appear as though they came from the trusted host. Shimomura's computer accepts these packets, and responds using the expected sequence number.
6. The automated utility used by Mitnik increments the incoming sequence number counter and sends ACK packets back to Shimomura's computer. The sequence numbers contained in the acknowledgements match, and Shimomura's computer is duped into believing that it is communicating with the trusted host.

Figure 3-11 illustrates the Mitnik Attack. For this attack to be successful, Shimomura's computer had to be idle. Each network connection would have incremented the ISN for subsequent connections, and a frequently changing ISN would have been very difficult for Mitnik to predict. Without the correct sequence number, the spoofed packets would not have been accepted by Shimomura's computer. Knowing this, Mitnik decided to launch his attack on Christmas day, when few people would be connecting to Shimomura's computer.

Countermeasures for Address-Spoofing and Session-Hijacking Attacks

Here are a number of suggestions for preventing address-spoofing and session-hijacking attacks:

- *Use a TCP/IP implementation that randomly increments ISNs.* This technique was originally suggested by Steven Bellovin (Bellovin, 1989). Bellovin later proposed a sequence number partitioning scheme that uniquely assigns ISNs based on a combination of source and destination addresses and ports of a connection (Bellovin, 1996).

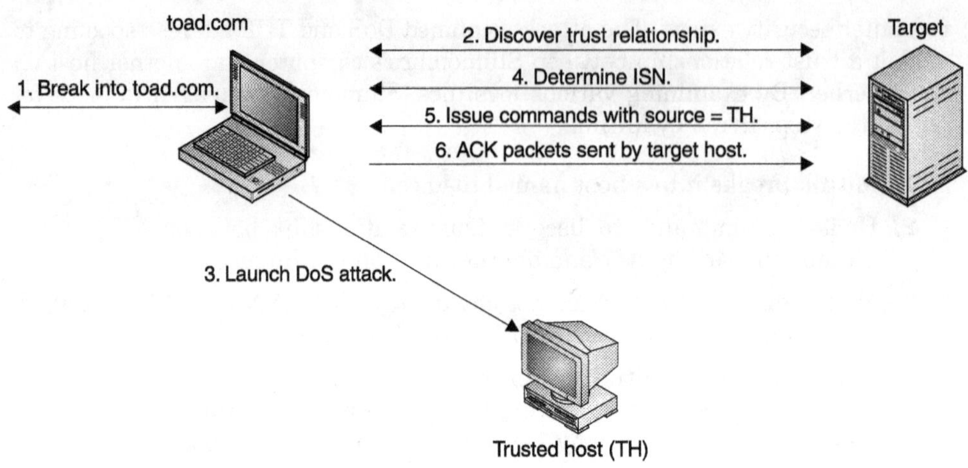

Figure 3-11 The Mitnik Attack combined DoS and TCP session hijacking to exploit a trust relationship between two hosts.

- *Use cryptography to conceal sequence numbers and protect IP header information.* IPSec can be used to encrypt upper-layer protocols, including TCP. This will effectively prevent an attacker from analyzing sequence numbers. IPSec also includes integrity mechanisms for protecting source address information contained within IP headers. (We will further discuss the benefits of IPSec in Chapters 6 and 7.)
- *Do not authenticate hosts based on source addresses.* Instead, use cryptographic authentication mechanisms, such as those included with IPSec, or user-based authentication.
- *Block external packets that have internal source addresses.* A router should drop any packets received on external network interfaces that contain source addresses from an internal network. Such packets almost always signify an address-spoofing attack.

TCP/IP Denial of Service

From its inception, TCP/IP was designed to be robust and reliable. Nonetheless, malicious users have shown no lack of imagination or creativity in devising DoS attacks to undermine the operation of TCP/IP. Basic attacks exist for exploiting TCP connection establishment, IP fragmentation and reassembly, and Internet Control Message Protocol (ICMP) messaging. Some make use of vulnerabilities specific to a particular operating system, while others target inherent weaknesses in network protocols. A few of the popular TCP/IP DoS attacks include TCP SYN flooding, Land.c, Smurf, Teardrop.c, and the Ping of Death.

A *TCP SYN* attack, also known as an *SYN flood,* exploits the behavior of the TCP three-way handshake. Recall from Figure 3-9 that a typical TCP connection requires three steps:

1. The source host sends an SYN request to the destination host.
2. The destination responds with an SYN/ACK, letting the source know that it has received the connection request.
3. The source host completes the connection by sending an ACK to the destination host.

For each incoming request, the destination host places an entry in a pending connection queue of limited size. The destination host uses this queue to track incoming connections waiting to be completed. Entries remain in the queue until the destination host receives the final ACK from the source, or until the entry times out (typically after 60 seconds). The destination usually receives acknowledgements quickly (within milliseconds), so the finite length of the queue rarely causes problems.

To exploit this behavior, an attacker can generate TCP SYN requests using arbitrary source addresses and send them to a target host. The target will respond to each SYN request with an SYN/ACK, and place an entry in its connection queue. If the target never receives a reply, the connection will never be completed, and the entry will remain in the queue until it expires. An attacker who quickly fills the connection queue with bogus requests can deny service to other users. Once the queue is filled, the target host will ignore further connection requests until an entry expires or a connection is completed. As long as the attacker continues to fill the queue, the target host will not accept incoming connections.

NOTE:

For TCP SYN flooding to work, each source address should belong to a nonexistent host. A host receiving an SYN/ACK for a connection it did not initiate will issue a connection reset, resulting in immediate removal of the connection from the target host's connection queue.

Land.c is another type of TCP/IP DoS attack that takes advantage of a weakness in some TCP implementations. An attacker can cause an operating system to hang by sending a connection request (TCP SYN packet) with both the source and destination addresses set to that of the target machine, and with matching source and destination ports. For this attack to be successful, a service must be running on the selected port.

While the SYN flooding and Land.c attacks focused on disabling a target host, the *Smurf* attack (see Figure 3-12) slows network performance by consuming available bandwidth. An attacker begins by sending an ICMP echo request to

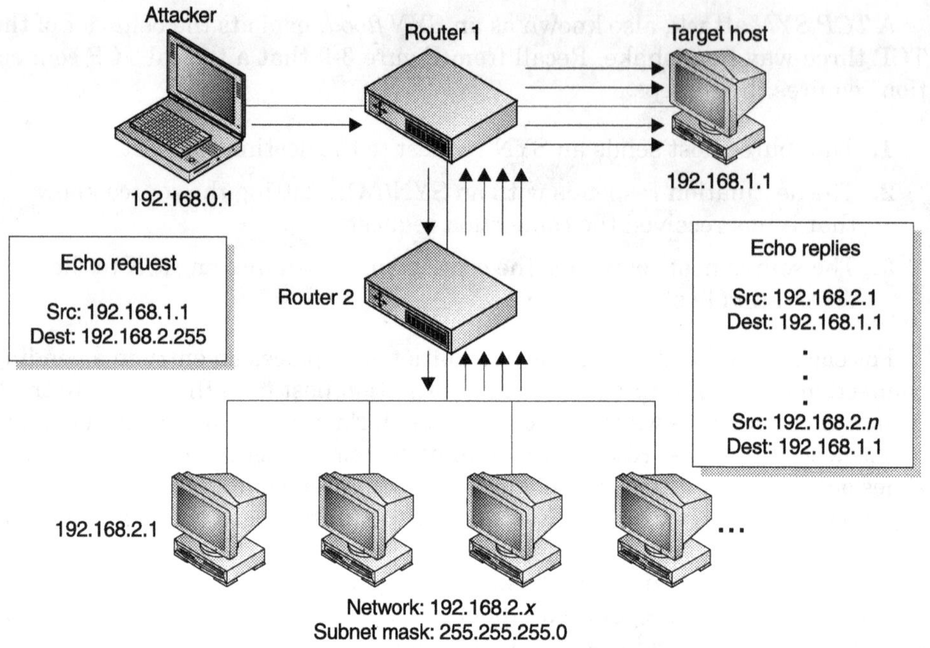

Figure 3-12 The Smurf attack results in excessive network traffic directed at a single target host. If the number of hosts responding to the attacker's request is large enough, the spoofed host will be overwhelmed.

the broadcast address of a remote network. The echo request packet contains a forged source address of the attacker's choosing. If the routing device servicing the remote network is configured to forward broadcast traffic, all hosts on the network will respond with echo replies directed at the spoofed source address. This attack is also referred to as an *ICMP flood* or a *ping flood*.

NOTE:
ICMP stands for Internet Control Message Protocol. *ICMP is used to exchange control messages and handle error conditions arising from abnormal or unexpected network behavior. Utilities such as the ping command use a particular ICMP message, an echo request, to determine whether another host is responding. A functioning host that receives an echo request will respond with an echo reply message. (See Appendix A for more information on ICMP and types of ICMP messages.)*

IP and ICMP Fragmentation

Fragmentation allows an IP datagram, also known as an IP packet, to traverse networks with a maximum transmission unit (MTU) smaller than the size of the datagram. For Ethernet, the maximum size of an IP datagram is 1500 bytes. If a router directs the datagram through a network segment that has an MTU smaller than the size of the packet, the router must fragment the datagram before placing it on the network. All fragments arrive independently at their destination, where they are reassembled by the destination host. In order for the destination host to reassemble the packet, each fragment must contain a fragment identifier and an offset. Every fragment from the same original IP datagram is assigned the same fragment ID. The offset indicates the fragment's position in the datagram. Table 3-4 lists MTUs for many popular network architectures.

In an attempt to bypass or evade security mechanisms, an attacker might purposely fragment an IP datagram before placing it on a network. For instance, some fragmented packets may be allowed to pass through packet-filtering devices and firewalls if they do not contain certain header information. When a packet that encapsulates TCP data is fragmented, the first fragment is likely to contain the entire TCP header. A device that filters traffic based on header information may block the fragment containing the TCP header, but allow the remaining fragments to pass onto the internal network.

The *Teardrop.c* DoS attack manipulates a weakness found in some protocol stacks. It is an older attack, but is still used for attacking hosts attached to the Internet. In this attack, an intruder splits a packet into two fragments, giving the second fragment an offset value that is less than the length of the first fragment. To interpret the fragment, the operating system must decrement the offset counter after processing the first fragment. Depending on how the IP stack implements integer arithmetic, a small negative offset may be incorrectly converted into a much larger positive value. (For those programmers in the audience, this is akin to assigning a signed value to an unsigned variable.) Notice that this type of attack is much more efficient than the brute force methods mentioned in the preceding sections; it requires only a single fragmented message.

Another attack that takes advantage of the unpredictable way in which some protocol stacks react to fragmented packets is the *Ping of Death*. The largest allowable length for an IP packet is 65,535 bytes. By fragmenting an ICMP echo request, an attacker might be able to generate an oversized message. A valid offset

Table 3-4 Maximum Transmission Units (MTUs) for Various Network Media

Media Type	MTU (in bytes)
Ethernet	1500
IEEE 802.3	1492
Token Ring	4464 (4 Mbps), 17,914 (16 Mbps)
X.25	576
FDDI	4352
Hyperchannel	65,535
PPP	1500
SLIP	1006 (logical restriction)

can be combined with an acceptable fragment size in the final fragment such that the entire echo request (IP header + ICMP header + ICMP data) is greater than 65,535 bytes. The receiving host will have no way of knowing that the message is invalid until it attempts to reassemble the message. In trying to handle the reassembled message, the host may overwrite data in memory that belongs to another process, resulting in a system crash (Northcutt and Novak, 2001).

Countermeasures for TCP/IP DoS Attacks

As already mentioned, it is very difficult to completely prevent certain DoS attacks. A large-scale DDoS attack may disable a host or network, regardless of the security mechanisms employed, if it generates enough traffic to overwhelm a router or flood a low-bandwidth Internet connection. Fortunately, the effects of this type of attack can be moderated, and many vendors release periodic updates to correct security vulnerabilities discovered in their operating systems and protocol stacks.

TCP SYN flood countermeasures include the following:

- *Increase the size of the pending connection queue.* This allows for more incoming connections. Be conservative when using this approach; maintaining too many incoming connections may deplete all available memory and cause a host to crash.
- *Decrease the connection timeout period.* This causes invalid connections to be removed more quickly from the pending connection queue. Be wary when reducing the timeout period for long-distance connections that traverse many network hops. Too short a timeout period may cause some valid connections to be dropped before the TCP handshaking session is completed.
- *Couple TCP with a network layer security protocol, such as IPSec.* IPSec can deny access by unauthenticated hosts before a single TCP packet has been exchanged. This successfully counters TCP SYN flooding and illicit connection resets, such as those used in session-hijacking attacks.

Neither of the first two solutions completely alleviates the possibility of SYN flooding, but they both increase the amount of work necessary to launch an attack. It may be possible to configure an appropriate connection queue size and timeout period such that an attacker using a single computer cannot generate enough SYN requests to fill the queue before previous connections time out. Unfortunately, this technique will eventually fail if there are enough computers participating in the attack. The third option eliminates TCP SYN flooding where an attacker uses spoofed source addresses. A downside to this approach, however, is that all hosts must support IPSec.

CAUTION:
While IPSec prevents many DoS attacks, the key-exchange protocol used in conjunction with IPSec—Internet Key Exchange (IKE)—may be susceptible to such attacks (Krywaniuk, 2001). (We discuss IKE in Chapter 6.)

To prevent fragmentation attacks, you can configure packet filters and firewalls to store and reassemble all fragmented packets. This requires additional processing time and memory, however, and may not be practical in some environments. Alternatively, all fragmented packets can be dropped at the risk of denying legitimate traffic from entering the network.

Additional DoS precautions include

- Employing countermeasures against address spoofing to prevent DoS attacks based on falsified source addresses
- Disabling directed broadcasts on routing devices
- Rate-limiting problematic message types (such as SYN requests and ICMP echo requests) through routing devices to avoid overwhelming the hosts on an internal network (Cisco Systems, 2000)

Attacks on Cryptography

Many of the countermeasures listed in the preceding sections employ cryptographic security mechanisms. While cryptography does provide considerable functionality, cryptographic mechanisms themselves may be subject to attack. Here are some examples of common cryptographic attacks:

- Exhaustive key searches (discussed in Chapter 2)
- Cryptanalysis
- Testing for weak keys
- Block replay
- Man-in-the-middle attacks

Cryptanalysis

Cryptanalytic attacks undermine the security of cryptographic algorithms by identifying implementation weaknesses. Regardless of the key size used, an algorithm that is highly susceptible to cryptanalysis is not considered secure. In Chapter 2's discussion of cryptographic algorithms, we introduced a number of security concerns and design decisions that can affect a cipher's resistance to cryptanalysis.

Early cryptographic algorithms demonstrated repetitive behavior by which an attacker could determine plaintext through statistical analysis. *Statistical analysis* associates the relative frequency of ciphertext characters to those of a written document. For example, in the English language, the letter "e" occurs most frequently, followed by the letter "t." The most common character pattern is "th." Using statistical analysis, an attacker may assume that the most common ciphertext character equates to the letter "e," while the most common pattern equates to "th." Fortunately, advancements in cryptography have greatly reduced the effectiveness of statistical analysis by ensuring that ciphertext patterns do not directly correlate to plaintext patterns.

Despite the resistance of modern-day cryptographic algorithms to statistical analysis, many cryptanalytic techniques do exist for effectively attacking symmetric- and public-key cryptosystems. Depending on the type of information available to an attacker, most cryptanalytic attacks fall into one of the following classifications (listed in order of decreasing difficulty): ciphertext-only, known-plaintext, chosen-plaintext, adaptive-chosen-plaintext, chosen-ciphertext, or adaptive-chosen-ciphertext. In a *ciphertext-only* attack, a cryptanalyst captures a ciphertext sample and attempts to determine the corresponding plaintext. Ciphertext can be obtained by eavesdropping on an encrypted communication channel. This type of attack is difficult and does not work well for small ciphertext samples. However, an intruder can capture a significant amount of ciphertext in a relatively short time by monitoring a busy broadband connection—which makes this type of attack plausible.

An attacker who has also obtained the corresponding plaintext can mount a *known-plaintext* attack. By comparing the compromised plaintext and ciphertext samples, an attacker may uncover a relationship between the encrypted and unencrypted data that he or she can use to partially decrypt future messages. In extreme cases, the attacker may be able to determine an encryption key, and then decrypt ciphertext at will. (Two specific examples of known-plaintext attacks are *differential cryptanalysis* and *linear cryptanalysis*. Discussion of these techniques is beyond the scope of this book.)

CAUTION:

For a stream cipher, XORing plaintext and ciphertext results in the keystream used to encrypt that portion of plaintext. If the keystream eventually repeats, the attacker can use the compromised portion of the keystream to decrypt any encrypted data. (In other words, the attacker can XOR the compromised keystream with ciphertext to produce plaintext.)

The cryptanalytic attacks discussed up to this point do not allow an attacker to select arbitrary plaintext or ciphertext for analysis. In a *chosen-plaintext* attack, a cryptanalyst has the ability to choose a plaintext message and then ascertain the corresponding ciphertext. The goal of this type of attack is to recover a decryption key, or to emulate the behavior of the encryption algorithm without knowing its operational details firsthand.

An *adaptive-chosen-plaintext* attack allows an attacker to interactively choose plaintext and then determine the associated ciphertext. This type of attack provides more flexibility than chosen-plaintext attacks, because the attacker can modify his or her choice of plaintext based on previous ciphertext outputs. Properly chosen plaintext samples may reveal patterns or other repetitive behavior.

In a *chosen-ciphertext* attack, a cryptanalyst can choose an arbitrary ciphertext sample and ascertain the corresponding plaintext. This type of attack mainly threatens public-key cryptosystems, where it can be used to identify private keys.

An *adaptive-chosen-ciphertext* attack is a special case of a chosen-plaintext attack in which a cryptanalyst can iteratively decrypt ciphertext to identify a relationship between the ciphertext and plaintext without knowing the decryption key. That relationship can then be used to derive the plaintext corresponding to a given ciphertext message. Using this type of attack is similar to accessing a black box that performs decryption. An attacker can decrypt messages at will, but cannot determine the decryption key (RSA Laboratories, 2000). An example of an adaptive-chosen-ciphertext attack has been described by Daniel Bleichenbacher (Bleichenbacher, 1998, and Schneier, 1998).

Testing for Weak Keys

A *weak key* is an encryption key that causes a block cipher to behave in a predictable or weakened state. For example, DES has four weak keys for which encryption and decryption operations are identical. For a given weak key k and plaintext data P, $E_k(E_k(P)) = P$. DES also has 12 semi-weak keys that exist in pairs ($k1$, $k2$), such that $E_{k1}(E_{k2}(M)) = M$. In general, weak keys comprise a very small fraction of the available keys, so the likelihood of randomly generating a weak key is quite low. Even so, a good key-generation program will check to ensure that the keys it is generating are not weak.

Block Replay

As noted in Chapter 2, the ECB mode of block cipher operation is prone to block-replay attacks. The absence of a chaining mechanism allows individual ciphertext blocks to be altered or replaced without affecting other blocks. (For more information, see the section entitled "Electronic Codebook Mode" in Chapter 2.)

Man-in-the-Middle Attacks

Public-key cryptosystems are vulnerable to man-in-the-middle attacks. An attacker who intercepts an initial public-key exchange may substitute his or her public key for that of a legitimate user without detection. Figure 3-13 illustrates a typical man-in-the-middle attack.

1. Before Alice and Bob can use public-key cryptography, they must exchange at least one of their public keys. Alice decides to send her public key to Bob.

2. An attacker, Mallory, captures the message sent to Bob containing Alice's public key and generates a new message replacing Alice's public key with his own.
3. Mallory then sends a message to Bob containing the altered key. Bob has no way of knowing that Alice's key has been replaced in the message he has just received.
4. Believing that he has the correct public key, Bob generates a session key, encrypts the session key using what he believes to be Alice's public key, and sends the encrypted key to Alice. Again, however, Mallory intercepts the message.
5. Mallory extracts the session key using his private key and reencrypts a new message, using Alice's public key. Upon receiving the message, Alice uses her private key to obtain the session key. Alice and Bob begin encrypted communication using the shared session key. Mallory can now listen to all communication between Alice and Bob.

Countermeasures for Attacks Against Cryptographic Mechanisms

The strength of cryptography depends greatly on how cryptographic mechanisms are implemented and deployed. All too frequently, intruders discover vulnerabilities not by attacking cryptographic algorithms directly, but by finding weaknesses in how the algorithms are used. In Chapter 2, we introduced many concepts fundamental to properly choosing and deploying PRNGs, ciphers, message

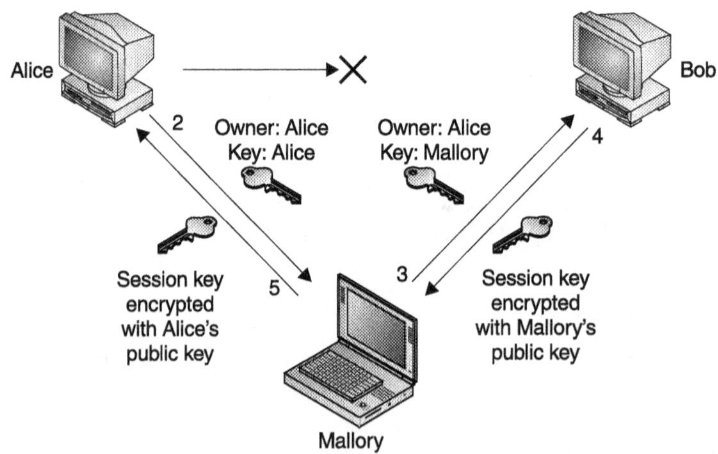

Figure 3-13 Public-key cryptosystems are susceptible to man-in-the-middle attacks, where an intruder substitutes his or her public key with that of a legitimate user during an initial key exchange.

digests, and other cryptographic mechanisms. The following five precautions relate directly to the attacks mentioned in this section.

- *Choose appropriate key lengths for the data being protected.* We have mentioned repeatedly the importance of choosing appropriate key lengths. Extremely valuable information, or data that must remain confidential for long periods, may call for the use of longer keys.

NOTE:
The less time required by an encryption operation, the more quickly a keyspace can be searched. As a result, fast encryption algorithms require longer key lengths than do slower algorithms to provide the same level of protection against brute force attacks. This is a subtle point that is frequently overlooked when choosing encryption algorithms.

- *Check for weak keys during key generation.* While the probability of generating a weak key or a set of weak keys is very low, searching for weak keys is one of the first things a cryptanalyst will try after discovering the encryption algorithm you are using. Testing for weak keys during key generation is a simple process. The number of weak keys for most cryptographic algorithms is very small, and the keys themselves are well published.
- *Employ cryptographic feedback mechanisms (chaining).* When using block ciphers, chaining mechanisms reduce the effectiveness of statistical analysis and other cryptanalytic techniques. Chaining conceals plaintext patterns and creates dependencies between successive blocks of ciphertext, making data-modification and block-replay attacks difficult.
- *Avoid encrypting large quantities of data with the same key.* The probability of successful cryptanalytic attack improves as the amount of data encrypted with a single key increases.
- *When using public-key cryptosystems, use digital certificates and third-party trust authorities to bind keying material to user identities.* A digital certificate securely associates a user's ID with his or her public key. To prevent man-in-the-middle attacks, users swap certificates during the initial public key exchange, and extract user IDs and corresponding public keys from the certificates. The certificates are digitally signed and issued by a certification authority (CA) that both parties trust. Before a CA issues a certificate, it verifies the identity of the user who has requested the certificate, thus ensuring that a malicious user can't obtain a certificate under the authentic user's identity. (Refer to Appendix B for details on digital certification and certificate authorities.)

NOTE:
The use of digital certificates and trusted third parties does not provide foolproof security. A CA that does not properly authenticate certificate requests can be tricked into issuing falsified certificates. For instance, early in 2001, Verisign, a highly regarded public CA, was tricked into issuing two Class 3 Software Publisher certificates to someone posing as a Microsoft employee. With these certificates, an attacker could have digitally signed rogue code as though it had originated from Microsoft.

Social Engineering and Dumpster Diving

No discussion of security threats would be complete without mentioning *social engineering* and *dumpster diving*. What benefit are complex security mechanisms when an attacker can quickly subvert them by simply asking unsuspecting users for their passwords? Smart attackers look for the simplest avenue possible, which often includes posing as an administrator to coax users into supplying passwords or executing rogue programs on their computers. They may also masquerade as maintenance personnel to gain physical access to secured areas. Attackers not opposed to getting a bit dirty may even search dumpsters or trash bins for documents or storage media (such as floppy disks and discarded hard drives) containing passwords or other information that might aid an attack. User education, physical security measures, and secure disposal of printed material and storage devices play the biggest role in thwarting social engineering and dumpster diving attacks.

References

Bellovin, Steven. April, 1989. "Security Problems in the TCP/IP Protocol Suite." *Computer Communication Review*.

Bellovin, Steven. May, 1996. *RFC 1948: Defending Against Sequence Number Attacks*. Internet Engineering Task Force.

Bleichenbacher, Daniel. 1998. "Chosen Ciphertext Attacks Against Protocols Based on RSA Encryption Standard PKCS #1." *Advances in Cryptology—CRYPTO'98*.

Braden, R. October, 1989. *RFC 1122: Requirements for Internet Hosts—Communication Layers*. Internet Engineering Task Force.

CERT Coordination Center. 6 August, 2001. *Home Network Security*. Carnegie Mellon University. 10 Oct., 2001 (http://www.cert.org/tech_tips/home_networks.html).

Chambers, Chris, Justin Dolske, and Jayaraman Iyer. 19 October, 2001. *TCP/IP Security*. Ohio State University (http://www.linuxsecurity.com/resource_files/documentation/tcpip-security.html).

Cisco Systems. 17 February, 2000. *Strategies to Protect Against Distributed Denial of Service (DDoS) Attacks*. 20 Oct., 2001 (http://www.cisco.com/warp/public/707/newsflash.html).

Dave Paras, and Nathan Moussa. 10 October, 2001. "TCP Connection Hijacking." *NIUNet*. Baylor University (http://cs.baylor.edu/~donahoo/NIUNet/hijack.html).

Dittrich, Dave. 9 December, 1999. *Session Hijack Script* (http://staff.washington.edu/dittrich/talks/agora/script.html).

Furlan, Fabiano. 23 October, 2001. "Securing RIP and OSPF Protocols." *SANS Institute Information Security Reading Room*. Sans Institute. 30 May, 2001 (http://www.sans.org/infosecFAQ/protocols/RIP.htm).

Joncheray, Laurent. 24 April, 1995. "A Simple Active Attack Against TCP." *Proceedings of the Fifth USENIX Unix Security Symposium.*

Kent, Stephen. September, 1977. "Encryption-Based Protection Protocols for Interactive User-Computer Communication." *Proceedings of the Fifth Data Communications Symposium*.

Krywaniuk, Andrew. 9 July, 2001. *Using Isakmp Message Ids for Replay Protection.* Internet Engineering Task Force.

Levy, Steven. 1994. *Hackers: Heroes of the Computer Revolution*. New York: Penguin Books.

Northcutt, Stephen, and Judy Novak. 2001. *Network Intrusion Detection: An Analyst's Handbook, Second Edition*. Indianapolis: New Riders.

Power, Richard. 2000. *Tangled Web: Tales of Digital Crime from the Shadows of Cyberspace*. Indianapolis: Que.

RSA Laboratories. May, 2000. *Frequently Asked Questions About Today's Cryptography, Version 4.1*.

Schneier, Bruce. 15 July, 1998. "Breaking RSA in PKCS1." *Crypto-Gram Newsletter* (http://www.counterpane.com/crypto-gram-9807.html).

Stevens, W. Richard. 1994. *TCP/IP Illustrated Volume 1: The Protocols*. Reading, MA: Addison-Wesley.

Chapter 4

Broadband Networking Technologies

When designing and implementing network security, it is important that we understand the network in terms of its layers, protocols, and applications. Before we can choose appropriate security mechanisms, we must familiarize ourselves with the function of each network layer as it relates to the overall operation of the network. We must also acquaint ourselves with the various communication and security protocols that run within each layer. Equally as important is an understanding of the applications we are trying to protect. Performance characteristics and service-level requirements vary greatly from one application to the next, and the security mechanisms adequate for one application may not be adequate for another. The multimedia applications of the future will increase the need for high-performance, cost-effective security implementations that augment—but do not interfere with—network operations. Without knowledge of the underlying network structure and the parameters used to gauge network performance and quality, designers, implementers, and administrators will have a difficult time providing this crucial security.

This chapter begins with a discussion of the Open Systems Interconnection (OSI) and TCP/IP network reference models used to describe the functional characteristics of the various network layers. We discuss many of the important protocols running at each layer, and provide a number of characteristics for classifying communication protocols. We then describe a typical service provider network in terms of backbone, distribution, and access components, and briefly discuss the most common access technologies. The chapter concludes with an introduction to quality of service (QoS) and the difficulties associated with ensuring QoS over IP-based networks. The intent of this chapter is to provide an understanding of the fundamental network concepts and terminology that we will apply in future chapters.

The Origins of Broadband

In the late 1970s and early 1980s, innovations in local area network (LAN) technology resulted in two competing communication technologies known as *baseband* and *broadband*. Baseband signaling uses *time-division multiplexing (TDM)* to transmit multiple signals simultaneously over the same wire. TDM divides the total available bandwidth into a number of lower-bandwidth signals. Each signal is periodically allotted a slice of time during which it can transmit data over the wire (see Figure 4-1). The amount of time allotted to each signal correlates directly to its available bandwidth. (Notice in Figure 4-1 that Signal 1 has twice the available bandwidth of Signal 2 or 3). The longer the time slice, or the more frequent the allotment, the greater the effective bandwidth available to the signal. In baseband transmission, digital signals are placed directly, onto the cable in the form of high and low voltages, which are represented by 1's and 0's. These signals require no *modulation*—change in amplitude, phase, or frequency—and have access to the full bandwidth of the cable during their respective time slices. Examples of baseband technologies include the Ethernet and Token Ring network topologies.

Token Ring is a classic case of TDM, where a host must possess a token to communicate over the network. Only the host with the token can transmit, and

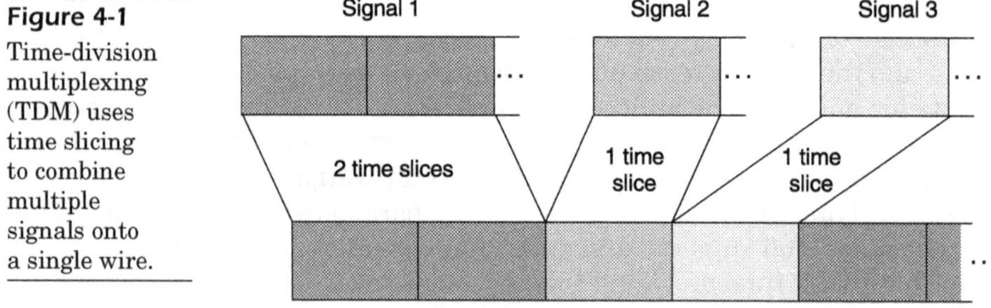

Figure 4-1 Time-division multiplexing (TDM) uses time slicing to combine multiple signals onto a single wire.

the host must relinquish control of the token following a brief transmission period. Ethernet, on the other hand, deserves further explanation. Although hosts communicating over an Ethernet network have complete control of the medium during transmission, they are not assigned slices of time during which to transmit. Instead, before transmitting data, a host checks that no other host is communicating on the network. If the host detects no traffic, it begins transmitting. On busy Ethernet networks, two hosts may begin communicating at the same time, which results in a collision. Following a collision, both hosts stop transmitting and wait a random delay interval before attempting retransmission. Each host repeats this operation until it receives exclusive control of the transmission medium.

The term *broadband* originally referred to analog communication systems. Broadband signaling techniques, such as those used by the cable television industry, transmit data in analog format. Any digital information must first be converted to an analog signal before being placed on the line. The device responsible for analog modulation is known as a *modem* (modulator/demodulator). Modems use *frequency-division multiplexing* (*FDM*) to simultaneously transmit multiple signals over the same wire. FDM works by assigning each signal a unique frequency range, or *carrier,* and transmitting all carriers in parallel (see Figure 4-2). The frequency ranges are separated enough that they do not interfere with one another during transmission. Signals with larger frequency ranges have more bandwidth available to them, and data is transmitted by modulating the carrier. Telephone companies were the first to use FDM in conjunction with high-speed digital transport services.

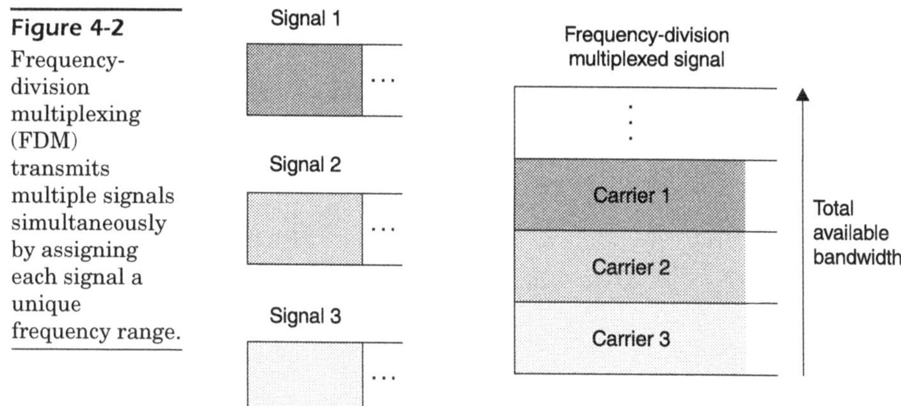

Figure 4-2 Frequency-division multiplexing (FDM) transmits multiple signals simultaneously by assigning each signal a unique frequency range.

The ISO/OSI Reference Model

The *Open Systems Interconnection (OSI) Reference Model* was developed in 1974 by the ISO (International Standards Organization). It provides a method for describing network protocols and distributed applications, and arose from the need for interoperability among communication products developed by different vendors. The model consists of seven layers and defines the functions performed by each layer. The actual hardware or software implementation, however, is left up to the vendor. A major benefit of the OSI model is that each layer can be developed independently. This autonomy allows vendors to implement a particular layer without concerning themselves with the details of the other layers. Table 4-1 lists the protocols that commonly run within each layer of the OSI model.

Layer Number	Layer Name	Common Protocols
7	Application	FTP, TFTP, Telnet, SMTP, POP3, IMAP, HTTP, NNTP, SNMP, NFS, FINGER, DHCP, X Windows
6	Presentation	ASCII, EBCDIC, JPEG, GIF, TIFF, MPEG, MIDI, CORBA IIOP, encryption, compression
5	Session	RPC, NFS, NetBIOS, SQL, LDAP, SAP, RTP
4	Transport	TCP, UDP, SPX
3	Network	IP, ARP, RARP, ICMP, IPX, RSVP, various routing protocols (RIP, BGP, IGMP, EGP, IGRP/EIGRP, OSPF)
2	Data Link	Ethernet, Token Ring, Frame Relay, ATM, FDDI, HDLC, SDLC, PPP, SLIP, X.25, V.42, V.42bis, ISDN
1	Physical	X.21, V.24, V.35, V.90, EIA/TIA-232, RJ45, 10Base2, 10BaseT, 100BaseT, HSSI, SOnNET

Table 4-1 Layers of the OSI Reference Model

Layer 7—Application

Layer 7, the highest level in the OSI model, defines the component of an application that communicates with applications running on other computers. This layer is responsible for determining whether appropriate system resources exist to establish a network connection, and whether the intended recipient is available. A user interacts directly with the Application layer when using a program. Typical examples of Application layer activities include file transfers, remote terminal access, e-mail, and database queries.

Layer 6—Presentation

The Presentation layer defines message formats for representing data. It is also responsible for *translating* (converting) data from one format to another. Encryption and compression are frequently applied at the Presentation layer, since they can be considered forms of data translation. However, both encryption and compression can be implemented within any of the OSI layers. The standard message syntax used by the Presentation layer is *Abstract Syntax Notation One (ASN.1)*. ASN.1 provides a mechanism for sharing information between computer architectures that have completely different internal representations of the transmitted data.

Layer 5—Session

The Session layer controls setup and termination of communication sessions. It identifies which data streams belong to a particular session, and describes three transfer modes: simplex, half-duplex, and full-duplex. *Simplex* mode allows data transmission in only one direction. *Half-duplex* mode supports two-way transmission, but allows only one communicating party at a time to transmit data. *Full-duplex* mode supports simultaneous two-way communication. The Session layer provides a “seamless” stream of data to the Presentation layer by performing a series of checks on incoming messages. If an error or failure occurs in a session, then only the data transmitted since the last successful check must be retransmitted.

Layer 4—Transport

Transport layer protocols establish a logical connection between two hosts. This connection is often referred to as a *virtual circuit*. The Transport layer multiplexes data streams from upper-layer applications into a single transport connection. Connection-oriented Layer 4 protocols, such as TCP, handle error recovery by acknowledging receipt of transmitted data. If the data contains an error, or if it is not received within a given period of time, it is retransmitted. The Transport layer reorders traffic arriving out of sequence (based on sequence numbers), and provides a level of flow control to prevent a sender from flooding a recipient with traffic. Connectionless Transport layer protocols (UDP, for example) do not perform error recovery. (We will compare and contrast connection-oriented and connectionless protocols later in this chapter, in the section entitled "Communication Protocol Characteristics.")

Layer 3—Network

The Network layer defines the logical addressing schemes used to provide end-to-end delivery of data over multiple network segments. It specifies how routing works via path determination and route discovery, and converts Layer 3 logical addresses to Layer 2 hardware addresses. Network layer protocols, such as IP, also handle fragmentation of packets as they move between networks that have different maximum transmission units (MTUs).

NOTE:

In Table 4-1, ICMP (Internet Control Message Protocol) *is listed as running at the Network layer. ICMP is an extension of the IP protocol, and is used exclusively to communicate Network layer status information and error codes relating to IP. However, some texts list ICMP as running at the Transport layer, since ICMP data is actually encapsulated within IP datagrams.*

Routers are Layer 3 network devices responsible for forwarding data packets from one network segment to another. To determine the appropriate forwarding behavior, a router analyzes the contents of a packet's Network layer protocol header. The router then references its *routing tables* to determine the most efficient path to the destination network.

Layer 2—Data Link

Data Link layer protocols are responsible for moving data across a single link in the network path between the source and destination. They provide physical addressing mechanisms, and are responsible for deciding when it is appropriate to access physical transmission mediums. Layer 2 protocols also perform error detection using a *frame check sequence (FCS)*. The FCS usually takes the form of *cyclic redundancy checksums (CRCs)*, which work as follows: The sending device calculates a checksum over the data contained within the Layer 2 frame, and appends the checksum to the end of the frame. The recipient verifies the integrity of the data by calculating a corresponding checksum and comparing it to the one contained in the frame. If the two checksums match, it is reasonable for the recipient to assume that no errors have occurred during transmission.

NOTE:

In general, a CRC represents the sum of bits contained within a message, and can only detect an odd number of bit errors. In the event that an even number of bits is flipped, the CRC may provide a false verification. Bit errors occur infrequently, however, so the chances of multiple bit errors in a single frame are slim.

Switches and bridges are Layer 2 network devices that forward traffic based on the hardware addresses contained in Layer 2 protocol headers. The Data Link layer consists of two sublayers:

- **Logical Link Control (LLC)** The LLC layer manages data link communication. It simplifies the task of the Network layer by abstracting the details of the underlying MAC layer. It also facilitates the transfer of data from the MAC layer to the Network layer once Layer 2 headers have been removed, and in some cases, provides flow control, sequencing, and timing.
- **Media Access Control (MAC)** For outgoing data, the MAC layer is responsible for physical addressing, calculation of the FCS, and conversion of upper-layer data into bits appropriate for the physical layer. For incoming data, the MAC layer creates frames from physical layer bits, checks CRCs, and examines the hardware address contained within the frame to determine whether the data is intended for the local host. The MAC layer is also responsible for determining the most appropriate time for media access. Media access methods include contention, token passing, and polling.

Layer 1—Physical

Layer 1, the lowest level in the OSI model, defines the physical characteristics of the transmission medium. Cabling, connectors, pin layouts, voltage and current levels, modulation techniques, and transmission speeds are all Layer 1 characteristics. The Physical layer describes the signaling mechanisms used to transmit bit streams between network interfaces. Signaling mechanisms take many forms, including audio tones, voltage transitions, and light pulses.

Transmission signals degrade as the distances increase between sending and receiving devices. Layer 1 devices, known as *repeaters,* boost weak signals to increase the distance over which they can be transmitted. Repeaters do not examine any addressing information; their functionality exists at the Physical layer only.

The following table lists a number of common network devices and indicates the layer (or layers) where each device functions:

Devices	Network Layers
Application-level gateways	Application, Presentation, Session, and Transport
Routers	Network
Bridges and switches	Data Link
Hubs, concentrators, and repeaters	Physical

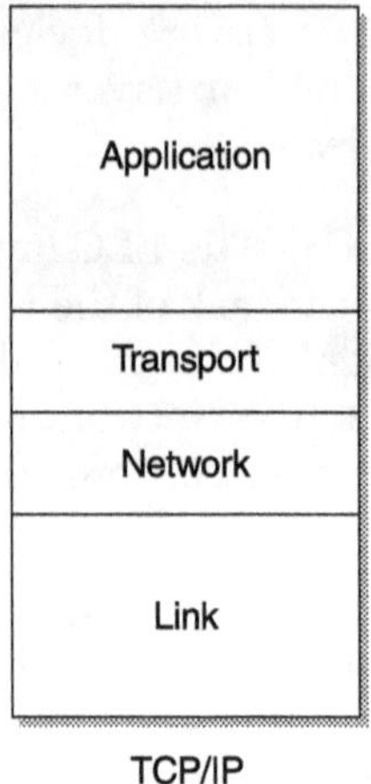

Figure 4-3 The TCP/IP Reference Model consists of four layers, which operate in much the same way as those of the OSI Reference Model.

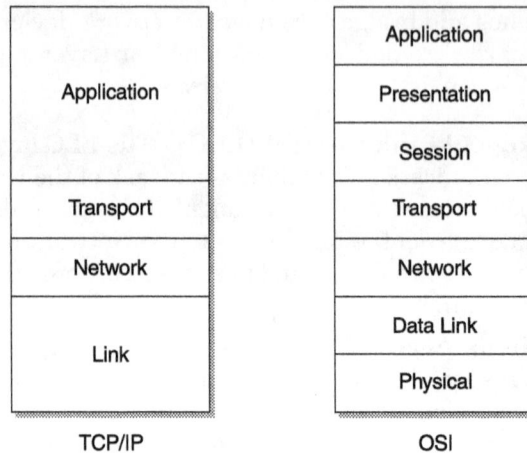

The TCP/IP Reference Model

The *TCP/IP Reference Model* is somewhat simpler than the OSI model. It consists of four layers—Link, Network, Transport, and Application—that function in much the same was as the layers of the OSI model (see Figure 4-3). However, the TCP/IP model pertains directly to the TCP/IP suite of protocols (see Figure 4-4). Notice that the Link layer encompasses Layers 1 and 2 of the OSI reference model, and that the Application layer corresponds to OSI Layers 5 through 7. Also, notice that the TCP/IP protocol suite, illustrated in Figure 4-4, contains only three layers. The Link layer is typically implemented within device drivers for network interface adapters, and not directly within the TCP/IP suite.

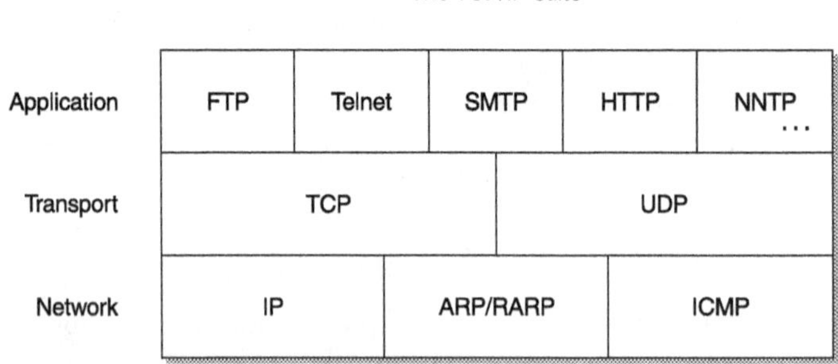

Figure 4-4 The TCP/IP suite consists of a number of applications and protocols running at different layers of the TCP/IP Reference Model. (Many more applications exist than are shown here.)

NOTE:
For convenience, and unless otherwise stated, we will use the term "application" to refer to Layers 5 through 7 of the OSI model, since the functions of these layers are implemented directly by the application, and not by a protocol stack.

Data Encapsulation

As user data moves through the protocol stack, each network layer wraps the data passed to it from upper-layer protocols in fields containing control information, a process known as *encapsulation*. Fields preceding the data are known as *headers,* and those following the data are known as *trailers*.

NOTE:
The Data Link layer is the only network layer that uses trailers (although the padding added by encryption at the Presentation layer might also be considered a trailer).

Headers and trailers provide the control information necessary for transmitting and interpreting data between applications. For example, Ethernet headers contain hardware addresses for identifying particular network adapters, and IP headers contain logical source and destination addresses essential for routing packets across multiple network segments. Both of these are necessary for data transmission over the Internet. Presentation layer headers, such as those for GIF images and MPEG video, contain information about encoding formats; receiving applications use these headers to interpret incoming data. An example may help to demonstrate the concept of encapsulation (here we are employing the naming conventions used by the OSI Reference Model):

1. Once an application (the combined Application, Presentation, and Session layers) receives user data, it adds header information to the data and passes it to the Transport layer.
2. The Transport layer creates a Layer 4 *segment* by adding the appropriate protocol header and passes the segment to the Network layer.
3. Once the Network layer receives the Transport layer segment, it adds a Network layer header to generate a Layer 3 *packet*, or *datagram*. The Network layer then passes the packet to the Data Link layer.
4. The Data Link layer adds a header and trailer to produce a Layer 2 *frame*, and it communicates the frame to the Physical layer as a sequence of *bits*. The Physical layer converts the bits to a signal (i.e., high and low

voltages) for transmission across a physical medium. The reverse occurs when a message reaches the intended recipient. Each network layer strips its associated header (and trailer) before passing the contents of the data field to the layer above it.

NOTE:
In a typical communication session, an application operating within Layers 5 through 7 of the OSI model adds headers only once during initialization. The lower-layer protocols (2 through 4), on the other hand, apply header information to each unit of data transmitted over the network.

You may sometimes hear segments, packets, and frames referred to as Layer 4, Layer 3, and Layer 2 *protocol data units (PDUs),* respectively. A PDU is a grouping of bits into autonomous units consisting of a header and data field. "Layer N PDU" is often written in shorthand notation as LNPDU. Figure 4-5 illustrates the encapsulation process for the TCP/IP protocol stack (notice now that we are using the terminology of the TCP/IP Reference Model, since we are referring specifically to the TCP/IP suite).

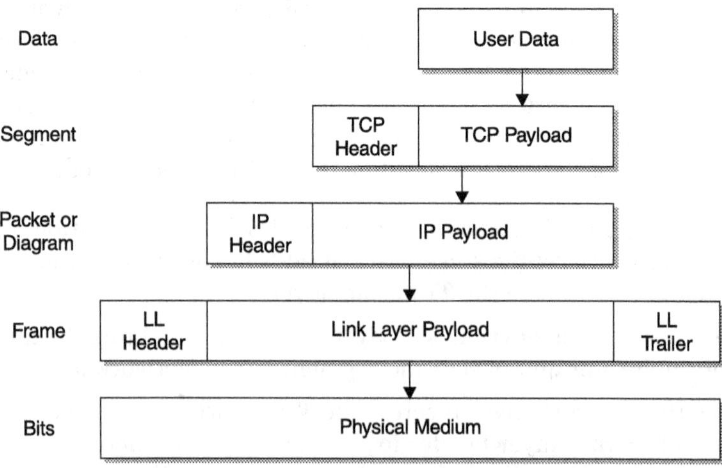

Figure 4-5 Each layer within the TCP/IP protocol stack encapsulates data from upper-layer protocols by adding a header and, in the case of the Link layer, a trailer.

Communication Protocol Characteristics

Communication protocols are generally classified as a combination of connection-oriented or connectionless, and reliable or unreliable. A *connection-oriented* protocol establishes a logical connection between communicating parties prior to any exchange of information. Such a connection is known as a *virtual circuit*. Establishing a connection may involve the exchange of initialization messages, such as in the TCP three-way handshake, or the existence of a predefined relationship between the communication endpoints (Odom, 2000). ATM and Frame Relay are connection-oriented protocols that make use of *permanent virtual circuits (PVCs)*. In a PVC, a logical connection always exists between the endpoints, so there is no need to exchange initialization messages to establish a session. Connection-oriented protocols also provide a mechanism for gracefully terminating a session. In the case of TCP, communicating peers exchange TCP FIN (finish) commands to signal that they are ready to tear down a virtual circuit.

Most connection-oriented protocols use sequence numbering to facilitate ordered data delivery. Prior to transmission, each packet is assigned a unique sequence number that identifies its location in the original data stream. By comparing the sequence numbers, the recipient can reconstruct a message from packets that arrive out of order (that is, in an order different from that in which they were originally transmitted). In addition, sequence numbering allows the recipient to detect duplicate packets. Any packet containing a sequence number that has been previously received is discarded.

A *connectionless* protocol lacks a mechanism for formally establishing a session, and, as a result, it does not create a virtual circuit. The sender doesn't verify that the destination host is operational before transmitting data. It simply addresses and transmits the data, and then forgets about it. In many cases, connectionless protocols are combined with higher-layer protocols that provide connection-oriented behavior.

A reliable protocol allows communicating parties to establish an ongoing dialogue for detecting and recovering from transmission errors. Characteristics commonly associated with reliable communication protocols include acknowledgment, error detection, and retransmission. To ensure successful data delivery, reliable protocols acknowledge receipt of every PDU. PDUs that are not received in a timely fashion, or that contain errors, are not acknowledged; these must be retransmitted by the sender. For instance, TCP uses checksums to detect bit errors during transmission, and acknowledges successful delivery by using a technique known as *forward acknowledgment*: After receiving an error-free packet, the recipient sends a TCP ACK message to the sender containing the sequence number of the next packet it expects to receive.

An *unreliable* protocol may detect errors, but it provides no means of retransmitting lost or corrupted data. Retransmission is left up to layers higher in the protocol stack. Reliability is often associated with connection-oriented protocols; however, not all connection-oriented protocols are reliable. Likewise, not all connectionless protocols are unreliable.

An analogy often used to describe reliable, connection-oriented protocols is a telephone conversation. A phone call typically begins with "Hello!" and "Is so-and-so there?" This is similar to a protocol handshake. Communicating parties maintain a dialogue throughout the entire call, and mutually agree to end the conversation. The call ends with "Goodbye" or "Talk to you later" (analogous to a TCP FIN message). Either side can quickly detect a problem if the other side stops responding.

Communicating over an unreliable, connectionless protocol is similar to sending a letter in the mail. You address the letter and drop it in a mailbox, but, typically, you do not expect to receive a return letter stating that your letter has been received. Depending on how busy the postal service happens to be, your letter may be lost or misrouted. However, you have no way of knowing this unless you use a connection-oriented method (a phone call, for instance) to contact the intended recipient.

Service Provider Networks

Recall from Chapter 1 that a typical service provider network is composed of a backbone network and multiple distribution and access networks (see Figure 4-6). A *backbone* network consists of high-capacity lines called *trunks* that are used to transmit large quantities of data over long distances. Backbones are the highest-bandwidth sections of a service provider's overall network, and they transmit data over the greatest distances. To achieve the capacity necessary for supporting a large number of consumers, backbone networks commonly employ fiber optic technologies and high-performance protocols such as Synchronous Optical Network (SONET) and Asynchronous Transfer Mode (ATM). (We'll discuss ATM in the section entitled "Quality of Service," later in the chapter.) A single backbone network typically services many geographically disperse distribution networks.

A *distribution* network sits between the backbone and access networks. It connects neighboring access networks and carries traffic between them, and it passes traffic destined for remote locations onto the backbone. Distribution networks typically span entire metropolitan areas, and utilize high-performance, high-bandwidth technologies similar to those used by backbone networks. A single distribution network may service many disparate access networks (e.g., voice, video, and data) and can potentially support multiple access network technologies (e.g., cable and digital subscriber line).

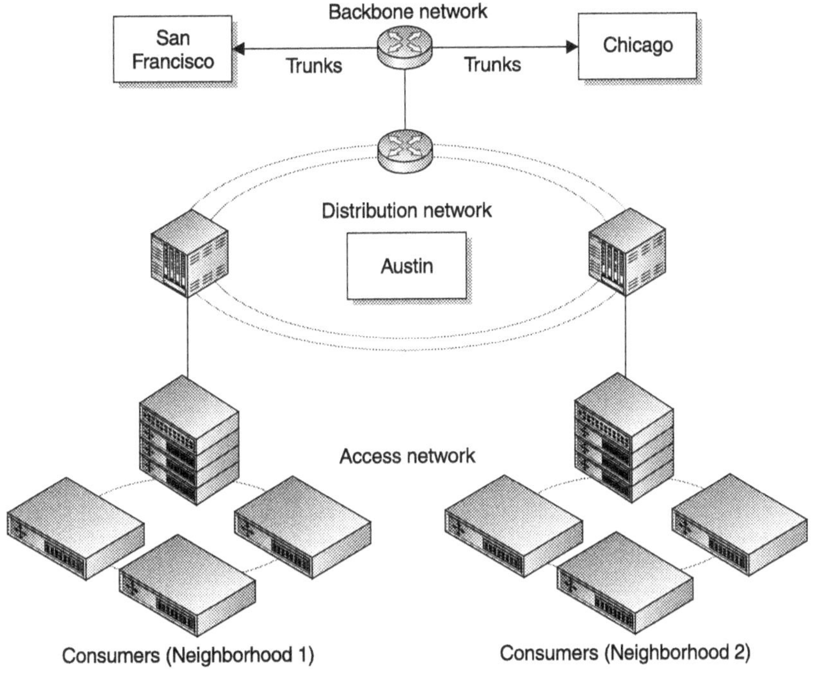

Figure 4-6 A typical service provider network consists of backbone, distribution, and access network components.

An *access* network carries traffic between a distribution network and the homes or offices of the end-using consumers (that is, the customers' premises). Due to their proximity to the customers' premises, access networks are commonly referred to as the *last mile.* The term *local loop* is also used to describe those access networks that employ coaxial cabling or UTP copper. Network technologies deployed in access networks vary greatly in terms of available bandwidth and transmission distances. The most common broadband access network technologies include cable, digital subscriber line, fixed wireless, and satellite.

Cable

Community Antenna Television (CATV) has been around since the early 1960s. The original intent of CATV was to provide television programming to consumers who, due to geographic restrictions or interference, could not receive over-the-air signals. Cable services have since expanded to include entertainment, voice communication, and Internet access. Early cable networks used a *tree architecture* (see Figure 4-7) to distribute CATV signals to consumers over coaxial cable. Each *headend* was equipped with satellites and directional antennas for receiving broadcast television signals. Following modulation, the headend placed the signal

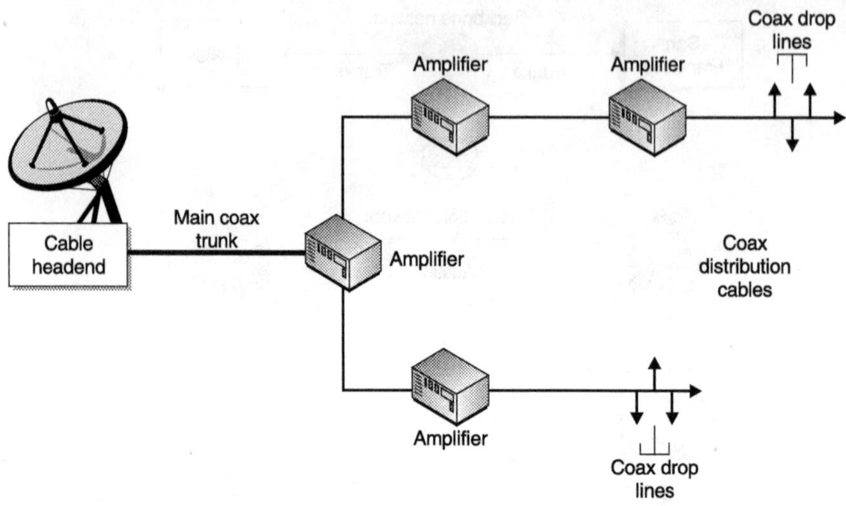

Figure 4-7 Early CATV networks used a basic tree architecture to distribute cable signals to consumers.

on the main *trunk* of the tree. The trunk carried the signal from the headend to each neighborhood, where *distribution cables* (branches) carried the signal past consumers' homes. Finally, *feeder lines* (also called *drop lines*) ran directly from the distribution cable into the homes, each one terminating at a set-top box or cable-ready television that demodulated the signal.

The basic tree architecture required amplifiers to periodically boost the strength of the signal as it traveled from the headend to the home. Unfortunately, the amplifiers boosted not only the signal strength, but noise and interference as well. In some cases, they caused unacceptable levels of signal distortion, resulting in poor picture quality. In the late 1980s, cable operators discovered that they could use fiber optics to increase available bandwidth in the distribution network while improving performance and signal quality (Bates, 2000). Fiber optic technologies are highly tolerant of environmental interference and are capable of transmission over long distances with minimal signal degradation. The use of fiber optics reduced the number of amplifiers needed to carry signals to consumers' homes greatly improving picture quality.

Current-day cable networks consist of a mixture of fiber optics and coaxial cabling in hybrid fiber-coax (HFC) architectures (see Figure 4-8). Within an HFC network, a *fiber loop* distributes the cable signal from the headend to a local *fiber node*. Coaxial cable carries the signal from the fiber node directly to the customer premises. Although coaxial cables still service the last leg of the network, the use of fiber significantly increases distribution capacity and moves the increased bandwidth closer to the consumer. Notice that the headend in Figure 4-8 also contains fiber connections back to the core network in support of data and other interactive services.

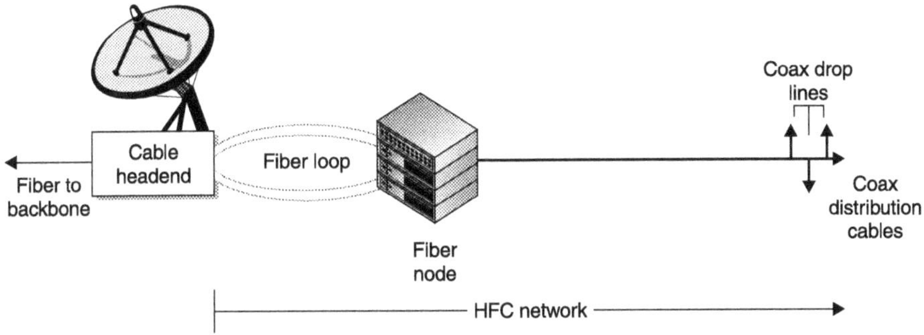

Figure 4-8 A hybrid fiber-coax (HFC) network combines fiber optics and coaxial cabling to carry cable signals to the customer premises.

NOTE:
Even with the increased bandwidth offered by fiber optics, the number of users serviced by a single node has declined steadily over the past decade. Numbers have dropped from 5,000–20,000 users per node to around 500 users (Bates, 2000). This demonstrates a growing trend in bandwidth consumption by end user services.

Cable Transmission Techniques

The analog cable spectrum is divided into a number of 6MHz channels. The effective broadcast spectrum for each channel actually occupies about 4.2 MHz, with the remainder forming buffers on either side of the 4.2MHz band to prevent interference with other channels. All channels are transmitted in parallel over the same cable using FDM. Common modulation techniques include *quadrature amplitude modulation (QAM), vestigial side band (VSB)* amplitude modulation, and *quadrature phase shift keying (QPSK).* The cable provider is responsible for apportioning the available cable spectrum into video, voice, and data channels. Figure 4-9 illustrates the current allocation of the spectrum into analog and digital video, voice, and data.

Until fairly recently, cable services had been limited to the transmission of analog sources. The advent of digital cable services carried with it the potential for many new and exciting interactive applications. However, digital services required new encoding techniques for modulating bit streams into phase and amplitude components appropriate for analog transmission mediums. The encoding and compression schemes used today make better use of available bandwidth than did traditional analog techniques, and they are more resistant to noise and interference.

Digital television (DTV) emerged in 1998 as one of the first digital services. Current MPEG-2 DTV encoding supports a video stream of 18.9 Mbps and a Dolby AC-3 5.1 surround-sound audio stream of 384 Kbps over a single 4.2MHz

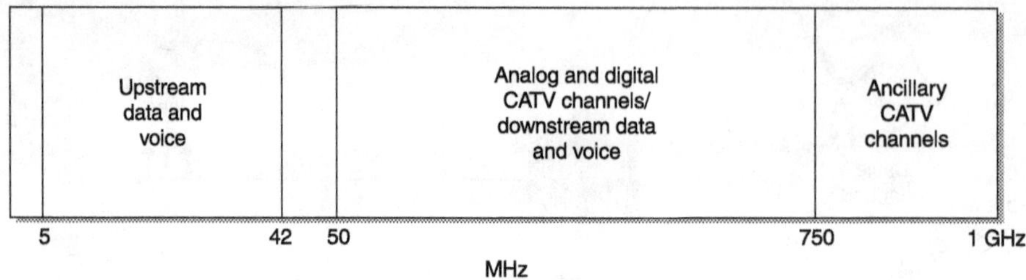

Figure 4-9 Cable providers divide the available cable spectrum into analog and digital video, voice, and data frequency bands.

cable channel. These are the requirements for a single high-definition television (HDTV) channel. Both signals are multiplexed into a single bit stream for transmission to consumers. Digital encoding and compression allow multiple standard-resolution video and audio streams to share the same 4.2MHz band; as a result, cable providers can now offer far more channels than ever before. This increase in available bandwidth also opens the door to interactive television, telephony, and high-speed data services, all offered by the same provider.

Cable Modems

Up until recently, cable networks provided one-way, downstream distribution of content from the cable operator to the consumer. Before a cable company could offer interactive services, it had to install an additional upstream channel for carrying consumer data back to its provider network. Despite the cost of this upgrade, most cable networks do now support two-way communication. One of the first practical interactive services to utilize the upstream channel has been high-speed Internet access.

Users of high-speed Internet access connect to the cable data network using cable modems. A cable modem converts digital data from a PC into an analog format compatible with the cable network, and vice versa. The digital signal must be modulated into amplitude and phase components before it is placed on the cable, and must be demodulated before it is sent to the PC. A subscriber attaches a PC to a cable modem—the customer premises equipment—using a 10 Mbps Ethernet connection.

NOTE:

Since cable providers currently limit data rates to well under 10 Mbps, there is no immediate need for a cable modem that would support Ethernet connections faster than 10 Mbps.

The other end of the cable modem attaches directly to the cable network via a coaxial drop line. This drop line taps into a distribution cable that carries data to

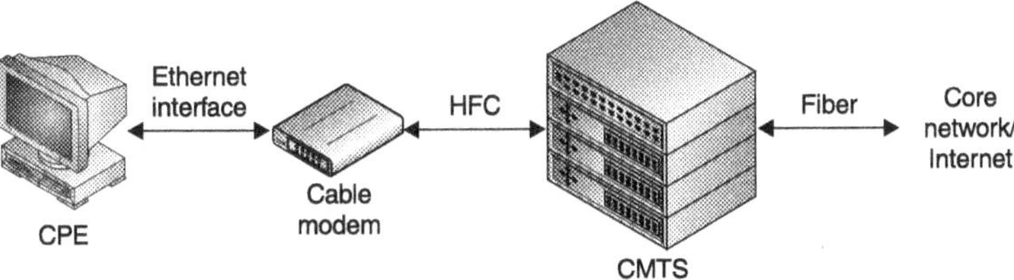

Figure 4-10 A cable modem exchanges data with a cable modem termination system (CTMS) via an HFC network. The CMTS forwards data to and from the provider's core network.

and from the fiber node servicing the area. On the other end of the fiber ring is a headend equipped with a *cable modem termination system (CMTS)*. The cable modem communicates directly with the CMTS, which is responsible for demodulating the modem's signal and passing the data onto the provider's network over the fiber backhaul connection. Figure 4-10 illustrates the components of a cable modem network. Although cable networks are capable of downstream bandwidths near 30 Mbps, downstream data rates for most cable modem services are currently 1.2–2 Mbps, with upstream rates of 128–512 Kbps.

Cable networks are shared media networks, which means that all cable modem subscribers on the same network compete for the bandwidth of a single cable. Bandwidth varies greatly depending on the number of concurrent users, and subscribers are likely to experience poor performance during peak hours. Cable service providers do not yet guarantee bandwidth availability. However, evolving industry standards are beginning to address bandwidth and service-level provisioning.

DOCSIS and PacketCable

Cable Television Laboratories, also known as CableLabs, is a research and development consortium of *multiple systems operators (MSOs)*—companies that operate multiple cable systems in different geographic locations—whose goal is to promote the advancement of cable technologies and vendor interoperability across the cable market. CableLabs are the developers of the *Data-Over-Cable Service Interface Specifications (DOCSIS)*, which were adopted by the International Telecommunication Union (ITU) in 1998 and which have since become international standards. DOCSIS consists of a number of specifications defining the various components of cable modem network architecture (such as CPE-to-cable modem interfaces, physical and MAC layer transmission, communication, security protocols, and operations support systems). DOCSIS standardization provides many benefits to consumers, providers, and hardware vendors (Kennebeck, Marcheck, and Mendelson, 2001):

- Consumers can be sure that cable modems and CMTSs from CableLabs-certified vendors will operate successfully with one another.

- Modem vendors do not have to sell through MSOs; they can sell directly to consumers. (However, most MSOs specify the modems they are willing to support within their networks, and distribute these modems directly to their customers.)
- Competition between vendors of cable modems and CMTS results in lower prices for MSOs and consumers.
- Cable operators are not limited to working with a single cable modem or CMTS vendor; they can reduce operating costs by purchasing from many competitive vendors.
- DOCSIS 1.1 includes quality-of-service provisioning to support real-time applications.
- The up-and-coming DOCSIS 2.0 series of specifications includes enhanced Physical layer modulation techniques that will increase upstream data rates from 10 to 30 Mbps.

Another CableLabs initiative, PacketCable, defines a series of specifications for delivering real-time multimedia services over packet-based cable networks running the Internet Protocol. It includes specifications for audio and video encoding, quality of service, and security. PacketCable provides end-to-end message transport and call signaling, and defines a number of operations relating to network management and billing. The first practical application for PacketCable networks will be IP telephony; however, other streaming multimedia applications—such as video conferencing and interactive gaming—will likely follow. We will discuss DOCSIS and PacketCable further in Chapters 9 and 10.

Digital Subscriber Line

Back in the good old days, consumers used analog modems to connect to the Internet. For years, modem technology kept pace with bandwidth demands. Traditional modem technology (see Figure 4-11) converts data from customer premise equipment to an analog "voice" signal that can be transmitted over the *public switched telephone network (PSTN).* The modulated analog signal occupies a 4KHz frequency band, which is all that is required for voice communication. However, this meager frequency range limits the raw throughput of analog modems to less than 56 Kbps. While this is sufficient for almost any static application consisting of text and simple graphics, today's multimedia applications have quickly outgrown this modest allotment. Furthermore, in order for a consumer to achieve data rates of nearly 56 Kbps, the copper wiring between the customer premise and the telephone company's central office (CO) has to be in good condition. Due to poor line quality, many consumers never connect at speeds above 33 Kbps.

In the early 1990s, telephone companies—looking to compete directly with cable and satellite providers—experimented with offering video services. However, they needed faster connectivity to support these services. The telephone companies had

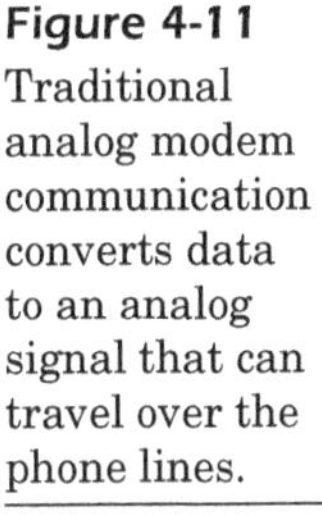
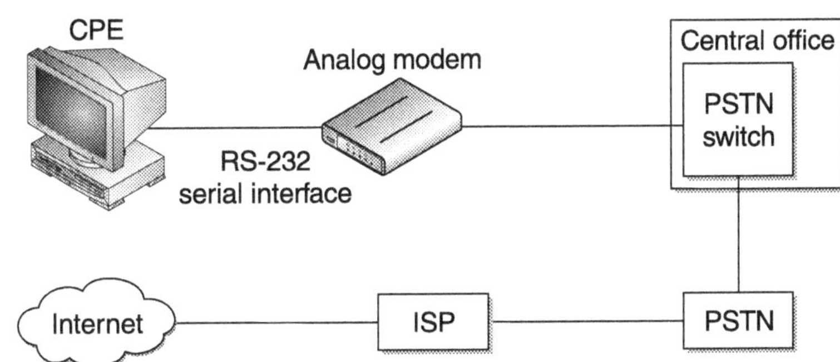

Figure 4-11 Traditional analog modem communication converts data to an analog signal that can travel over the phone lines.

already invested significantly in *unshielded twisted pair (UTP)* copper wiring, and needed a high-speed solution that could take advantage of the existing wire plant. *Digital Subscriber Line (DSL)* proved to be a likely fit, but not for what the telephone companies had originally intended. Initial video services were unsuccessful, and telephone companies quickly turned instead toward offering high-speed Internet access. Internet access remains the primary application for DSL today.

DSL and voice communication coexist on the same copper wire. Voice signals occupy the frequency range between 0 and 4 KHz, leaving a considerable range of higher frequencies untouched. DSL signals occupy these higher frequencies, allowing the transmission of data without interference with voice signals. Whereas voice communication travels from the consumer's home or office through the telephone company's CO and onto the PSTN, DSL communication runs only the length of the copper wire between the customer premises and the CO. CPE data enters the DSL modem through an Ethernet connection and passes through a *splitter,* where it is combined with voice traffic. The copper wire then carries the combined data/voice signal to the CO. At the CO, another splitter separates the voice and data frequencies. The voice signal passes through to the normal switched telephone network, and the data signal enters a receiving DSL modem contained within a *digital subscriber line access multiplexer (DSLAM).* The modem converts the frequencies back into a data stream that is carried, via any number of LAN/WAN networking technologies, to an *Internet service provider (ISP)* or a *network service provider (NSP).* This process is illustrated in Figure 4-12.

Analog telephones can be connected to the line using active or passive splitters. Passive splitters draw power directly from the phone line, whereas active splitters have their own power supplies. The configuration shown in Figure 4-12 uses a passive splitter.

Unlike cable, DSL is not a shared-bandwidth service. A dedicated wire runs directly from the consumer's location to the CO. This ensures that bandwidth remains constant, and allows telephone companies to guarantee upstream and downstream bandwidths.

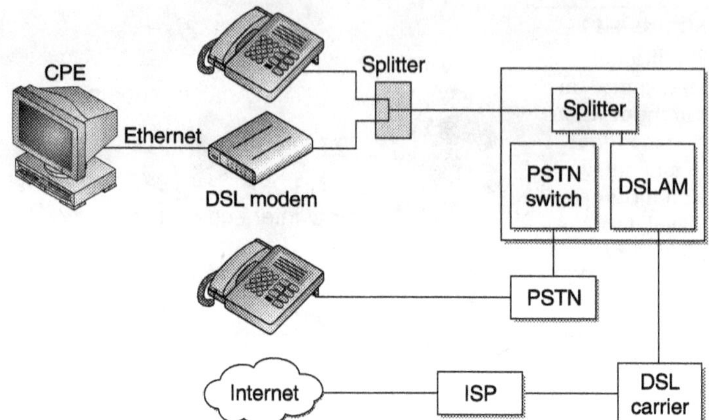

Figure 4-12 On DSL networks, voice and data signals are transmitted over the same copper wire to the telephone company's central office.

DSL technology actually consists of a family of DSL variants. The term x*DSL* is used to describe the entire family. The transmission rates and distances for members of the *x*DSL family vary greatly. Table 4-2 lists many common *x*DSL variants, and Table 4-3 lists the maximum transmission distances from the most common DSL—Asymmetric DSL (ADSL).

NOTE:
The data rates and distances provided in Tables 4-2 and 4-3 are theoretical limits based on copper wiring that is in good condition. In many cases, poor quality wiring or environmental factors, such as water saturation, greatly reduce the capacity of x*DSL.*

sxDSL Variant	Typical Downstream Data Rates	Typical Upstream Data Rates
Asymmetric DSL (ADSL)	1.544–6.144 Mbps	128–640 Kbps
High bit rate DSL (HDSL)	1.544–2.048 Mbps	1.544–2.048 Mbps
Rate Adaptive DSL (RADSL)	64 Kbps–8.192 Mbps	128–768 Kbps
Very High-Speed DSL (VDSL)	13–55 Mbps[1]	1.6–2.3 Mbps and 19.2 Mbps
Symmetric DSL (SDSL)	1.544–2.048 Mbps	1.544–2.048 Mbps

[1] High data rates require very short transmission distances (on the order of 1000–4000 feet).

Table 4-2 Upstream and Downstream Data Rates for Common *x*DSL Variants

Current Data Rate	Wire Gauge[2]	Distance in Thousands of Feet	Distance in Miles/ Kilometers
1.5–2.048 Mbps	24	18	3.4/5.5
1.5–2.048 Mbps	26	15	2.8/4.6
6.144 Mbps	24	12	2.3/3.7
6.144 Mbps	26	9	1.7/2.7

[2] *Wire gauge* refers to the diameter of a wire. The thicker the wire, the lower the gauge. Common gauges for telephone wiring are 24 and 26.

Table 4-3 Maximum Transmission Distances for ADSL (Bates, 2000).

Although DSL deployments use a handful of modulation techniques, the industry as a whole has settled on a single technique known as *discrete multitone modulation (DMT)*. DMT is a form of frequency division multiplexing that breaks up the input frequency spectrum into a large number of narrow carrier bands, called *subchannels*. Each subchannel is independently modulated by a *carrier frequency* located in the middle of the band. All subchannels are of equal bandwidth, and they are processed in parallel. The ANSI (American National Standards Institute) standard for DMT in conjunction with ADSL uses 256 subchannels, each with a bandwidth of 4.3125 KHz. A total of 1.1 MHz is subdivided into ranges, with the 0–4 KHz range reserved for voice, 26–133 KHz reserved for upstream traffic, and 142 KHz–1.1 MHz reserved for downstream traffic. The ranges are sufficiently separated to prevent interference between voice, upstream, and downstream data signals. A benefit of DMT is that it is capable of maximizing throughput in the presence of noise and interference by dynamically allocating data to each channel; channels with little interference carry more data than noisy channels.

DSL suffers from the following deployment hurdles:

- The maximum bandwidth of UTP copper wiring is limited. The current theoretical limitation for DSL is just over 8 Mbps.
- Signal attenuation causes performance to degrade over distance. Thus, the bandwidth available to subscribers located farther away from the CO will not be as great as that available to customers in close proximity to the CO.
- Even at lower data rates, subscribers must be within 18,000 feet of the CO in order to receive service at all.

Fixed Wireless Technology

Fixed wireless technologies exist in *point-to-point* (dedicated) and *point-to-multipoint* (broadcast) configurations. Dedicated point-to-point links typically service organizations that require large amounts of bandwidth (see Figure 4-13).

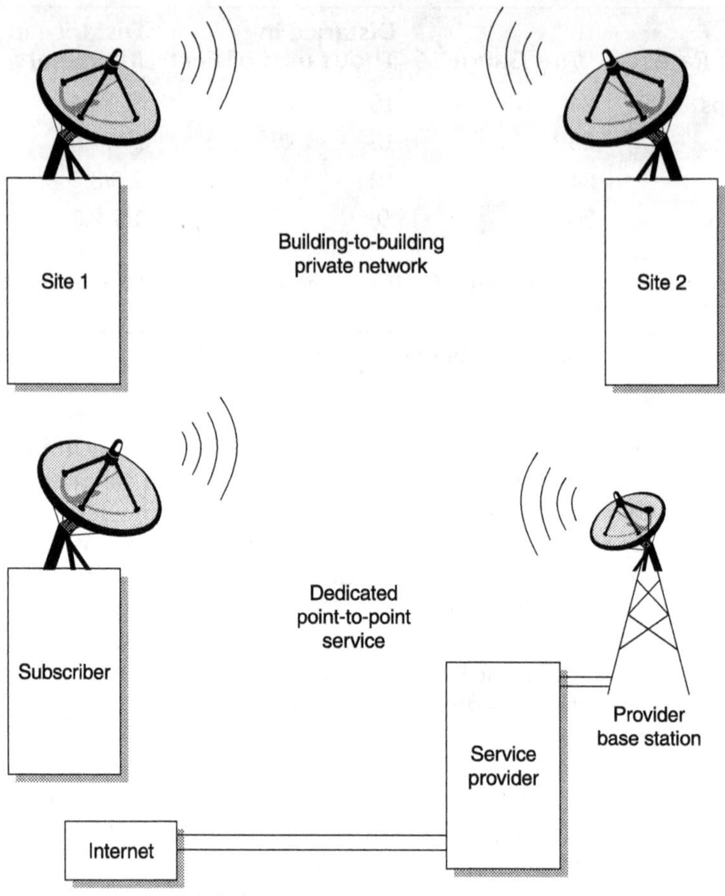

Figure 4-13 In fixed wireless point-to-point communication, a dedicated link exists between the transmitter and receiver.

A company can purchase its own roof-mounted transmitter tower to communicate information to remote offices across town. With a dedicated connection, the owner or subscriber has complete access to the full bandwidth of the link.

In a point-to-multipoint environment, a single transmitter tower may service a large number of subscribers within close proximity to the tower (see Figure 4-14). The bandwidth provided by each tower is shared among its subscribers; as the number of subscribers increases, the available bandwidth decreases. To receive fixed wireless signals, subscribers must install antennas at their home or office. For most fixed wireless technologies, the antenna must lie in direct line of sight of the transmitter tower. Environmental obstructions, such as hills, mountains, or large buildings, may disable service.

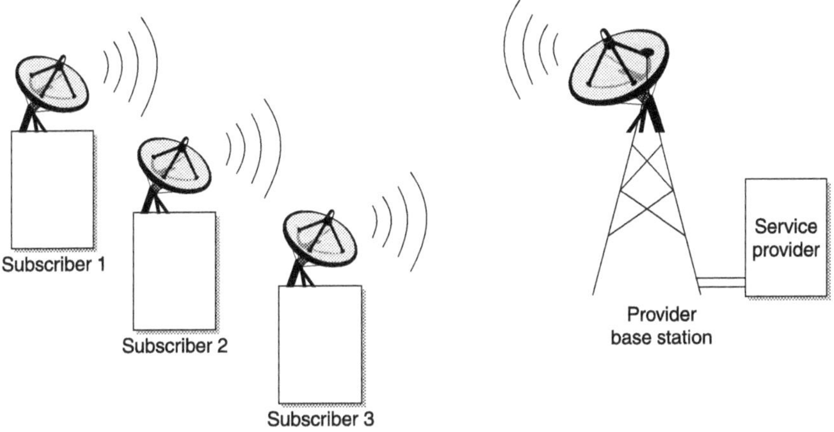

Figure 4-14 Fixed wireless point-to-multipoint communication allows a provider to service many subscribers at once.

Fixed wireless technologies offer some key advantages over traditional wired approaches:

- Fixed wireless technologies work in areas where cable and UTP copper do not exist.
- Microwave transmissions are very clear, and much less susceptible to electrical noise and interference than transmissions made over copper wire. They are ideal for carrying voice, video, and data traffic.
- Wireless solutions are often quicker to deploy, and simpler to operate and maintain, than their wired counterparts.

The two most common forms of commercial fixed wireless are *Multichannel Multipoint Distribution Services (MMDS)* and *Local Multipoint Distribution Services (LMDS)*. Both of these occupy frequency bands that must be licensed from the FCC (Federal Communications Commission), which means that there can be only one fixed wireless provider in a given geographical region. Unlicensed frequency bands, including the Unlicensed National Information Infrastructure (U-NII) and the Industrial, Scientific, and Medical band (ISM) are freely available for public use; anyone with the appropriate equipment who wishes to transmit in these ranges can do so without a license.

In general, higher transmission frequencies are capable of greater bandwidths. However, microwave signals attenuate more quickly at higher frequencies, reducing the transmission range. MMDS, which runs at 2.15–2.68 GHz, is capable of transmission over distances of 30 miles or more. LMDS, which runs at 27.5–40 GHz, is practical only over short distances of 1–3 miles. Atmospheric water absorption—the absorption of microwave energy by water in the air—affects the range of fixed wireless devices. The effects of water absorption are multiplied during rainy weather, which further reduces transmission distances. This reduction is known as *rain fade*. MMDS and other lower-frequency bands are less susceptible to rain fade than is LMDS. Table 4-4 compares transmission distances and bandwidths for common fixed wireless services.

Two-Way Satellite Communication

Satellite technology has been around for many years. In fact, the first satellites began orbiting the earth in 1957. Satellite technology has long been used for telecommunication and broadcast television. It wasn't until recently that its role was expanded to include high-speed data access. Satellite access networks are relatively straightforward. *Digital broadcast satellite (DBS)* uses a series of *geosynchronous earth orbit (GEO)* satellites to relay information from the service provider's network to consumers (see Figure 4-15). The satellites rotate the earth at a distance of 22,300 miles, and their period of rotation is exactly equal to the rotational speed of the earth. As a result, they appear stationary from the earth's surface. DBS service is available to almost anyone with a clear view of the sky. All the customer needs is a small satellite dish and receiver. As with fixed wireless, the absence of wires simplifies deployment, making satellite service available to those consumers not serviced by cable or DSL. Unlike fixed wireless transmissions, however, satellite signals are capable of traveling enormous distances, so subscribers need not be near a transmitter tower.

GEO satellite technology sounds great, but it has severe drawbacks that limit its use for real-time multimedia services. Latency-intolerant applications, such as telephony, video conferencing, and online gaming, operate poorly over services

Service	Frequency Ranges	Typical Transmission Distance	Typical Bandwidth (Downstream and Upstream)
LMDS	27.5–40 GHz[3]	1–3 miles	10–100 Mbps
MMDS	2.15–2.68 GHz	30–35 miles	1.5–10 Mbps
ISM	902–928 MHz	5 miles	< 1 Mbps
U-NII	5.15–5.25 GHz (indoor use only), 5.25–5.35 GHz, and 5.725–5.825 GHz	5–15 miles	> 1 Mbps[4]

[3] LMDS does not occupy the entire range of frequencies between 27.5 GHz and 40 GHz.

[4] The throughput of the unlicensed ISM and U-NII bands varies depending on the number of users transmitting over a given frequency. Even though they run at a higher frequency, the U-NII bands are not likely to match the throughput of licensed MMDS services.

Table 4-4 Frequency Ranges, Transmission Distances, and Bandwidths for Common Fixed Wireless Services

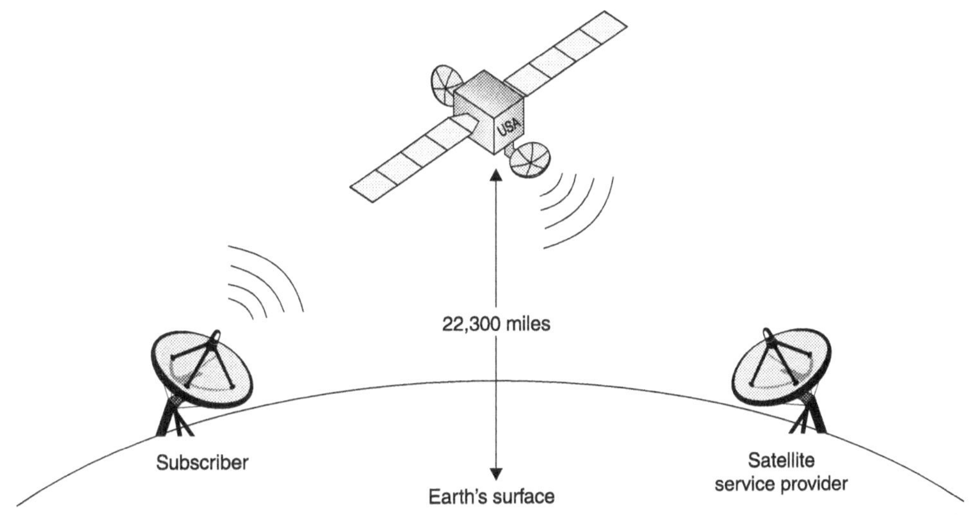

Figure 4-15 Satellite communication uses GEO satellites orbiting the earth at a distance of 22,300 miles to relay data between a service provider's network operations center (NOC) and the consumer's location.

employing GEO satellites, due to the distance that the signal must travel to and from the satellites (see Figure 4-16). It takes a minimum of 240 milliseconds for a satellite signal traveling at the speed of light to traverse the path from a point on the earth's surface to a satellite and back to the earth's surface. This constitutes a roundtrip time of nearly half a second to cover a single network hop between subscriber and provider. Most communication will not end at the service provider's network operations center (NOC), and additional latency is incurred as the signal is routed to its final destination. In contrast, roundtrip times for terrestrial links spanning the entire United States and more than a dozen network hops typically fall within 100–150 milliseconds. You can test this for yourself using the *ping* and *traceroute* (*tracert* on Microsoft Windows hosts) utilities included with almost every TCP/IP stack implementation on the market.

NOTE:
A latency of 400 ms or greater is considered intolerable for interactive applications.

A second drawback of two-way satellite is signal quality. The long transmission distances required by satellite communication result in relatively high bit error rates. Although reliable protocols, such as TCP, are capable of retransmitting corrupted data, retransmission reduces the effective throughput of a satellite link, and further increases delay.

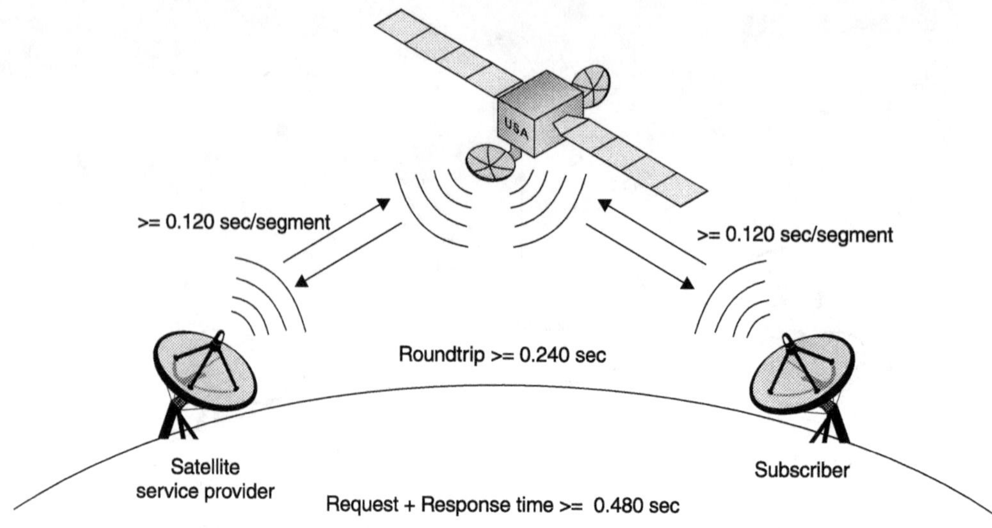

Figure 4-16 The roundtrip transmission time for satellite signals is undesirable for interactive applications, such as online gaming.

NOTE:
Low earth orbit (LEO) *satellite systems currently under development consist of satellites orbiting close to the earth's surface (often at distances less than 500 miles). Roundtrip times for these satellites are much less than those for GEO satellites, making them better suited for interactive applications. The decreased distance also reduces error rates.*

As we discussed in Chapter 1, first-generation DBS provided one-way downstream communication. The upstream channel consisted of an analog dial-up line. Two-way DBS offers communication in both the upstream and downstream directions via the satellite link. Typically, the return path for two-way satellite runs at 128–150 Kbps. The downstream data rate for one- and two-way satellite is much greater, varying from a few hundred Kbps to more than 1 Mbps. Service providers commonly limit the downstream bandwidth to somewhere between 400 and 500 Kbps. The transmission frequencies used by two-way satellite are as follows:

Band	Frequency range
C-band	4–8 GHz
Ku-band	11–17 GHz
Ka-band	20–30 GHz

Quality of Service

The explosive growth of the Internet and the introduction of numerous multimedia applications have created significant demand for responsive, high-bandwidth networks. Future demand is likely to grow as more resource-intensive applications come to market. These applications, especially those involving human interaction and entertainment, have strict bandwidth and timing requirements that greatly affect the user experience. Furthermore, service providers now deliver a multitude of applications and services directly to consumers' homes and offices via a single network interface. A service provider may offer video conferencing, audio-on-demand, telephony, and basic Internet access, all of which compete for shared network resources. As the number and variety of applications increase, the need for resource management quickly becomes apparent. To remain competitively priced while still providing high-quality services, network service providers must strive to minimize operating costs by utilizing network resources to the fullest and most efficient extent possible.

Some proponents of fiber optic and gigabit Ethernet argue that all problems relating to resource allocation can be solved simply by adding bandwidth. While this addresses issues stemming from a lack of bandwidth, it oversimplifies the needs of many applications sensitive to signal delay and variation. Such delay and variation often results from packet processing in switches and routers along the transmission path, regardless of the available bandwidth. Resource requirements vary greatly from one application to the next. E-mail, for example, requires little more than a very modest level of bandwidth. Video conferencing, on the other hand, is extremely demanding in terms of bandwidth and other resources necessary for maintaining signal quality.

Typical network-based applications fall into one of two broad categories: *real-time* or *elastic*. Real-time applications can be further categorized as *rigid* or *rate-adaptive,* and elastic applications can be categorized as *interactive burst-traffic, interactive bulk-traffic,* or *asynchronous bulk-traffic*. Table 4-5 lists examples of each type of application. In general, real-time applications are much more sensitive than elastic applications to signal delay and variation,

Applications	Traffic Type	Tolerance to Signal Delay and Variation
Video conferencing and telephony	Real-time/rigid	Very low
Audio- and video-on-demand (MPEG)	Real-time/rate adaptive	Low

Table 4-5 A Comparison of Various Real-Time and Elastic Applications

Applications	Traffic Type	Tolerance to Signal Delay and Variation
Terminal emulation (Telnet) and online gaming	Elastic/interactive burst traffic	Moderate
File transfer (FTP) and web browsing	Elastic/interactive bulk traffic	High
E-mail	Elastic/asynchronous bulk traffic	Very high

Table 4-5 A Comparison of Various Real-Time and Elastic Applications *(continued)*

with real-time/rigid application being the least tolerant and elastic/asynchronous bulk traffic being the most tolerant.

The needs of these various applications can be met through *quality of service (QoS)* assurance. QoS refers to the ability of a network or network element to provide a guaranteed level of service to its users on a per-application basis. QoS mechanisms ensure the timely and predictable delivery of data, and QoS-enabled devices determine how to handle traffic based on parameters defined by each individual application.

Guaranteed, end-to-end QoS requires cooperation among all network layers and devices between the sending and receiving applications. If one network layer or device in the path between the sender and recipient does not provide guaranteed QoS, then the overall QoS cannot be guaranteed. Guaranteed QoS also requires resource allocation on a per-data-stream basis (Quality of Service Forum, 1999). This is something that is difficult for connectionless protocols such as IP to provide, since their protocol headers rarely carry session state information.

QoS Parameters

In order to ensure QoS, service providers must have some quantitative means of measuring network performance. The primary parameters used to gauge QoS are bandwidth, latency, jitter, packet loss, and availability. Table 4-6 provides a summary of these parameters.

Bandwidth

Bandwidth is a measure of the capacity of a communication channel in terms of the maximum amount of data it can transfer in a given time. In most cases, bandwidth refers to communication speeds at the Physical layer, and is usually expressed in bits per second (bps). Most often, bandwidth describes the maximum *sustainable* transfer rate achievable by the network, and not the maximum *burst*

Parameter	Alternate Name	Description
Bandwidth	Throughput	The maximum data transfer capacity of a network expressed in bits per second (bps); also used to measure the rate at which an application transmits data.
Latency	Delay	The interval between the times when a signal is encoded by the sender and when it is decoded by the recipient. *Network latency* refers to the time data spends traveling over a network from sender to recipient.
Jitter	Delay variation	The variation in interpacket delay times caused by signal encoding/decoding and queuing times in network routing devices.
Packet loss	Drop rate	The number of packets dropped before a transmission reaches its final destination. Usually expressed as a percentage of the total number of transmitted packets.
Availability	Uptime	The amount of time that a service is available for use by customers or other end users. Usually expressed as a percentage of the total time for which a service is offered.

Table 4-6 Parameters Used to Measure Quality of Service

rate. For shared media networks, the burst rate will almost always be higher than the sustainable rate due to the likelihood of collisions. When many applications share the bandwidth of a single cable, steps must be taken to ensure that high-priority applications maintain minimum bandwidth requirements while not starving lower-priority applications of bandwidth. In general, multimedia applications are the most bandwidth intensive, with video requiring the highest data transfer rate of all.

Latency

Latency, also known as *delay,* generally describes the interval between the times when a network node transmits a message and when it is received at its destination. This definition includes delays introduced by network components on the path between the sender and recipient; however, it ignores delays associated with signal processing before and after transmission. For instance, when using multimedia applications, end users will experience delay while waiting for the sending and receiving hosts to process voice and video signals. With this in mind, we will expand our definition of latency to include

- Signal processing delays (encoding and decoding) by the sending and receiving hosts
- Packetization of data before it is placed on the network
- Propagation delays associated with the transmission medium
- Queuing delays within network routing and switching devices

When all communication occurs on the same network segment, there is usually very little latency aside from that associated with network propagation and host signal processing. If all devices use high-speed digital signal processors, then the overall latency is negligible. However, latency becomes an issue when signals must traverse multiple network segments. Each routing device on the path between sender and recipient introduces further delay due to packet processing. As a rule of thumb, latency increases as the number of hops between sender and recipient grows. Minimizing the number of hops a signal takes between source and destination, combined with the use of advanced Layer 3 switching technologies such as *IP switching* and *multiprotocol label switching (MPLS),* greatly reduces network latency.

REMEMBER:

Latency of more than 400 ms is considered intolerable for most interactive applications (video conferencing and telephony, for example).

Jitter

Not all data packets traversing a network experience the same latency; some packets remain in routing queues longer than others, or may follow a different route to their destination. The resulting variation in interpacket delay times is known as *jitter*. For multimedia applications, jitter greatly affects signal quality, and it results in jerky playback of audio and video. In addition to network latency, overburdened processors may introduce jitter during signal processing on end devices.

NOTE:
Jitter affects signal quality more than latency does. Delay may hinder interactivity, but jitter results in signal degradation.

Buffering helps to reduce jitter by collecting and storing packets until they can be played in the correct order and at the proper timing. Buffering allows the recipient to reorder out-of-sequence packets, and ensures that slower packets are received well before they are needed. Higher-bandwidth applications usually require larger buffers in order to minimize the effects of jitter. For instance, a 64 Kbps *Pulse Code Modulated (PCM)* audio signal fills a 100KB buffer in 12.5 seconds. At a rate of nearly 20 Mbps, an HDTV signal fills the same buffer in 0.04 seconds.

CAUTION:
Buffering minimizes jitter but increases overall delay; the larger the buffer, the greater the delay. Applications with stringent delay requirements greatly restrict the use of buffering. This means that for video conferencing, telephony, and other rigid, real-time multimedia applications, the benefits of buffering are significantly reduced.

Packet Loss

For applications such as web browsing and file transfer, lost packets can be retransmitted with no significant impact on the user experience. In fact, these applications require that lost packets be retransmitted. Due to the demanding performance of most real-time applications, the retransmission of lost packets often results in unacceptable delay. This is especially true with streaming audio and video applications. For these applications, it is better to forget about lost packets than try to retransmit them, and service providers must guarantee that the network drops lower-priority packets before it drops those of higher priority.

NOTE:
Most users find a packet loss rate of greater than five percent intolerable for interactive, real-time applications, —especially for telephony, which lacks a visual context. (Source: TF-NGN Task Force.)

Availability

The final QoS parameter, *availability,* measures the amount of time that a network is operational, and is usually represented as a percentage of the combined uptime and downtime of the system. Most network service providers have availability targets of 99.9 percent or higher.

Table 4-7 compares many common applications in terms of their QoS requirements.

Application	Bandwidth	Latency (End-to-End)	Jitter (End-to-End)	Packet Loss
Video conferencing and low-quality video applications	64 Kbps–2 Mbps; varies greatly depending on image quality	100–150 ms target; > 400 ms becomes intolerable	20–50 ms (Benefits of signal buffering are minimal.)	Far less than 5%
Telephony	8–64 Kbps; varies greatly depending on encoding method (64 Kbps uses no compression.)	100–150 ms target; > 300 ms becomes intolerable	10–20 ms (Benefits of signal buffering are minimal.)	Far less than 5%
Real-time control and monitoring systems with high sampling rates	Varies greatly from one application to the next	< 100 ms	10–15 ms	Far less than 5%
Streaming video	MPEG-1: up to 1.5 Mbps MPEG-2: up to 20 Mbps (HDTV)	< 500 ms (e.g., response time for a control operation, such as play, stop, or pause)	Very tolerant to jitter when using large receive buffers	Far less than 5%
Streaming audio	MPEG-1: 32–448 Kbps MPEG-2: 8 Kbps – 1 Mbps	< 500 ms (e.g., response time for a control operation, such as play, stop, or pause)	Very tolerant to jitter when using large receive buffers	Far less than 5%

Table 4-7 Typical QoS Requirements for Various Network Applications

Application	Bandwidth	Latency (End-to-End)	Jitter (End-to-End)	Packet Loss
Interactive online gaming	Moderate	< 250 ms	N/A	Uses retransmission
Web browsing	Minimal	< 5 sec (in the second range)	N/A	Uses retransmission
File transfer	Minimal	N/A	N/A	Uses retransmission
E-mail	Minimal	N/A	N/A	Uses retransmission

Table 4-7 Typical QoS Requirements for Various Network Applications *(continued)*

Table 4-8 lists a number of audio sampling rates and their corresponding bit rates when MPEG-1 or MPEG-2 compression is used.

NOTE:
The optimal compression ratio for MPEG audio is between 1:6 and 1:7. This level of compression results in the greatest reduction in signal size while maintaining a high level of quality (Berkeley Multimedia Research Center, 2000).

Sampling Rate	Bandwidth	Relative Quality	Example
8-bit at 16 KHz (monophonic)	21.3 Kbps	Low	Speech
16-bit at 24 KHz (stereophonic)	64 Kbps	Moderate	Radio
16-bit at 44.1 KHz (stereophonic)	117.6 Kbps	High	CD-ROM
24-bit at 96 KHz (multichannel surround)	384 Kbps and higher	Very High	DVD audio

Table 4-8 Bandwith Required for Common Audio Sampling Rates with a 1:6 MPEG Compression Ratio

Degrees of QoS

Varying degrees of QoS exist, including guaranteed/reservation-based, differential/priority-based, and best effort (Agarwal, 2000, and Quality of Service Forum, 1999):

- **Guaranteed/reservation-based** Resources are explicitly assigned to all packets belonging to a particular traffic flow. For convenience, we will define a *traffic flow* as the collection of all packets from a single instance of an application that require the same level of service. Once assigned, reservations exist for the lifetime of a session. Guaranteed service implies an assured bandwidth, bounded end-to-end delay, and no queuing losses resulting from forwarding activities of routing devices.
- **Differentiated/priority-based** Provides a relatively coarse level of classification for each packet traversing a network. Unlike reservation-based services, there is no firm resource allocation, and no notion of traffic flow. Traffic is grouped into classes of packets, all receiving similar levels of service. Forwarding policy is applied to individual packets on a per-hop basis, according to the information contained within each packet.
- **Best effort** Offers no resource allocation for QoS provisioning. Packets are queued on a first-in, first-out (FIFO) basis, with no one packet receiving preferential treatment over another.

We will discuss these models further during our discussion of IP integrated and differentiated services later in the chapter.

The Great Debate: Cell-Relay vs. Standard Packet Switching

There has been much debate over the past few years as to the WAN technology most appropriate for the delivery of high-bandwidth and low-delay, low-jitter applications. The two common approaches are cell relay and standard packet switching. *Cell-relay* protocols, such as Asynchronous Transfer Mode (ATM), transmit data within fixed-length units known as *cells*. *Standard packet-switching* protocols, such as IP, use variable-length packets. For ATM, the cell length is 53 bytes—5 bytes for the header and 48 bytes for payload—which is considerably smaller than the maximum packet size of 1500 bytes for IP-over-Ethernet. When variable-length packets are used, it is difficult to determine ahead of time how long it will take to process each packet. Smaller packets may be delayed if they are received behind larger packets, due to the greater overhead required to process the larger packets. Fixed-length cells allow strict control over jitter because all packets require the same amount of time for processing. For switches and routers running at line speeds, the maximum queuing delay is the time required to process a single 53-byte cell. ATM also transmits cells at fixed intervals, which allows firm control over both bandwidth usage and delay for a particular data stream. ATM provides the following suite of QoS guarantees:

- Cell delay variation (reduces jitter)
- Maximum cell transfer delay (reduces latency)
- Cell loss ratio (reduces packet loss)
- Cell error ratio
- Severely errored cell block ratio
- Cell misinsertion rate

Only the first three parameters are in practical use today (Lambarelli, 2001). Standard packet-switched protocols provide little in the way of QoS provisioning. Packets are treated as autonomous units that are individually addressed and transmitted to their destinations. They are not associated with a data stream until they are received by an upper-layer protocol. For instance, IP is a connectionless network-layer protocol that offers no guarantee that packets will be received or received in a particular order. As a result, IP is often termed a "best-effort" protocol that relies on TCP for reordering and retransmission. IP efficiently uses available network bandwidth, and is well suited for bursty traffic (the term "bursty" implies the transmission of varying amounts of data at irregular intervals). For non-real-time applications, such as e-mail and file transfer, IP has proven sufficient. These applications are generally not overly sensitive to delays introduced by routing queues and the retransmission of lost packets. However, the best-effort characteristics of IP make it inappropriate for the delivery of real-time applications. IP does provide limited support for prioritizing traffic, through the *Type of Service* and *Class of Service* fields contained within IPv4 and IPv6 headers, respectively (see Figure 4-17), but these fields are rarely used today.

NOTE:
To avoid introducing additional delay, lost packets are generally not retransmitted in delay-sensitive applications.

While ATM does provide guaranteed QoS, it can be costly in terms of *header-to-payload ratios*. Regardless of the size of the data set to be transmitted, the most data that each ATM cell can hold is 48 bytes. This results in a very large number of cells when transmitting large data sets. The bandwidth consumed by the 5-byte ATM headers quickly becomes substantial. When transmitting the same data over IP, you can break the data up into many packets, each equal in size to the MTU of the underlying Data Link layer minus the space required for the IP header. Let's look at an example.

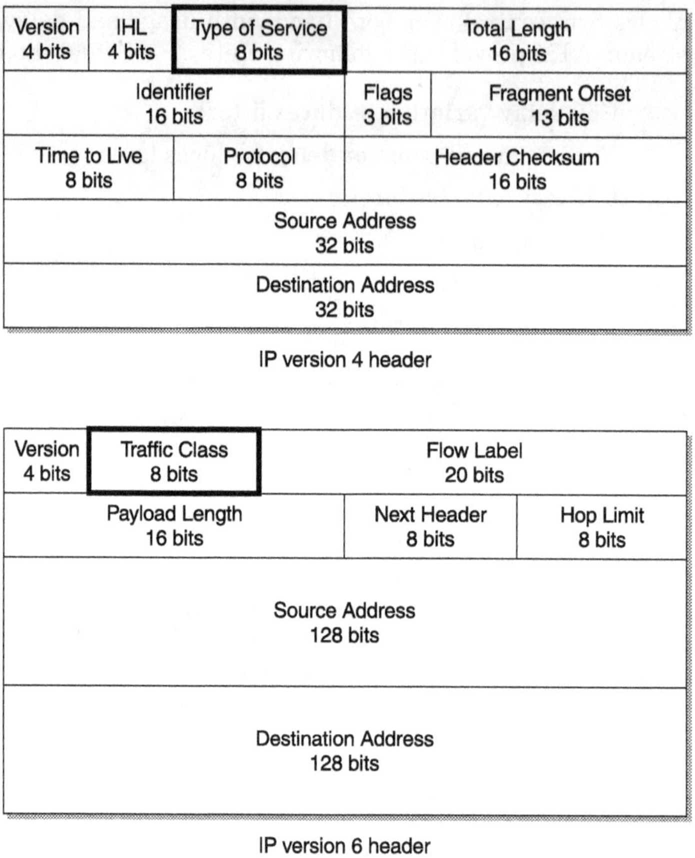

Figure 4-17 IPv4 and IPv6 headers contain Type of Service and Traffic Class fields for use in priority-based QoS environments.

Version 4 bits	IHL 4 bits	Type of Service 8 bits	Total Length 16 bits	
Identifier 16 bits			Flags 3 bits	Fragment Offset 13 bits
Time to Live 8 bits		Protocol 8 bits	Header Checksum 16 bits	
Source Address 32 bits				
Destination Address 32 bits				

IP version 4 header

Version 4 bits	Traffic Class 8 bits	Flow Label 20 bits	
Payload Length 16 bits		Next Header 8 bits	Hop Limit 8 bits
Source Address 128 bits			
Destination Address 128 bits			

IP version 6 header

The MTU for Ethernet is 1500 bytes. An IP packet running over Ethernet is limited to 1480 bytes of payload (1500-byte Ethernet MTU and 20-byte IP header with no options). If we want to send a 1MB (1024KB) file using IP-over-Ethernet, the total number of packets required is 692. This corresponds to 1,050,456 total bytes of data. Sending the same file over ATM requires 21,334 cells, and results in 1,130,702 total bytes of data. The difference is 80,246 bytes, or 6.42 milliseconds of wasted bandwidth on a 100 Mbps network. This wasted bandwidth quickly adds up!

ATM-native applications that run directly above Layer 2 can take full advantage of the QoS functionality provided by ATM. Since ATM runs at Layer 2 and IP at Layer 3, it is possible to run IP over ATM. This technique has been used in the past, with limited success, to add advanced QoS provisioning features to IP-based applications. Unfortunately, there is a lack of communication facilities between TCP/IP and ATM, and most IP-based applications cannot take full advantage of the QoS functionality built into ATM. In addition, the combination of

IP and ATM often results in undue cost and complexity. Due to the number of applications written for IP, and the relative simplicity and cost-effectiveness of IP-based networks, significant effort is now being placed into adding support for Integrated and Differentiated Services architectures to IP. Table 4-9 compares the protocol characteristics and QoS functionality of ATM and IP.

Models for QoS over IP Networks

Over-provisioning—making sure that you have more resources on hand than you anticipate needing—is a costly solution for maintaining quality of service, and does not make for efficient use of network resources. While it does ensure that the bandwidth requirements for all applications are met, it ignores the

Feature	ATM	IP
Data transport format	Fixed-length cells	Variable-length packets
Protocol characteristics	Connection-oriented. Establishes virtual circuit before communication begins.	Connectionless. Requires upper-layer protocol (e.g., TCP) to create virtual circuit.
Independent delivery of PDUs (Layer 3 addressing)	No. Requires establishment of switched virtual circuits.	Yes. Packets can be independently addressed and delivered.
Support for QoS	Resources reserved for entire session. Excellent built-in QoS provisioning. Can guarantee QoS levels.	Best-effort delivery, but supports packet classification. QoS must be applied individually to each packet (prioritization). Detailed QoS support requires additional protocols (e.g., IntServ with RSVP).
Efficient use of shared bandwidth	Cells must contain fixed-length payload, or padding must be added. Short cell length increases header-to-payload ratio.	Variable packet length reduces header-to-payload ratio, and facilitates bursty traffic.
Flexibility	Limited. Lack of supported applications.	High. Most Internet applications written for IP.
Economy	Hardware is complex and costly to develop due to added QoS functionality.	Simple and inexpensive.

Table 4-9 A Comparison of ATM and IP in Terms of Protocol Characteristics and QoS

unique QoS requirements of each individual application. A major disadvantage of this model is that unused or idle resources cannot be reclaimed. Most service providers opt for more flexible methods of resource management. Two popular models for offering QoS over IP-based networks are the Integrated Services and Differentiated Service architectures. These architectures efficiently balance network resources between high- and low-priority applications, while using the fewest resources possible.

IP Integrated Services Architecture

The Integrated Services (IntServ) architecture proposed by the IETF in RFC 1633 (Braden and Clark et al., 1994) supports both best-effort and real-time IP traffic, and consists of three service classes: guaranteed, controlled-load, and best-effort. As discussed earlier, *guaranteed* service provides absolute resource allocation for real-time applications with strict delay and jitter requirements. Best-effort services offer no resource allocation for QoS provisioning; all data packets are treated as having the same priority. *Controlled-load* service provides no firm QoS guarantees; instead, it treats traffic as best-effort on a lightly loaded network. Unlike true best-effort delivery, however, the performance of controlled-load service does not noticeably degrade when the network becomes congested. In both the guaranteed and controlled-load service models, resource allocation requests are made before the data is sent. Each application reserves resources for a particular traffic flow using a dynamic signaling protocol, such as RSVP, which we discuss in the following section. The primary components of the IntServ architecture (and other guaranteed service models) are

- **Admission control** Determines whether to accept a resource request based on the level of service purchased by a customer, the network resources available at the time of the request, and other policy considerations.
- **Policy control** Verifies that incoming traffic conforms to a predefined traffic specification; traffic that does not conform is treated as best-effort. Policy execution occurs at network borders.
- **Packet classification** Determines the QoS class for each packet, and associates the packet with a particular traffic flow.
- **Packet scheduling** Schedules the transmission of incoming packets, based on their QoS classes and associated traffic flow. Queue management mechanisms ensure minimal delay and loss for high-priority flows, and *traffic shaping* can be applied to smooth out bursty traffic.
- **Reservation setup protocol** Used to negotiate resource reservations between host applications and routing devices, and to maintain session

state information related to traffic flows. The application defines upper and lower boundaries for bandwidth, latency, and jitter, which are converted to bandwidth and buffer space within each routing device.

Resource Reservation Protocol The *Resource Reservation Protocol (RSVP)* allows host applications to reserve network resources within routers in IntServ environments (Braden and Zheng et al., 1997). RSVP provides guaranteed and controlled-load QoS. It is responsible for negotiating connection parameters between routers and maintaining session state information for each traffic flow. Within the context of RSVP, a traffic flow is identified by a destination address, a destination port number, and a transport-layer protocol. Reservations are path specific, meaning that they exist only along a defined route between the sender and recipient. If the traffic follows a different path, or does not conform to negotiated traffic characteristics, it is treated as best-effort. RSVP reservations create *soft states* within routers that must be periodically refreshed or they will be dropped.

RSVP supports allocation of resources for both unicast and multicast data streams via two message types: PATH and RESV. PATH messages originate from the sender. They are used to establish a reverse route back to the sender and to convey resource requirements to the recipient(s). The recipient uses the traffic specification (Tspec) information contained within the PATH message to determine the resources required by the application. The recipient is then responsible for requesting the necessary resources from an adjacent router, using an RESV message. The reservation propagates in the upstream direction until it is received by a router adjacent to the sender, or until it meets another reservation for the same source stream (see Figure 4-18). RSVP runs on top of IP, but is not a routing protocol. (For more information on RSVP, see Braden and Zheng et al., 1997 and White, 1997.)

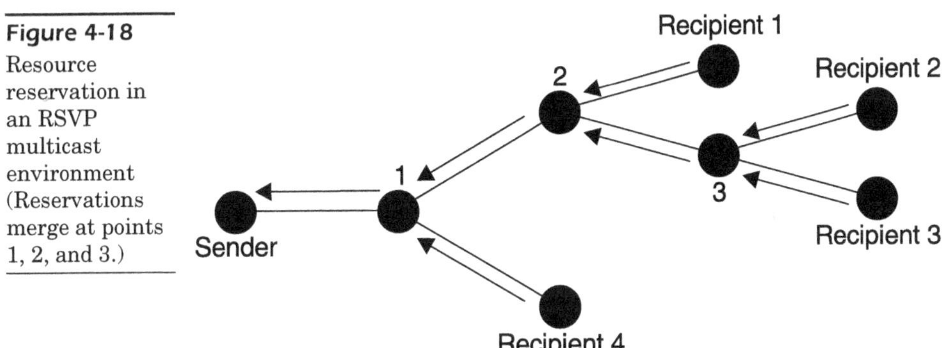

Figure 4-18 Resource reservation in an RSVP multicast environment (Reservations merge at points 1, 2, and 3.)

RTP and RTCP The *Real-time Transport Protocol (RTP)* provides transport services for real-time multimedia applications in both unicast and multicast environments (Schulzrinne et al., 1996). RTP is not a reliable transport protocol, nor does it natively support flow control. However, it does supply hooks, such as sequence numbering, for adding this functionality. RTP runs on top of UDP. UDP was chosen over TCP for two primary reasons (Liu, 1999):

- RTP was designed for multicast usage. The session-oriented nature of TCP does not scale well in such environments.
- Within real-time applications, timely delivery is more important than reliability. In most cases, retransmission is undesirable for real-time applications because it results in added delay.

NOTE:
UDP provides less protocol overhead than TCP (8 bytes for UDP, and a minimum of 20 bytes for TCP).

RTP includes timestamping and sequence numbering, so that the receiving application can place incoming packets in the correct order and play them back with the proper timing. When reliable transmission is necessary, sequence numbers can also be used to detect lost packets. RTP offers no mechanisms for establishing QoS. Instead, it relies on RSVP for resource allocation.

An RTP header includes a payload type field describing the format of the data encapsulated within the RTP message. The sender uses this field to inform the recipient of the method (PCM, MPEG-1, MPEG-2, H.261 video, and so on.) that has been used to encode the data. The contents of this field may change during a communication session. For example, the sender may wish to adjust parameters related to encoding or compression to compensate for changing network conditions.

The *Real-Time Control Protocol (RTCP)* is the control protocol used in conjunction with RTP. Using RTCP, communicating parties periodically exchange information associated with the delivery quality of the current session. This allows the sender to dynamically adjust the transmission based on feedback from the recipient.

IP Differentiated Services Architecture

The IntServ architecture requires that all end-user systems and intermediate routing devices support RSVP. It is a complex architecture that does not scale well to very large networks (Agarwal, 2000). The Differentiated Services Architecture (DiffServ) provides a simple, scalable mechanism for achieving QoS (Blakely et al., 1998). DiffServ offers coarse control of QoS parameters through the aggregation of packets into *buckets*. Packets within the same bucket all receive the same level of treatment. No formal resource allocation requests are made by the application before sending traffic. Instead, policy is applied to individual packets at every node between sender and recipient. There exists no notion of a traffic flow, so there is no need to maintain state information relating to traffic flows. Forwarding behavior is based on the contents of certain fields included in the IP header of each packet. IPv4 uses a *Type of Service* field, and IPv6 uses a *Traffic Class* field. The DiffServ architecture actually modifies these fields to represent a greater number of packet classifications. Packet classification, marking, and policing occur only at the edges of the network. The only operation performed in the core network is forwarding, based on the priority of each packet. DiffServ is a scalable model that is easy to implement, but it does not provide guaranteed allocation of resources.

References

Agarwal, Anjali. Fall, 2000. "Quality of Service (QoS) in the New Public Network Architecture." *IEEE Canadian Review*: 22-25.

Bates, Regis J. 2000. *Broadband Telecommunications Handbook*. New York: McGraw-Hill.

Berkeley Multimedia Research Center. 2000. *MPEG-1 FAQ*. 5 November, 2001 (http://bmrc.berkeley.edu/frame/research/mpeg/mpegfaq.html).

Blakely, S., D. Black, M. Carlson, E. Davies, Z. Wang, and W. Weiss. December, 1998. *RFC 2475: An Architecture for Differentiated Services*. Internet Engineering Task Force.

Braden, R., D. Clark, and S. Shenker. June, 1994. *RFC 1633: Integrated Services in the Internet Architecture: An Overview*. Internet Engineering Task Force.

Braden, R., ed., L. Zhang, S. Berson, S. Herzog, and S. Jamin. September, 1997. *RFC 2205: Resource ReSerVation Protocol (RSVP)—Version 1 Functional Specification*. Internet Engeneering Task Force.

Kennebeck, Keith, Jason Marcheck, and James S. Mendelson. 2001. *Residential High-Speed Internet: Cable Modems, DSL, and Fixed Wireless*. Washington D.C.: The Strategis Group.

Lambarelli, Livio, ed. 3 October 2001. *ATM Services Categories: The Benefits to the User*. The ATM Forum. 8 November 2001 (http://www.atmforum.com/pages/library/whitepapers/6.html).

Liu, Chunlei. 1999. Multimedia Over IP: RSVP, RTP, RTCP, RTSP. In *Handbook of Communication Technologies: The Next Decade*. Rafael Osso, ed. Boca Raton: CRC Press. 3 November 2001. Ohio State University (http://www.cis.ohio-state.edu/~cliu/ipmultimedia/)

Odom, Wendell. 2000. *Cisco CCNA Exam #640-507 Certification Guide*. Indianapolis, IN: Cisco Press.

Quality of Service Forum. 1999. *The IP QoS FAQ*. Campbell, CS: Stardust Forums.

Schulzrinne, H., S. Casner, R. Frederick, and V. Jacobson. January, 1996. *RFC 1889: RTP: A Transport Protocol for Real-Time Applications*. Internet Engineering Task Force.

White, Paul P. May, 1997. "RSVP and Integrated Services in the Internet: A Tutorial." *IEEE Communications Magazine*.

Chapter 5

A Survey of Existing Broadband Security Standards and Specifications

With broadband entertainment and communication services quickly becoming commonplace in the home and office, hardware and application vendors are racing to take advantage of the growing market. Leading market analysts expect the number of broadband services, applications, and devices to grow considerably in the near future as more and more consumers embrace the technology. Along with this growth come many challenges. How can we ensure the interoperability of products from different vendors? What standards exist for securing current broadband applications? Who is responsible for developing security for future applications? In this chapter, we present the organizations responsible for developing and promoting broadband security standards and practices. We also introduce the most influential security standards and specifications in existence today, many of which we will discuss further in Chapters 9–11. The chapter concludes with a case study of the 802.11 Wired Equivalent Privacy (WEP) standard. This case study serves to highlight the shortcomings in WEP, which resulted from a lack of public scrutiny during development of the standard.

Standards Bodies and the Role of Standardization

Standards bodies and research consortiums play many important roles in the development and exchange of technology. They provide a forum for industry-leading technology manufacturers to share their knowledge and expertise, while co-developing products and services for the future. The standards and specifications produced by these organizations help ensure interoperability among products from different vendors by reducing the number of proprietary implementations. Organizations such as the European Telecommunications Standards Institute and Cable Television Laboratories offer testing and certification of products to guarantee interoperability and adherence to published specifications. As a result, service providers and consumers can purchase standards-compliant products, knowing that these products will work for their intended purposes. Standards bodies and research consortiums also promote public examination by industry experts. This scrutiny greatly reduces the likelihood of poor design decisions making their way to market.

Standards place all manufacturers, large and small alike, on even ground when it comes to producing products that will be accepted by the marketplace. Vendors may still distinguish their products from others on the market through value-added features, but it is difficult for a single large vendor to lock out competition by being the first to offer a proprietary implementation. Consumers benefit directly from standardization as well. Standardization enhances competition by improving the availability of products, thereby lowering their prices. The remainder of this section introduces many active standards bodies and research consortiums integral to the development of broadband network and security technologies.

ANSI (American National Standards Institute)

ANSI is a nonprofit organization responsible for administering and coordinating the U.S. voluntary standardization and conformity assessment system. While ANSI produces no standards of its own, it facilitates the development of American National Standards by promoting consensus among qualified organizations. ANSI's scope encompasses many aspects of life in the U.S.: agriculture, finance, technology (aerospace, telecommunications, information systems, and so on), safety and health, and much more.

ANSI's website can be found at http://www.ansi.org

The BWIF (Broadband Wireless Internet Forum)

The BWIF is a nonprofit association of broadband wireless vendors promoting cost-effective, broadband wireless access technologies and solutions using Vector

Orthogonal Frequency Division Multiplexing (VOFDM). The BWIF is a program of the IEEE-ISTO (the IEEE Industry Standards Organization).

The BWIF's website can be found at http://www.bwif.org.

Cable Television Laboratories

Cable Television Laboratories is a nonprofit research consortium consisting of North American and South American cable systems operators. Also known as CableLabs®, this organization is dedicated to the advancements and interoperability of cable technologies for high-speed data, streaming multimedia, and interactive cable television applications. CableLabs consists of numerous projects, including Data-Over-Cable Service Interface Specifications (DOCSIS), PacketCable, OpenCable, and CableHome. CableLabs does not produce standards. Instead, they submit specifications for adoption by standards bodies.

The CableLabs website can be found at http://www.cablelabs.com.

The DVB (Digital Video Broadcasting) Project

The DVB Project of ETSI is a consortium of broadcasters, network operators, hardware and software manufacturers, and regulatory bodies devoted to the development of international standards for delivering digital television and data services. According to the DVB Project, "DVB is the point-to-multipoint data delivery mechanism with guaranteed quality of service." It produces standards relating to transmission, encoding, interfacing, conditional access, and code-execution platforms within DVB environments.

The website of the DVB Project can be found at http://www.dvb.org.

The DSL Forum

The DSL Forum is a corporation that promotes the DSL market and the development of interoperable DSL-based network components. Its membership comprises network hardware/software vendors and nonprofit/educational institutions. The DSL Forum manages a number of working groups addressing architecture and transport, autoconfiguration, operations and network management, testing and interoperability, and voice-over-DSL (VoDSL).

The DSL forum's website can be found at http://www.adsl.com.

ETSI (European Telecommunications Standards Institute)

ETSI is a nonprofit organization dedicated to producing telecommunications, broadcasting, and information technology standards for use in Europe and the rest of the world. ETSI's membership consists of administrators, network operators, manufacturers, service providers, research bodies, and users from around the world. Current efforts and partnerships focus on third-generation mobile systems, broadband mobile devices, and global interoperability testing.

ETSI's website can be found at http://www.etsi.org.

The IETF (Internet Engineering Task Force)

The IETF is an international organization composed of network designers, operators, vendors, and researchers committed to the advancement of Internet standards and architectures. Most well known for its extensive list of RFCs and Internet draft documents, the IETF consists of numerous working groups, which focus on topics such as network applications, Internet protocol development, operations and management infrastructures, routing, security, data transport, and user services.

The IETF's website can be found at http://www.ietf.org.

The ITU (International Telecommunication Union)

The ITU is an international organization focusing on the development and standardization of telecommunication and information services. The ITU promotes global connectivity and management of telecommunication resources. It consists of three subdivisions: ITU-R (Radio Communication), ITU-T (Telecom Standardization), and ITU-D (Telecom Development). Current initiatives focus on broadband technologies, third-generation mobile devices, and building trust in network infrastructures.

The ITU's website can be found at http://www.itu.org.

The IEEE (Institute of Electrical and Electronics Engineers)

Endorsed as "the world's largest technical professional society," the IEEE (pronounced "eye-triple-E") is a nonprofit corporation devoted to the advancement of electrical and information technologies and sciences. The IEEE consists of many councils and working groups focusing on a wide variety of topics, including broadcast technology, communications, information technology, measurements, and power and energy. One of the most well-known working groups within the IEEE is the 802 LAN/MAN Standards Committee.

The IEEE's website can be found at http://www.ieee.org.

The ISO (International Standards Organization)

The ISO is a global federation of standards bodies dedicated to the development of international standards for promoting the exchange of goods and services, as well as knowledge relating to science, technology, and economics. Responsible for standardization in most aspects of daily life, the ISO's activities include auditing, telecommunications, healthcare, metallurgy, the clothing industry, and much more. (We discussed the ISO-OSI Network Reference model in Chapter 4.)

The ISO's website can be found at http://www.iso.org.

Current Broadband Security Standards and Specifications

Today, there exist broadband security standards and specifications for protecting high-speed data services, voice- and videoconferencing, interactive television, and copyrighted content distribution. The network environments, performance characteristics, and security requirements of these broadband applications vary greatly. No one standard applies equally well to all environments, nor can a single standard address the needs of all current and future broadband applications. For example, security mechanisms that work well for protecting Internet traffic crossing a single network segment may fail miserably when applied to streaming multimedia signals traversing multiple access, distribution, and provider networks. Each class of application carries with it its own unique QoS requirements for bandwidth, latency, jitter, and packet loss, which dictate the use of cryptography and other security mechanisms. As we will soon see, existing standards implement security at various network layers and employ different security protocols and mechanisms depending on the application(s) they are protecting.

Despite their differences, these standards and specifications share a number of common traits. First and foremost, security cannot interfere with usage. Transparent security is easy for end users to accept. Most users remain content just knowing that security mechanisms are in place, without ever having to touch them. The more complexity security adds to the user experience, the more likely users are to find ways to circumvent it. They may even decide not to use a service if they view security as too cumbersome. In addition to ease-of use, service providers grapple with maintaining performance and service availability. Network resource management is a constant juggling and balancing act with little room for sluggish security. (Imagine yourself crossing a burning high-wire while tossing around a few large bowling balls, and you'll get the picture.) Therefore, high-performance security is a necessity. Limited processing power and available memory, as well as other restrictions inherent to embedded devices, call for creative and efficient design by manufacturers. Finally, most security standards employ widely available, standards-based components in order to reduce costs and avoid favoring a particular vendor.

In the following sections, we discuss many pivotal standards and specifications developed by CableLabs, the ITU, and the IETF. All of these organizations have been instrumental in promoting security in broadband applications and services.

The DOCSIS 1.0 Baseline Privacy Interface

DOCSIS 1.0 specifies physical and MAC layer protocols for secure transmission of data over the Hybrid Fiber-Coax (HFC) segment of a cable network. Included within the DOCSIS 1.0 series of specifications is a Data Link layer security protocol known as the Baseline Privacy Interface (BPI) (CableLabs, 2001c). BPI protects against unauthorized access to best-effort IP data services by enforcing traffic encryption. It provides security comparable to or greater than dedicated line services by encrypting traffic between cable modems and cable modem termination systems. DOCSIS networks are shared-media networks, and without encryption, subscribers sharing the same DOCSIS network segment could eavesdrop on one another's communication. BPI also includes a protocol for the periodic exchange of keying material between cable modems and cable modem termination systems. While BPI does provide a very basic level of protection for cable providers against theft of service, its primary goal is data privacy.

The characteristics of DOCSIS 1.0 BPI are outlined in Table 5-1.

Characteristic	Description
Name	DOCSIS 1.0 Baseline Privacy Interface (BPI)
Organization	Cable Television Laboratories
Description	BPI is a Link layer security protocol that provides data privacy over the HFC segment of a cable network running between a cable modem (CM) and a cable modem termination system (CMTS).
Applications	Best-effort, high-speed data services. The CableLabs BPI specification describes these as "Internet access-like" services (CableLabs, 2001c).
QoS requirements	DOCSIS 1.0 includes no built-in support for QoS, and may not be suitable for the transmission of traffic highly sensitive to signal variation. Consequently, BPI was designed with performance in mind, but not guaranteed QoS.
Security services and mechanisms	*Weakly* authenticates each CM based on a hardware (MAC) address; cannot protect against CM cloning. Controls access to protected cable network resources and services. Provides for secure exchange of encryption keys, and performs traffic encryption across a *single* network segment. Ensures the integrity of message types, including key requests and replies exchanged between the CM and CMTS, but does not verify the integrity of encrypted user data.
Network security protocols	N/A

Table 5-1 Characteristics of DOCSIS 1.0 BPI

Characteristic	Description
Public-key encryption algorithms	Uses a 768-bit RSA public/private key pair to encrypt and decrypt long-term key-encryption keys (KEKs).
Symmetric-key encryption algorithms	Uses 56-bit DES (ECB mode) KEKs for encrypting short-term traffic-encryption keys (TEKs). Uses 40-bit and 56-bit DES (CBC mode) TEKs for encrypting the contents of Layer 2 PDUs in the upstream and downstream directions.
Message integrity algorithms	Uses HMAC SHA-1 to ensure the authenticity and integrity of key request and reply messages. Uses MD5 for CM message integrity checks (MICs).
Use of digital certificates and PKI	N/A

Table 5-1 Characteristics of DOCSIS 1.0 BPI *(continued)*

NOTE:
Functionally, 40-bit and 56-bit DES are identical. With 40-bit DES, however, 16 bits of the key are masked with known, fixed values.

The DOCSIS 1.1 Baseline Privacy Plus Interface

The environment for DOCSIS 1.1 and the Baseline Privacy Plus Interface (BPI+) is nearly identical to that of DOCSIS 1.0 and the BPI (CableLabs 2001d). However, DOCSIS 1.1 includes QoS provisioning absent in DOCSIS 1.0. From a security perspective, BPI+ improves upon the BPI specification by enhancing cable modem authentication, employing stronger encryption algorithms, and enabling secure software upgrades in the field. The characteristics of DOCSIS 1.1 BPI+ are outlined in Table 5-2 (BPI+ is discussed in more detail in Chapter 9.).

Characteristic	Description
Name	DOCSIS 1.1 Baseline Privacy Plus Interface (BPI+)
Organization	Cable Television Laboratories
Description	BPI+ is a Link layer security protocol[1] providing both data privacy and cable modem authentication over the HFC segment of cable networks running between a cable modem (CM) and a cable modem termination system (CMTS).

Table 5-2 Characteristics of DOCSIS 1.1 BPI+

Characteristic	Description
Applications	Best-effort, high-speed data services and QoS-assured constant bit rate traffic.
QoS requirements	DOCSIS 1.1 includes built-in support for QoS. Consequently, BPI+ must support the QoS requirement of upper-layer protocols and applications, such as those defined by PacketCable. (See the next section, "The PacketCable Security Specification.")
Security services and mechanisms	*Strongly* authenticates CMs via an RSA public/private key pair and an X.509v3 digital certificate; protects against CM cloning. Controls access to protected cable resources and services. Provides for secure exchange of encryption keys, and performs traffic encryption across a *single* network segment.[2] Ensures the integrity of important message types, including key requests and replies and network management messages. Does not verify the integrity of encrypted user data. Implements secure software upgrades within CMs deployed in the field.
Network security protocols	Simple Network Management Protocol (SNMP) v3 security to protect network management messages exchanged between CMs and operations support systems.
Public-key encryption algorithms	Supports both 768-bit and 1024-bit RSA public/private key pairs for encrypting and decrypting long-term key-encryption keys (KEKs). Includes support for signed PKCS #7 software upgrade files;[3] signatures are created via RSA and SHA-1.
Symmetric-key encryption algorithms	Uses two-key 3DES (EDE-ECB mode)[4] KEKs for encrypting short-term TEKs. Uses 40-bit and 56-bit DES (CBC mode) traffic-encryption keys (TEKs) for encrypting the contents of Layer 2 PDUs in the upstream and downstream directions. (56-bit TEKs must be used when supported by both the CM and CMTS.) Uses DES (CBC mode) in conjunction with SNMPv3 security.

Table 5-2 Characteristics of DOCSIS 1.1 BPI+ *(continued)*

Characteristic	Description
Message integrity algorithms	Uses HMAC SHA-1 to ensure the authenticity and integrity of key request messages. Uses MD5 for CM message integrity checks (MICs). Uses HMAC versions of MD5 and SHA-1 to ensure the authenticity and integrity of SNMPv3 network management messages.
Use of digital certificates and PKI	DOCSIS 1.1 BPI+ specifies a three-tier PKI hierarchy consisting of a CableLabs-chosen DOCSIS Root Certification Authority (CA), subordinate manufacturer CAs, and cable modems, along with all applicable certificate formats. The DOCSIS Root CA assigns each CM manufacturer a 2048-bit RSA X.509v3 digital certificate for use in signing CM certificates during manufacturing. Each CM contains a 1024-bit RSA X.509v3 digital certificate and a corresponding RSA key pair. The DOCSIS Root CA assigns each CM manufacturer and MSO[5] a 1024-bit, 1536-bit, or 2048-bit X.509v3 code verification certificate (CVC) for use in signing software upgrade files.
Special characteristics	CMs and CMTSs should meet the minimum requirements of FIPS 140-1 Security Level 1 for protection of keying material (RSA key pairs) stored within hardware devices. Write-once memory modules containing RSA key pairs are not field-upgradeable.

[1] The DOCSIS 1.1 series of specifications (including BPI+) are being considered as the basis for Link layer protocols being developed by the BWIF and the IEEE 802.16 Working Group.

[2] DOCSIS 1.1 BPI+ includes a security capabilities selection feature that allows the CM and CMTS to negotiate a common cryptographic suite (a combination of a data-encryption and a data-authentication algorithm) prior to exchanging encrypted traffic. This functionality, which does not exist in DOCSIS 1.0 BPI, permits future upgrades to algorithms such as AES without altering the BPI+ specification.

[3] While DOCSIS 1.0 BPI supports remote software upgrades, it includes no mechanisms for authenticating the source or integrity of a code file. The signed PKCS #7 message format employed by DOCSIS 1.1 BPI+ accomplishes both data origin authentication and message integrity.

[4] EDE stands for encrypt-decrypt-encrypt. A typical 3DES encryption or decryption consists of three DES encryption/decryption operations, with the output from the first operation acting as the input to the second, and so on. Each individual DES encrypt/decrypt may use its own key, hence the one-key, two-key, and three-key nomenclature.

[5] A *multiple-systems operator* (MSO) is a large cable television provider operating multiple cable systems.

Table 5-2 Characteristics of DOCSIS 1.1 BPI+ *(continued)*

The PacketCable Security Specification

PacketCable operates at or above the Network layer, and it interfaces with DOCSIS 1.1 networks for QoS. PacketCable uses the existing cable television network infrastructure to deliver packet-based multimedia services to subscribers. IP telephony will serve as the first practical application for PacketCable networks. PacketCable environments consist of many components, including media terminal adapters (MTAs), call management servers, media and signaling gateways (including gateways for interfacing to PSTN networks), and operations support systems. PacketCable employs numerous communication protocols, such as Media Gateway Control Protocol (MGCP), Trivial File Transfer Protocol (TFTP), RTP/RTCP, and SNMP, for call signaling and data transfer. QoS-aware packet-based networks provide a cost-effective alternative to ATM-based networks for the delivery of telephony, videoconferencing, and other real-time multimedia applications. Since PacketCable must support these real-time applications, high performance is a key requirement for security. The PacketCable Security Specification implements a number of services for authenticating subscriber devices (such as MTAs), protecting subscriber data, and controlling access to MSO resources and services (CableLabs, 2002). The characteristics of PacketCable Security are outlined in Table 5-3. PacketCable Security is discussed in further detail in Chapter 10.

NOTE:

MTAs come in two flavors: embedded and standalone. An embedded MTA is combined with a DOCSIS 1.1 cable modem (one device contains both components). A standalone MTA is physically separate from the cable modem, but must be used in conjunction with a DOCSIS 1.1 cable modem to access the cable network. As of this writing, the PacketCable series of specifications defines only embedded MTAs.

The H.235 Security Standard

The H.235 standard provides security for H-Series terminals in H.323 networks (International Telecommunication Union, 2001). A product of the ITU, H.323 was originally designed to support videoconferencing over local area networks. H.323 has since been adopted for point-to-point and multipoint voice, video, and data conferencing over packet-based networks (LANs and WANs) without built-in guaranteed QoS. Voice telephony and packet-based services are likely to drive future need for H.323 networks. A typical H.323 environment is composed of H-Series terminals (that is, end-user devices), multipoint control units (MCUs), gatekeepers, and PSTN gateways.

NOTE:
A multipoint control unit (MCU) *enables conferencing between more than two endpoints. A gatekeeper is a server used for tracking active users and the nodes (that is, the IP addresses) to which they are connected.*

Characteristic	Description
Name	PacketCable Security Specification.
Organization	Cable Television Laboratories.
Description	The PacketCable Security Specification provides end-to-end authentication and encryption for multimedia applications over cable networks; it implements security services at or above the Network layer.
Applications	IP telephony, voice- and videoconferencing, and other real-time streaming multimedia applications
QoS requirements	Real-time streaming multimedia applications have stringent QoS requirements for bandwidth, latency, jitter, and packet loss: High-quality video conferencing may require bandwidth of 2 Mbps or more. IP telephony requires that end-to-end latency be much less than 400 ms (100–150 ms is acceptable), and that jitter not exceed the 10–20 ms range. Packet loss rates for real-time multimedia applications should not exceed five percent. (Less than two percent is desirable.)
Security services and mechanisms	*Strongly* authenticates each MTA via an RSA public/private key pair and an X.509v3 digital certificate— protects against MTA cloning. Controls access to protected cable network resources and services. Provides for secure exchange of encryption keys and performs end-to-end encryption (within the cable provider's network only) of user traffic contained within the payload of RTP messages. Ensures the integrity of key exchanges, network management and signaling messages, and user data. Uses digital signatures to provide authentication and message integrity; nonrepudiation is not supported. Leverages DOCSIS 1.1 secure software upgrades for updating embedded MTAs deployed in the field. Uses IPsec to provide protection against replay attacks.

Table 5-3 Characteristics of the PacketCable Security Specification

Characteristic	Description
Network security protocols	Kerberos/PKINIT/PKCROSS. IPSec Encapsulating Security Payload (ESP) in Transport mode with Internet Key Exchange (IKE) and Kerberos/ PKINIT/PKCROSS for negotiating encryption and authentication keys. SNMPv3 security. RADIUS authenticators. RSVP integrity objects. RTP and RTCP Application layer security protocols for negotiating cipher suites, exchanging keying information, and encrypting media streams between two MTAs and between MTAs and other PacketCable components.
Public-key encryption algorithms	Uses ephemeral-ephemeral Diffie-Hellman (DH) for key agreement in conjunction with Kerberos/PKINIT. Uses DH for key agreement within SNMPv3. Uses RSA digital signatures to provide data origin authentication and integrity in conjunction with IKE and Kerberos/PKINIT. Standalone MTAs are likely to include support for verifying signed PKCS #7 software upgrade files.
Symmetric-key encryption algorithms	Uses 3DES, RC5, IDEA, CAST, Blowfish, and AES (all in CBC mode) for traffic encryption in conjunction with IPSec. (3DES must be supported.) Uses 128-bit AES (CBC mode), XDESX/DESX (CBC mode), DES (CBC mode), 3DES (EDE-CBC mode), and 128-bit RC4 for encryption of RTP messages exchanged between MTAs. (AES must be supported.) Uses DES in CBC mode to secure SNMP network management messages (SNMPv3 security).
Message integrity algorithms	Uses HMAC versions of MD5 and SHA-1 to ensure the authenticity and integrity of: IPSec keying material and encrypted traffic. SNMPv3 network management messages. (HMAC MD5 must be supported.) RTCP control messages. Employs the Multilinear Modular Hash (MMH) MAC, in both 2-byte and 4-byte versions, to ensure the integrity of RTP messages. Uses SHA-1 to generate and verify digital signatures.

Table 5-3 Characteristics of the PacketCable Security Specification *(continued)*

Characteristic	Description
Use of digital certificates and PKI	Like DOCSIS 1.1 BPI+, PacketCable specifies a three-tier PKI hierarchy consisting of a CableLabs-chosen PacketCable MTA Root (CA), subordinate MTA manufacturer CAs, and MTAs, along with all applicable certificate formats. PacketCable also defines another four-tier PKI hierarchy consisting of the CableLabs Service Provider Root CA, service provider CAs, local system operator CAs, and various authentication and electronic surveillance servers. The PacketCable MTA Root CA assigns each MTA manufacturer a 2048-bit RSA X.509v3 certificate for use in signing MTA certificates during manufacturing. Each MTA contains a 1024-bit RSA X.509v3 certificate and a corresponding RSA key pair.
Special characteristics	An MTA should meet the minimum requirements of FIPS 140-1 Security Level 1 for protection of keying material (RSA key pairs, IPSec encryption and authentication keys, and Kerberos session keys) stored within hardware devices. Write-once memory modules containing RSA key pairs are not field-upgradeable.

Table 5-3 Characteristics of the PacketCable Security Specification *(continued)*

H.323 communication requires a handful of protocols for call management and data transmission, the most important being

- **RAS** Required for call registration, admission, and status; manages billing and gatekeeper-related processes.
- **H.225** Required for call setup and termination.
- **H.245** Required for call control and selection of capabilities (for example, audio and video codecs and cryptographic mechanisms).
- **RTP/RTCP** Required for transmission of real-time audio and video data.

H.235 includes services for authenticating end-user terminals, providing end-to-end communication privacy, and controlling access to multimedia conferencing environments and resources.

The characteristics of H.235 are outlined in Table 5-4.

Characteristic	Description
Name	H.235
Organization	International Telecommunication Union
Description	H.235 provides end-to-end authentication and encryption for multimedia applications over H.323 (and other H.245-based) networks; it implements security at or above the Network layer.
Applications	Multimedia (voice, video, and data) conferencing over packet-based networks that do not provide built-in guaranteed QoS. Services include real-time IP telephony and videoconferencing in both point-to-point and multipoint environments.
QoS requirements	Real-time streaming multimedia applications have stringent QoS requirements for bandwidth, latency, jitter, and packet loss: High-quality videoconferencing may require bandwidth of 2 Mbps or more. IP telephony requires that end-to-end latency be much less than 400 ms (100–150 ms is acceptable), and that jitter not exceed the 10–20 ms range. Packet loss rates for real-time multimedia applications should not exceed five percent. (Less than two percent is desirable.)
Security services and mechanisms	Provides multiple options for end-to-end authentication of H-Series terminals to gateways, gatekeepers, MCUs, and back-end systems: "Subscription-based" challenge/response authentication using shared secret keys Certificate-based authentication with digital signatures Authentication built into other security protocols (e.g., TLS, IPSec,[6] and RADIUS) Controls access to H.323 multimedia conferences. Supports key exchanges/agreement and the negotiation of encryption algorithms. Performs end-to-end traffic encryption of user data contained within the payload of RTP messages. Ensures the authenticity and integrity of key exchanges, signaling messages, and user data. Uses digital signatures to provide data origin authentication and integrity during terminal authentication. Includes anti-spamming[7] protection for RTP media streams. Employs trusted third parties (TTPs) for recovery of lost media-encryption keys.

Table 5-4 Characteristics of H.235

Characteristic	Description
Network security protocols	IPSec[6] for Network layer authentication and encryption of call connection and control channels; IPSec is not suggested, however, for use with RTP media streams.[8] TLS[6] to secure call connections (for example, TSAP[9] protocol messages) and control channels. RADIUS for challenge/response authentication to back-end systems
Public-key encryption algorithms	Uses 512- and 1024-bit DH with optional digital-signature–based authentication to generate RTP session-encryption keys Includes a *signature security profile*[10] specifying RSA signatures using MD5 and SHA-1 for ensuring the authenticity and integrity of key-exchange messages and signaling protocols (e.g., RAS, H.225, and H.245 messages). Incorporates the Elliptic Curve Diffie-Hellman algorithm and the Elliptic Curve Digital Signature Algorithm (ECDSA) for key management and digital signatures, respectively.
Symmetric-key encryption algorithms	RTP media stream encryption uses the following algorithms: 56-bit DES[11] in CBC 168-bit 3DES[11] in outer-CBC 56-bit "RC-2 compatible"
Message integrity algorithms	Uses 64-bit DES-MAC, 168-bit 3DES-MAC, and 96-bit HMAC SHA-1-96 to ensure the authenticity and integrity of RTP media streams. Uses 96-bit HMAC-SHA1-96 to ensure the authenticity and integrity of call connection and control messages. (e.g., RAS, H.225, and H.245). Uses MD5 and SHA-1 in conjunction with the generation of RSA digital signatures.

Table 5-4 Characteristics of H.235 *(continued)*

Characteristic	Description
Use of digital certificates and PKI	H.235 specifies a protocol for exchanging certificates, but not how the certificates are verified or trusted. H.235 does not specify certificate formats beyond what is required for the authentication protocols, and includes no PKI hierarchy.

[6] Additional public-key, symmetric-key, and message digest algorithms may be used when TLS or IPSec is employed within H.235.

[7] Spamming attacks attempt to crash a host by consuming critical resources, such as storage space used for virtual memory.

[8] Neither PacketCable nor H.235 suggest using IPSec for protecting real-time RTP media streams. In most cases, IPSec is too complex for real-time data channels, and is more applicable to securing call connection and control channels.

[9] TSAP (Transport Layer Service Access Point) is used to multiplex Transport layer services that share a single IP address.

[10] The signature security profile is geared at large-scale installations where the distribution of shared secrets would be prohibitively difficult to manage.

[11] Future versions of H.235 will likely support AES as a suitable replacement for DES. The current draft for H.235v3 also references the DES and 3DES algorithms running in OFB mode; this mode may or may not appear in the finalized standard.

Table 5-4 Characteristics of H.235 *(continued)*

The DVB Multimedia Home Platform

The DVB Multimedia Home Platform (MHP) defines a programming interface between interactive applications and their execution environment (Digital Video Broadcasting Project, 2001). This interface abstracts the details of the underlying hardware and operating system on which the applications run. This abstraction allows software developers to create non-platform-specific applications that run without modification on a wide variety of end-user devices. MHP is intended to extend the interactive functionality of DVB to all transmission networks, including cable, wireless, and satellite. MHP security authenticates the source of application code and other files downloaded to set-top boxes, televisions with integrated DVB receivers, and multimedia PCs. It also secures the return channel used to communicate subscriber information back to the service provider. The characteristics of the DVB Multimedia Home Platform are outlined in Table 5-5. DVB MHP Security is discussed in further detail in Chapter 11.

NOTE:
All DVB applications run within an execution environment known as DVB-J. DVB-J includes a Java virtual machine for executing Java applications, and supports the use of many popular Java classes from Sun Microsystems.

The OpenCable Copy Protection System

OpenCable defines standards for the hardware and software platforms required to support interactive television services over cable networks. In the OpenCable environment, the subscriber purchases the host device of his or her choice (an OpenCable-compliant set-top box or television, for instance). The cable provider supplies a *point-of-deployment (POD)* module that the subscriber attaches to the host device. The POD allows the host device to decrypt scrambled MPEG video and audio signals coming from a cable television head end. Before sending copy-protected content to the host device, the POD is responsible for re-encrypting the signal. The OpenCable Copy Protection System (CPS) defines the process for encrypting and decrypting the MPEG video and audio flowing across the POD-to-host device interface (CableLabs, 2001b). Its purpose is to protect "high-value" content from attackers tapping into the circuitry of the device.

The characteristics of the OpenCable Copy Protection System are outlined in Table 5-6.

Characteristic	Description
Name	DVB Multimedia Home Platform (MHP)
Organization	The Digital Video Broadcasting Project of ETSI
Description	MHP security authenticates the source and verifies the integrity of interactive digital video applications downloaded to and executed on DVB-compliant subscriber devices; it implements security within the Transport and Application layers.
Applications	MHP supports a wide range of interactive applications,[12] including electronic program guides, information services, synchronized application and TV content, e-commerce (for example, home shopping and online banking), online gaming, and other applications requiring security (DVB, 2001).
QoS requirements	Performance is important because it affects the load time of an application. Hash codes and signatures must be checked each time an application is loaded.

Table 5-5 Characteristics of the DVB Multimedia Home Platform

Characteristic	Description
Security services and mechanisms	Each MHP-enabled terminal authenticates applications using a combination of three message types: **Hash code** Ensures integrity by summarizing data contained within a directory (files and subdirectories). **Signature** Resides in the root of the application "tree"; associates all hash code files with a single application, ensures integrity of the application as a whole, and authenticates the source of the application. **Certificate** Contains a list of certificates required to verify the signature files. Allows multiple signers of an application. Provides the following types of access control using signed permission files:[13] Local resource and file access (including storage) Interapplication communication Service selection Return channel access Remote host access Application lifecycle control (for example, pause, stop, and resume functions) Smart card access Implements data privacy on the return channel from the subscriber to the service provider.
Network security protocols	TLS secures user data on the return channel.
Public-key encryption algorithms	Uses the RSA algorithm for Generation and verification of digital certificates and digital signatures Authentication and key exchanges in conjunction with TLS
Symmetric-key encryption algorithms	TLS uses the following algorithms[14] to encrypt user data on the return channel: 40-bit and 56-bit DES (CBC mode) 2-key 3DES (in EDE-CBC mode) 3-key 3DES (in EDE-CBC mode)

Table 5-5 Characteristics of the DVB Multimedia Home Platform *(continued)*

Characteristic	Description
Message integrity algorithms	Uses MD5 and SHA-1 message digests To create and verify hash code and signature files To create and verify signatures contained within digital certificates In conjunction with TLS on the return channel for message integrity
Use of digital certificates and PKI	Uses X.509v3 digital certificates for key exchanges and processing of digital signatures; MHP-compliant devices must support certificates with up to 4096-bit key lengths. Uses a TTP certification authority to provide trust between MHP terminals and application providers. Allows an *arbitrary-level* PKI hierarchy with a common trusted Root CA. Specifies formats for certificates and certificate files, but not for the PKI hierarchy. Specifies policies for the management, distribution, and processing of CRLs[15] and root certificates.
Special characteristics	The DVB-J runtime environment includes support for the java.security package from Sun Microsystems.[16]

[12] MHP provides the basis for the OpenCable Application Platform (OCAP). OCAP is the middleware software component of OpenCable interactive set-top boxes and televisions. As with MHP, OCAP offers a runtime environment for the execution of interactive digital video applications.

[13] Permission files allow application providers to request additional resources on MHP-enabled terminals.

[14] DVB also uses a proprietary content-encryption algorithm called the Common Scrambling Algorithm (CSA). This algorithm encrypts digital video content from the service provider to the subscriber. In cable television networks, the CSA would scramble cable signals sent from a head end to a set-top box or digital cable-ready television. ETSI maintains the specification for the CSA, and its details are only available after signing a nondisclosure agreement.

[15] A *certificate revocation list (CRL)* contains a list of digital certificates revoked due to key compromise, employee termination, or any other condition requiring premature invalidation of a certificate (see Appendix B).

[16] The java.security package contains many Java classes for implementing various security-related tasks within Java applications.

Table 5-5 Characteristics of the DVB Multimedia Home Platform *(continued)*

Characteristic	Description
Name	OpenCable Copy Protection System (CPS)
Organization	Cable Television Laboratories
Description	The OpenCable CPS secures the transfer of copy-protected content, such as pay-per-view or video-on-demand, between a POD and an end-user host device; this prevents unauthorized reproduction.
Applications	The OpenCable system supports interactive television (iTV) services, including interactive program guides, sports, and games shows; online games; impulse pay-per-view; video-on-demand; Internet access; and e-commerce (CableLabs, 2001a).
QoS requirements	MPEG audio and video is real-time, rate-adaptive content, requiring high-performance cryptographic operations. However, large receive buffers greatly reduce the effects of signal latency and jitter on streaming multimedia content.
Security services and mechanisms	A POD authenticates a host device via an X.509v3 digital certificate and shared long-term authentication keys; the POD will not descramble copy-protected content unless the host device contains a valid certificate. Supports DH key agreement for the exchange of keying material. POD modules report any (detected) non-OpenCable-compliant or compromised host devices to the service provider. Host devices use a special-purpose "authenticated tunneling protocol" to verify the delivery of valid control messages from POD modules. *Does not* specify built-in integrity checking for MPEG signals flowing across the POD-to-host device interface.
Network security protocols	N/A
Public-key encryption algorithms	Uses 1024-bit DH key agreement to generate a shared secret key, which is then used to generate long-term authentication and copy-protection keys.
Symmetric-key encryption algorithms	Uses DES[17] in ECB mode for encryption of MPEG content.

Table 5-6 Characteristics of the OpenCable Copy Protection System

Characteristic	Description
Message integrity algorithms	Uses SHA-1 In conjunction with proprietary DFAST algorithm[17] to generate copy-protection keys For generating and verifying certificate signatures
Use of digital certificates and PKI	Each host device and POD contains an RSA X.509v3 digital certificate for authentication and initial key exchange. Each head end maintains a CRL containing the certificates of revoked or compromised hosts.

[17] The POD and host device generate DES encryption keys using the proprietary and patented DFAST technology, which converts a 128-bit input to a 56-bit output. According to its creators, DFAST generates keystreams with long periods and nonlinear statistical properties (CableLabs, 2001b).

Table 5-6 Characteristics of the OpenCable Copy Protection System *(continued)*

Security Gone Wrong—A Case Study of 802.11 WEP Encryption

The IEEE 802.11 series of standards (802.11, 802.11b, and 802.11a) define Physical and Data Link layer protocols for wireless communication over local area networks. *Wireless LANs (WLANs)* provide cost-effective wireless connectivity for both homes and offices where wired infrastructures are infeasible or inconvenient. A typical WLAN consists of an *access point (AP)* and one or more wireless stations (wireless NICs). Figure 5-1 illustrates an 802.11 wireless network. The AP coordinates communication between wireless stations, and may act as a bridge for exchanging data with a wired network. Wireless NICs come in many forms, including PC cards, PCI expansion boards, and external USB adapters. Currently capable of transmission speeds of up to 11 Mbps, 802.11b wireless networks provide performance competitive with 10 Mbps wired Ethernet networks.

To be truly competitive, however, wireless networks need to be as secure as—or more secure than—their wired counterparts. To prevent eavesdropping on WLANs, the 802.11b working group selected a Link layer encryption protocol known as *Wired Equivalent Privacy (WEP)*. Encryption has the added benefit of

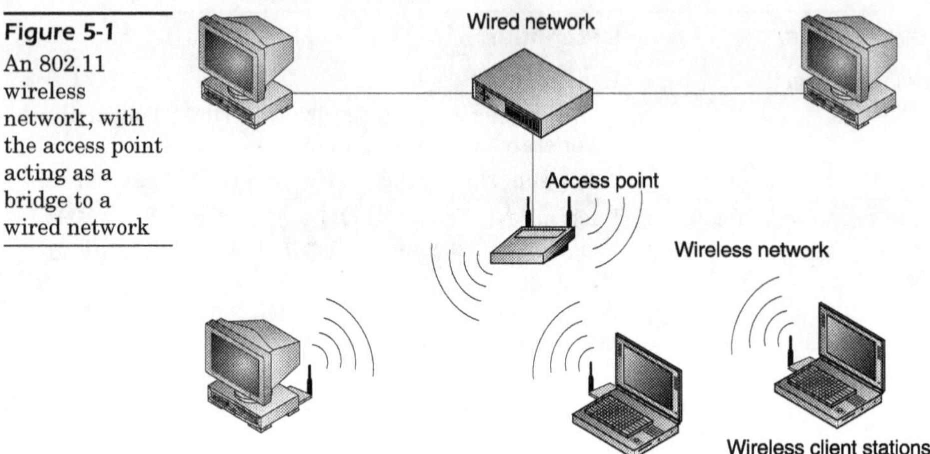

Figure 5-1 An 802.11 wireless network, with the access point acting as a bridge to a wired network

controlling access to wireless networks. Only those stations sharing a secret key with the AP may communicate on the network.

Despite its intentions, WEP is inherently flawed. Shortly following its adoption, researchers began discovering weaknesses in WEP, making it far less secure than its "wired equivalent." A number of poor design decisions—decisions that might have been avoided through open public debate—made their way into the specification for WEP.

WEP uses RC4 to encrypt the payload of each MAC layer frame. This in itself poses no problems; practically all SSL implementations in existence today use RC4 for traffic encryption, and the algorithm provides high resistance to cryptanalysis when used correctly. Unfortunately, WEP does not use RC4 correctly. Borisov, Goldberg, and Wagner discuss two very important weaknesses in WEP that make it vulnerable to a number of possible passive and active attacks (Borisov, Goldberg, and Wagner, 2001):

- The method used by WEP to ensure that two ciphertext messages are not encrypted with the same keystream is highly susceptible to cryptanalysis. This weakness facilitates passive eavesdropping attacks.
- The message integrity mechanism employed by WEP is relatively easy to subvert without detection. This weakness can lead to active data-modification attacks.

Recall from our discussion in Chapter 2 that RC4 is a stream cipher that encrypts data by XORing it with a very long keystream. Also, recall that stream ciphers are very susceptible to cryptanalysis when the keystream repeats, or when a single key is used to encrypt multiple data sets. WEP attempts to subvert cryptanalytic attacks by combining the secret key shared by the AP and each wireless NIC with a 24-bit initialization vector. The purpose of the *initialization vector (IV)* is to make the encryption key for each frame unique. Most

APs set the IV counter to 0 at the start of communication, and increment the IV by 1 for each successive packet transmitted. However, this is a very predictable method, and the 24-bit IV limits the keyspace to a total of 2^{24} (16,777,216) unique keys. In cryptographic terms, this is a *very* small number. Even without knowing the IV sequence, an attacker has a relatively high probability of retrieving a portion of the plaintext or the corresponding keystream after capturing a comparatively small amount of transmitted data. Borisov, Goldberg, and Wagner calculated that the keystream generated by a busy 802.11b AP sending 1500-byte packets at 11 Mbps begins to repeat after a mere five hours.

To prevent an attacker from actively modifying transmitted data, WEP attaches a CRC-32 checksum to each 802.11b MAC frame (see Figure 5-2). A special-purpose field, known as an *integrity check (IC)* field, contains the CRC, and is part of the encrypted payload. One characteristic of stream ciphers is that they are highly resistant to error propagation; a single ciphertext bit error results in a single plaintext error during decryption. With some work, an intruder could alter bits in the payload of a frame along with bits in the corresponding checksum, and thus produce data that decrypts and verifies correctly. This would fool the receiving device into believing that it had obtained a valid message.

In August 2001, Fluhrer, Mantin, and Shamir described a passive ciphertext-only attack against WEP's usage of RC4 (Fluhrer, Mantin, and Shamir, 2001). The attack took advantage of the way that WEP uses the RC4 key-scheduling algorithm to generate a unique keystream for each frame. By analyzing the first word of pseudorandom output from the keystream generator, they discovered a method for deriving information about a single byte of the shared secret key. This type of attack is especially effective when the first few bytes of plaintext input are predictable (an IP protocol header, for instance). Stubblefield, Ioannidis, and Rubin successfully implemented the attack in August of 2001 (Stubblefield, Ioannidis, and Rubin, 2001). They were able to retrieve a complete 128-bit secret key from a WEP-enabled AP.

Ron Rivest, the creator of RC4, suggested a simple solution to this attack (Rivest, 2001). He concluded that if WEP had discarded the first 256 bytes produced by the keystream generator before RC4 combined the keystream with plaintext, the attack would not have been possible. This technique is incompatible

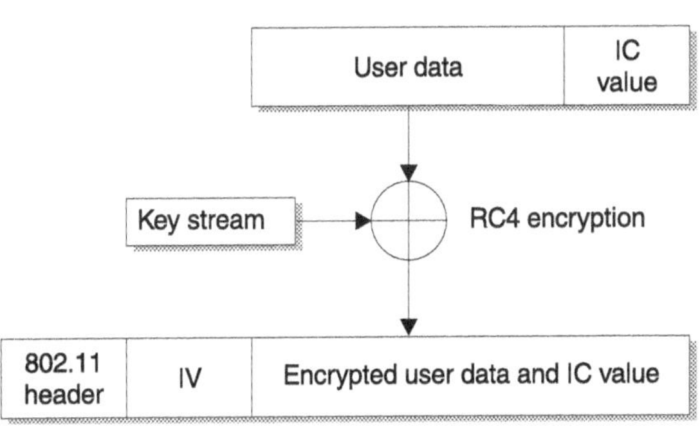

Figure 5-2 The WEP packet encapsulation process

with the current implementation of WEP, but had it been included in the original design, it would have effectively prevented the attack. This situation highlights the need for public scrutiny during the process of standards adoption.

References

Borisov, Nikita, Ian Goldberg, and David Wagner. 2001. "Intercepting Mobile Communications: The Insecurities of 802.11." *Seventh Annual International Conference on Mobile Computing and Networking*. Special Interest Group on Mobility of Systems, Users, Data, and Computing.

CableLabs, 2001. *CableLabs Project Primer: OpenCable*. Cable Television Laboratories. 18 December, 2001 (http://www.opencable.com/opencableprimer.html).

—15 May, 2001. *IS-POD-CP-INT05-010515: OpenCable Copy Protection System*. Cable Television Laboratories.

—16 January, 2002. *PKT-SP-SEC-I05-020116: PacketCable Security Specification*. Cable Television Laboratories.

—29 August, 2001. *SP-BPI-I03-010829: Data-Over-Cable Service Interface Specifications: Baseline Privacy Interface Specification*. Cable Television Laboratories.

—29 August, 2001. *SP-BPI+-I07-010829: Data-Over-Cable Service Interface Specifications: Baseline Privacy Plus Interface Specification*. Cable Television Laboratories.

Digital Video Broadcasting Project. 13 May, 2001. *Digital Video Broadcasting (DVB) Multimedia Home Platform (MHP) Specification 1.1*. European Telecommunications Standards Institute.

Fluhrer, Scott, Itsik Mantin, and Adi Shamir. August, 2001. "Weaknesses in the Key Scheduling Algorithm of RC4." *Eighth Annual Workshop on Selected Areas in Cryptography*.

International Telecommunication Union. 2001. *H.235: Security and Encryption for H-Series (H.323 and other H.245-based) Multimedia Terminals Version 3*.

Rivest, Ron. December, 2001. "RSA Security Response to Weaknesses in Key Scheduling Algorithm of RC4." *RSA Laboratories Tech Notes*. RSA Laboratories. 18 December, 2001 (http://www.rsasecurity.com/rsalabs/technotes/wep.html).

Stubblefield, Adam, John Ioannidis, and Aviel R. Rubin. 21 August, 2001. "Using the Fluhrer, Mantin, and Shamir Attack to Break WEP—Revision 2." *AT&T Labs Technical Report TD-4ZCPZZ*. New Jersey: AT&T Labs.

Part II

Broadband Security Design Considerations

Chapter 6

Existing Network Security Protocols

When designing network security infrastructures, you have many tools at your disposal. In addition to cryptographic mechanisms, there are a number of existing protocols for implementing security within various network layers (see Table 6-1). One of the decisions you will have to make when designing a security infrastructure is whether to use existing network security protocols. In Chapter 2, we discussed security services and explored how to implement many of the services using cryptography. In this chapter, we'll take the idea one step further by demonstrating how many of the popular network security protocols combine security mechanisms (both cryptographic and noncryptographic) to secure communication between communicating parties. While many protocols implement authentication, key exchange, and traffic encryption, they each do so in different ways. For example, two protocols may use similar cryptographic mechanisms for authentication and key exchange, but may define unique procedures and syntaxes for exchanging authentication and keying information. As we will discuss in Chapter 7, some protocols apply more directly to certain network environments than others do. In this chapter, we focus on three network security protocols: IPSec, SSL/TLS, and Kerberos.

OSI Network Layer	Security Protocols and Mechanisms
Application	Kerberos and Secure Multipurpose Internet Mail Extensions (S/MIME)
Transport	Secure Sockets Layer/Transport Layer Security and Wireless Transport Layer Security (WTLS)
Network	Internet Protocol Security (IPSec), Simple Key-Management for Internet Protocols (SKIP), and swIPe
Data Link	Baseline Privacy Interface (BPI), BPI+, Wired Equivalent Privacy (WEP), and Microsoft Point-to-Point Tunneling Protocol (MS PPTP)
Physical	Line or bulk encryption

Table 6-1 Common Security Protocols and Mechanisms Running at Various Network Layers

NOTE:
In Table 6-1, the Application layer protocols operate within Layers 5–7 of the OSI Reference Model.

IPSec

The Internet Protocol (IP) was not designed to be secure; as a result, it contains many vulnerabilities. IP is susceptible to address spoofing, replay attacks, session hijacking, and numerous DoS attacks. It also uses a weak checksum feature for calculating message integrity, and includes no confidentiality mechanism for preventing eavesdropping. In anticipation of the need for a robust network security protocol capable of protecting IP-based Internet traffic, the IETF designed *Internet Protocol Security (IPSec)* to counter many of the security shortcomings of IP. IPSec runs at the Network layer and consists of a suite of protocols for encryption, message authentication, and key management. The IPSec Working Group within the IETF is responsible for IPSec development, and it has defined the following RFCs specifying IPSec:

- RFC 1828: IP Authentication Using Keyed MD5
- RFC 1829: The ESP DES-CBC Transform
- RFC 2085: HMAC-MD5 IP Authentication with Replay Protection
- RFC 2401: Security Architecture for the Internet Protocol
- RFC 2402: IP Authentication Header (AH)
- RFC 2403: The Use of HMAC-MD5-96 Within ESP and AH

- RFC 2404: The Use of HMAC-SHA-1-96 Within ESP and AH
- RFC 2405: The ESP DES-CBC Cipher Algorithm with Explicit IV
- RFC 2406: IP Encapsulating Security Payload (ESP)
- RFC 2407: The Internet IP Security Domain of Interpretation for ISAKMP
- RFC 2408: Internet Security Associations and Key Management Protocol (ISAKMP)
- RFC 2409: The Internet Key Exchange (IKE)
- RFC 2410: The NULL Encryption Algorithm and Its Use with IPSec
- RFC 2411: IP Security Document Roadmap
- RFC 2412: The Oakley Key Determination Protocol
- RFC 2451: The ESP CBC-Mode Cipher Algorithms
- RFC 2857: The Use of HMAC-RIPEMD-160-96 Within ESP and AH

IPSec provides a multitude of security services for protecting IPv4 and IPv6 traffic, including data origin authentication, connectionless integrity, data privacy, limited traffic flow confidentiality, access control, and replay protection. The IPSec protocols responsible for implementing these services are the *Authentication Header (AH)* and the *Encapsulating Security Payload (ESP)*. AH ensures the integrity of upper-layer headers and data, along with many components of the IP header. ESP protects the integrity of upper-layer data encapsulated within IP headers, but not the headers themselves. However, ESP includes encryption mechanisms for providing confidentiality. For key management, IPSec employs IKE, the *Internet Key Exchange* protocol. (We'll discuss each of these protocols in detail in the following sections.) Table 6-2 compares the AH and ESP protocols in terms of the security services they support.

Table 6-2
A Comparison of Security Services Provided by the AH and ESP Protocols

Security Service	Supported by AH?	Supported by ESP?
Data origin authentication	Yes	Yes
Connectionless integrity[1]	Yes	Yes
Anti-replay protection	Yes	Yes
Data privacy	No	Yes
Limited traffic flow confidentiality	No	Yes

[1] The levels of integrity offered by AH and ESP differ. We'll discuss the differences in the sections entitled "Authentication Header" and "Encapsulating Security Payload," later in this chapter.

Transport and Tunnel Modes

The AH and ESP protocols operate in two modes: transport and tunnel. *Transport mode* provides end-to-end security in environments where the security endpoints are the same as the communication endpoints (see Figure 6-1). The sending host applies authentication, integrity, and—in the case of ESP—encryption mechanisms to TCP or UDP data prior to transmitting each IP packet. The associated AH and ESP formatting remains intact and unaltered as routing devices forward each packet across the network. Upon receiving a packet, the receiving host strips the IP header and any AH and ESP headers and trailers, and then decrypts and verifies the authenticity and integrity of the encapsulated data; for AH, this includes parts of the IP header. When using transport mode, both communication endpoints must support IPSec. This usually entails modification to the IP stack running on each host.

CAUTION:

Transport mode operation fails in the presence of network address translation (NAT). *One reason for the failure is that a NAT gateway masks internal network addresses by converting the source address of outgoing packets to that of the gateway. This conceals internal IP addresses and makes all packets look as though they have originated from the same host. Since IPSec maps authentication information to an IP source address, the modified address causes integrity checks to fail during processing by the recipient.*

Figure 6-2 illustrates the encapsulation process for transport mode. The sending host first applies the appropriate transforms to the upper-layer TCP or UDP data. It then attaches an AH header, ESP header, or both to the transformed data. Finally, the IP layer affixes an IP header to the packet.

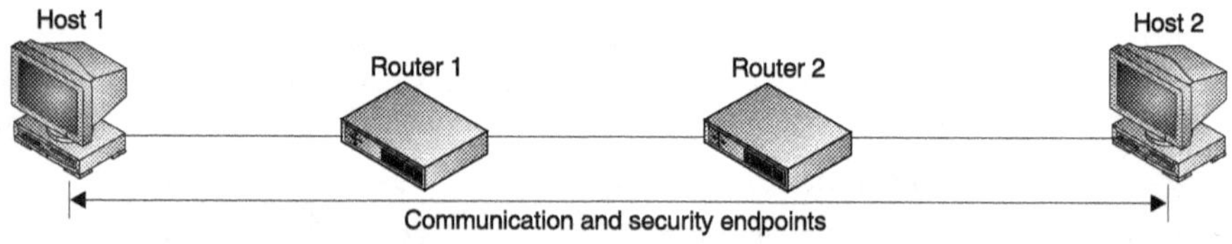

Figure 6-1 Transport mode provides end-to-end security when the communication and security endpoints are the same.

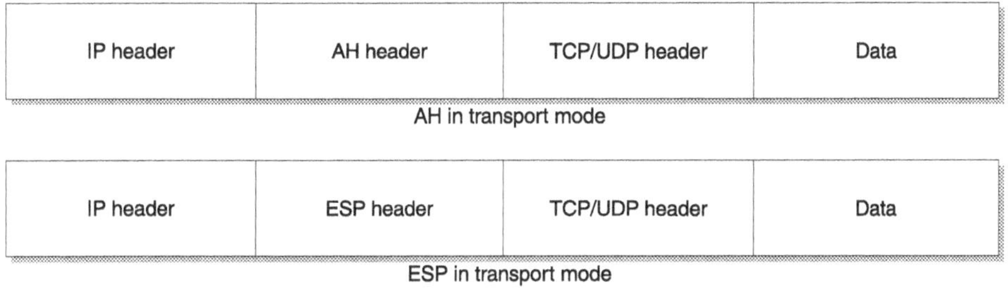

Figure 6-2 In transport mode, IPSec inserts AH and ESP headers between the IP header and the encapsulated upper-layer payload.

AH and ESP modes can be combined, as shown here, for privacy of the encapsulated payload and protection of the IP header:

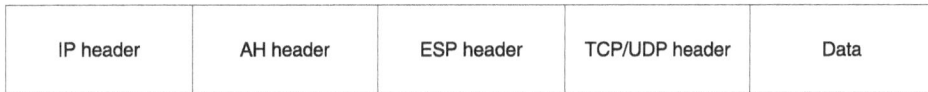

IP header	AH header	ESP header	TCP/UDP header	Data

This process is known as *transport adjacency*. Transport adjacency addresses only a single level of nesting; no added benefits arise from further nesting (Kent and Atkinson, 1998).

CAUTION:

Remember: When using both AH and ESP in transport mode, apply ESP before AH for maximum data integrity protection (Doraswamy and Harkins, 2000).

In *tunnel mode,* the security and communication endpoints usually differ (see Figure 6-3). This scenario is common in dedicated virtual private networks (VPNs) where a device other than the originating host provides security, or

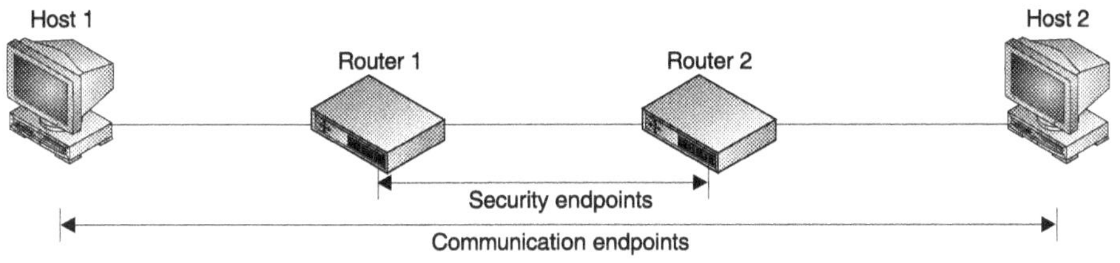

Figure 6-3 Tunnel mode is applicable to environments where the security endpoints and communication endpoints differ.

when security services do not extend all the way to the destination host. In a VPN, an IPSec-enabled router or other VPN device protects all traffic flowing between two or more remote locations by applying cryptographic transforms to every packet passing through the router. The VPN transparently applies encryption and integrity protection to end-user traffic—in other words, the endpoints do not need to be capable of the IPSec protocol. Before forwarding the data on to its final destination, the receiving router strips the headers, decrypts the data, and verifies that the data has not changed in transit. Tunnels can be arbitrarily nested as long as the endpoints of the inner tunnels do not overlap those of the outer tunnels; one or both endpoints may be the same, however. Figure 6-4 shows examples of valid scenarios for nested tunneling. For more information on nested tunneling, refer to RFC 2401 (Kent and Atkinson, 1998).

The encapsulation process for tunnel mode differs from that of transport mode in that it uses an additional IP header (see Figure 6-5). The sending host encapsulates upper-layer data within an IP header, and transmits the packet to

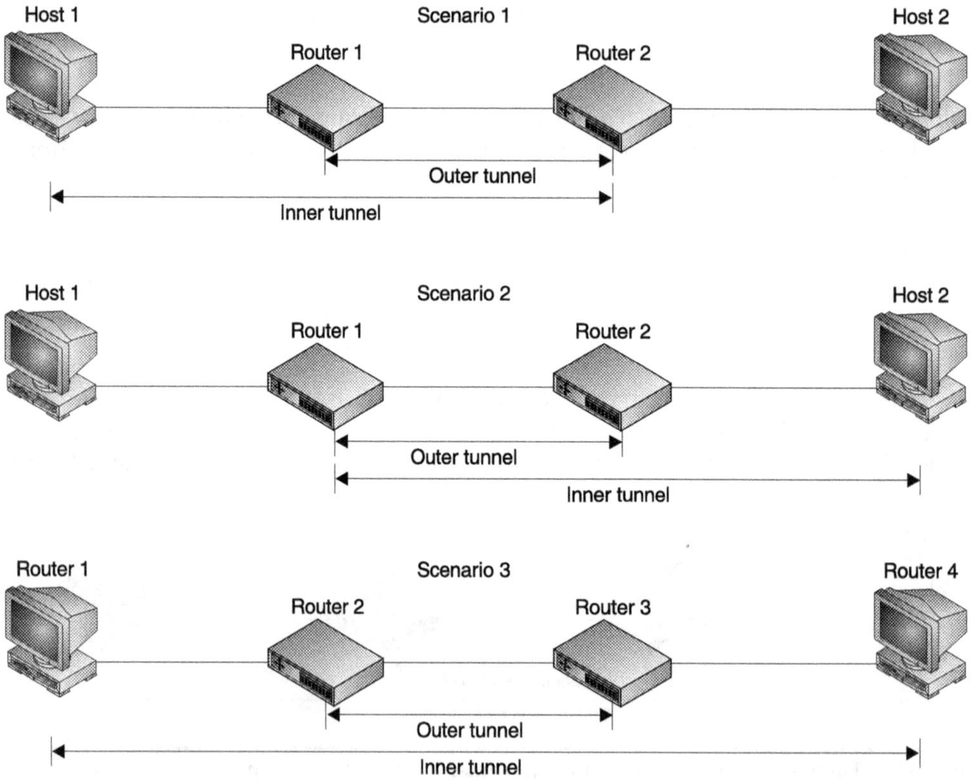

Figure 6-4 Tunnels can be nested as long as the endpoints of the inner tunnels do not overlap those of the outer tunnels.

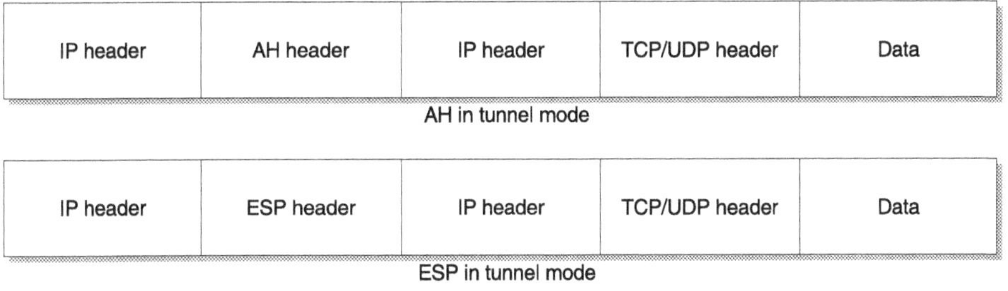

Figure 6-5 Tunnel mode encapsulation uses an inner and an outer IP header.

a VPN router (or whatever device is acting as the security endpoint). The router applies cryptographic transforms to both the upper-layer data and the original IP header. It then adds corresponding AH and ESP headers along with an outer IP header. The inner IP header, TCP/UDP header, and upper-layer data comprise the payload of the AH or ESP packet.

AH and ESP headers can be nested in tunnel mode, a process known as *iterated tunneling* (see Figure 6-6). Table 6-3 compares IPSec transport and tunnel modes.

NOTE:
While both transport and tunnel modes can be used to provide end-to-end security, transport mode is better suited for this task, as it does not include additional IP headers. The IP header nesting performed by tunnel mode increases the header-to-payload ratio.

Security Associations

Before IPSec can determine how to handle an IP packet, a relationship must exist between two communicating parties. *Security associations (SAs)* define this relationship in terms of the IPSec protocols, modes, cryptographic transforms, keying information, and key lifetimes associated with an IPSec session. SAs,

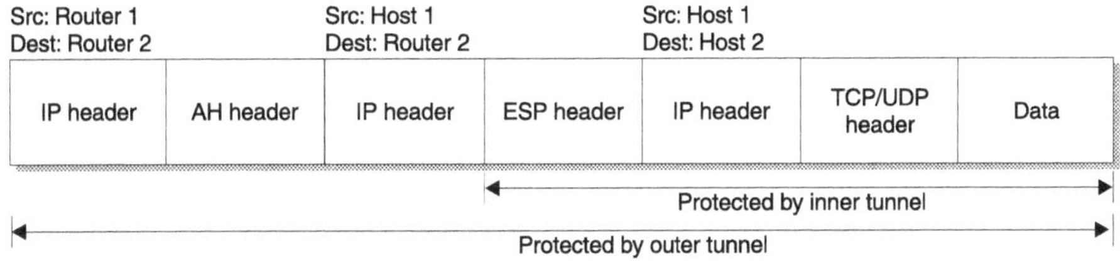

Figure 6-6 AH and ESP tunnel mode headers can be nested. (The encapsulation shown here corresponds to Scenario 1 in Figure 6-4.)

Transport	Tunnel
Security and communication endpoints are the same (end-to-end security).	Security and communication endpoints generally differ.
Adds a header for AH, ESP, or both.	Adds a header for AH, ESP, or both, plus an additional IP header.
Usually implemented on end-user hosts.	Typically implemented on routers and VPN devices.

Table 6-3 A Comparison of the IPSec Transport and Tunnel Modes

combined with IPSec policy, determine the steps taken by the sending and receiving hosts to process a packet. All traffic belonging to the same SA receives similar processing. A single SA applies to traffic flowing in one direction; bidirectional communication requires that each host maintain a pair of SAs—one for inbound traffic and the other for outbound traffic. We refer to these as $SA_{inbound}$ and $SA_{outbound}$, respectively (see Figure 6-7). Each outbound SA on one host corresponds to an inbound SA on another host, and vice versa. AH and ESP protocols use separate SAs, such that a single connection protected by both protocols requires multiple SA pairs. The collection of all SAs related to a particular one-way connection is referred to as an *SA bundle*. Security associations are identified by the following:

- Security parameter index (SPI)
- IPv4 or IPv6 destination IP address
- IPSec protocol (AH or ESP)

A *security parameter index (SPI)* is a unique, 32-bit integer value that, in conjunction with the destination IP address and security protocol (AH or ESP), uniquely identifies an SA. The recipient of a packet uses the SPI to determine what IPSec transforms have been applied to the packet. Since AH and ESP require separate SAs, both protocols must use their own SPI range (that is, an AH and an ESP SA cannot share the same SPI).

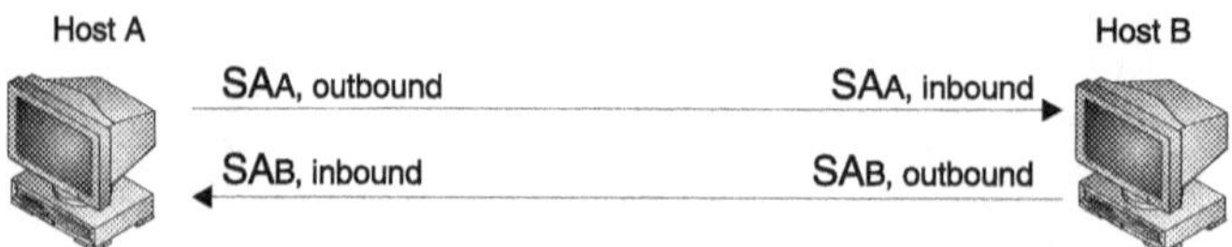

Figure 6-7 Two-way IPSec communication requires that each host maintain at least one pair of security associations—one for inbound traffic and the other for outbound traffic.

Before beginning secure communication, communicating parties must establish SAs. An SA can be created manually or automatically. One of the functions of IKE is the establishment and maintenance of IPSec SAs. IKE negotiates SAs automatically when a request occurs for a connection that requires security, but for which no SA yet exists. Once an SA has been created, it is added to the Security Association Database (SAD). (We'll discuss IKE in more detail in the section entitled "Internet Key Exchange," and we'll discuss the SAD in the following section.)

To limit the exposure of traffic in the event of a compromised key, SAs have limited lifetimes. They must be periodically refreshed or renewed to update their keying information. When an SA is renewed or refreshed, the lifetime of the existing and newly created SAs must overlap. If they do not overlap, there will be a period where no SA exists for the connection. As a result, the hosts will close the session and require negotiation of a new SA before communication can continue. When an SA's lifetime expires, IPSec removes the SA entry from the SAD, and frees its SPI for use by another SA.

SA deletion occurs for three primary reasons:

- The lifetime for the SA has expired.
- A predefined number of bytes have been processed.
- One of the communicating parties requests to have the SA deleted (in the event of key compromise, for instance).

Security Policy Database

Along with security associations, IPSec requires a security policy that defines protocols, modes, and transforms for use in securing a communication channel. Every IPSec implementation must offer an interface where users and administrators can define their own IPSec policies on a per-packet basis. The details of this interface are left to the particular IPSec implementation. IPSec stores policy information in a *Security Policy Database (SPD)*. Each SPD contains a list of policy entries used for processing inbound and outbound IP packets, and every IPSec-enabled interface employs separate SPDs for inbound and outbound traffic. Each entry in the SPD includes one or more *selectors,* along with pointers to SAs within the SAD. IPSec uses the selectors to match packets to policy entries by comparing selector values to the parameters contained within the IP header of each packet. Selectors provide varying levels of processing granularity, depending on the values chosen by the user. IPSec defines the following selectors for categorizing packets:

- **Destination address** An IPv4 or IPv6 destination address; can be a single IP address, a range of addresses, an address plus a subnet mask, or a wildcard address.

- **Source address** An IPv4 or IPv6 source address; can be a single IP address, range of addresses, address plus a subnet mask, or a wildcard address.
- **Source and destination ports** A TCP or UDP port; can be a single port or wildcard value.
- **User or system name** A fully qualified DNS name or name string, or an X.500 distinguished name (must be mapped to an IP address during SA creation).
- **Transport layer protocol** The Transport layer protocol contained within the IPv4 Protocol field or the IPv6 Next Header field.

Based on the policy settings, three options exist for handling packets:

- **Discard** Drop the packet with no further processing.
- **Process** Apply the necessary cryptographic transforms as specified by the inbound or outbound policy settings.
- **Bypass** Allow the packet to pass unaltered.

IPSec uses the first matching entry found in the SPD for processing. Those packets not matching a policy entry in the SPD are automatically discarded without further processing.

Security Associations Database

The Security Association Database (SAD) maintains the list of all inbound or outbound SAs for a particular host. As with the SPD, each interface maintains separate SADs for inbound and outbound traffic. Each entry in the SAD defines parameters associated with a single SA. In addition to the SPI, destination address, and IPSec protocol, entries within the SAD contain the following fields:

- **Sequence Number Counter** A 32-bit value used for creating the sequence numbers contained within the AH or ESP headers of outbound packets.
- **Sequence Counter Overflow** A flag used for determining whether a sequence number overflow has occurred, and whether the SA is still valid following an overflow condition.
- **Anti-Replay Window** A 32-bit counter used for determining whether an inbound AH or ESP packet is part of a replay attack.
- **AH Authentication Algorithm and Associated Key**
- **ESP Encryption Algorithm and Associated Key, IV, and IV Mode**

- **ESP Authentication Algorithm and Associated Key**
- **SA Lifetime** A field indicating the validity period of an SA in terms of the number of bytes processed using the SA and/or the amount of time that has passed since the creation of the SA.
- **Protocol Mode** A field identifying the mode (transport or tunnel) used by the AH and ESP protocols to secure a connection.
- **Path Maximum Transfer Unit (PMTU)** A field containing the maximum observed transmission unit for the path spanned by an IPSec tunnel; fragmentation occurs when packets are transmitted that are larger than the PMTU.

To determine the SA used by a packet, the sender or recipient uses the SPI, destination IP address, and IPSec protocol contained within the packet headers to index into the appropriate SAD. When a matching entry is located, the parameters of the SA are compared with those contained within the AH and ESP headers. If the values match, IPSec processes the packet; otherwise, the packet is discarded. If the SAD contains no SA entry describing the packet, one of two possible outcomes occurs:

- The packets will be discarded (incoming packets).
- A new SA /SA bundle will be created (outgoing packets).

Authentication Header

Authentication Header (AH), described in RFC 2402 (Kent and Atkinson, 1998b) provides data origin authentication, connectionless integrity, and optional anti-replay protection. AH offers no confidentiality services, but can be combined with ESP for data privacy. AH authenticates upper-layer protocol data and as much of the IP header as possible using a keyed MAC or HMAC. IP header authentication is not provided by ESP, so there are scenarios where it is desirable to use a combination of AH and ESP. For integrity calculations, keyed MACs and HMACs require that the sender know the values for all header and data fields prior to transmitting a packet. Some fields within IPv4 and IPv6 headers change during transit, and, in many cases, the sender cannot predict their final values. These fields, termed *mutable* fields, cannot be authenticated. Table 6-4 and Table 6-5 list the mutable and immutable header fields for IPv4 and IPv6, respectively. (See Appendix A for descriptions of these fields.) The AH protocol sets the value of mutable fields to 0 before calculating the authentication data. The authentication algorithms that must be supported by any RFC 2402–compliant IPSec implementation are

- HMAC using MD5
- HMAC using SHA-1

Table 6-4 Mutable and Immutable IPv4 Header Fields

IPv4 Field	Mutable or Immutable?
Version	Immutable
Internet Header Length (IHL)	Immutable
Type of Service (TOS)	Mutable
Total Length	Immutable
Identification	Immutable
Flags	Mutable
Fragment Offset	Mutable
Time to Live (TTL)	Mutable
Protocol	Immutable
Header Checksum	Mutable
Source Address	Immutable
Destination Address	Immutable/Mutable but predictable
Options	Immutable/Mutable but predictable/Mutable

NOTE:

In the case of source routing, the destination address may be considered "mutable but predictable." Most IP implementations ignore source-routed packets, as they are rarely used for legitimate purposes.

Since the predictability of IPv4 options varies from one option to the next, they may be classified as immutable, mutable but predictable, or mutable. (See Appendix A of RFC 2402 for details.)

Table 6-5 Mutable and Immutable IPv6 Header Fields

IPv6 Field	Mutable or Immutable?
Version	Immutable
Priority	Mutable
Flow Label	Mutable
Payload Length	Immutable
Next Header	Immutable
Hop Limit	Mutable
Source Address	Immutable
Destination Address	Immutable
Extension Header and Options	Immutable/Mutable but predictable/Mutable

NOTE:
Appendix A of RFC 2402 provides an explicit list of the IPv6 extension header fields, and classifies them as immutable, mutable but predictable, or mutable. According to RFC 2402, if an IPv6 extension header contains any mutable options (in other words, options whose values may change in transit), the entire "Options Data" field must be considered mutable, and is filled with zeros.

The IPSec Anti-Replay Service

The IPSec anti-replay service assigns a sequence number to each transmitted packet. The receiving host tracks the sequence numbers of all incoming packets and compares them with a list of valid sequence numbers. If the recipient receives a packet containing a sequence number that has already been received, or that is outside a window of acceptable values, it flags and rejects the packet. Although this service is optional, there is no way for the sender to know ahead of time whether the recipient will be tracking sequence numbers. As a result, the sender must always include a sequence number within each AH and ESP header.

AH Header Format

In transport mode, the AH header, illustrated in Figure 6-8, follows the IP header and immediately precedes the upper-layer protocol information. In tunnel mode, the AH header is placed between the original and new IP headers—we refer to these as the inner and outer IP headers, respectively.

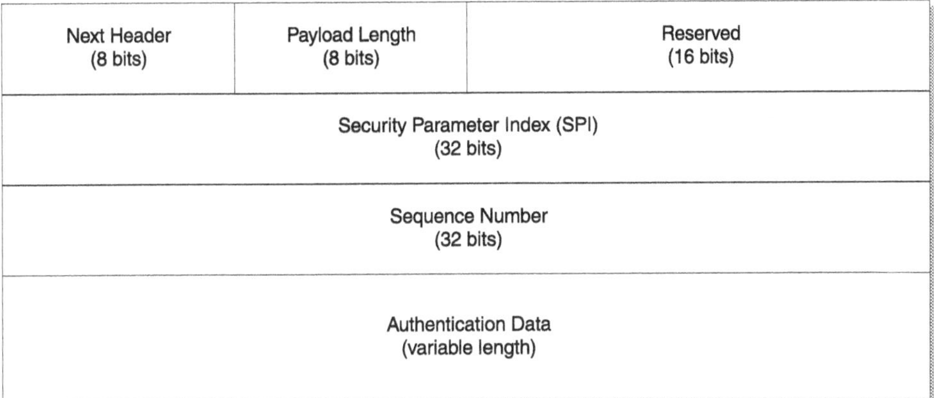

Figure 6-8 An AH header

NOTE:
The placement of AH headers when using IPv6 varies somewhat from the IPv4 case previously mentioned. In the case of IPv6, the AH header is placed after any extension headers (for instance, hop-by-hop, routing, and fragmentation extension headers).

An AH header consists of the following fields:

- **Next Header** An 8-bit field identifying the type of payload (for instance, upper-layer protocol information) following the AH header. RFC 1700 (Reynolds and Postel, 1994) contains a list of IP protocol numbers assigned by the IANA (Internet Assigned Numbers Authority).
- **Payload Length** An 8-bit field containing the length of the AH header in 32-bit words minus 2. When used in conjunction with IPv6, AH is considered an extension header. (See Appendix A for a description of extension headers.) The length of IPv6 extension headers is calculated by subtracting one 64-bit word from the header length. Since, AH encodes the header length in multiples of 32-bit words (two 32-bit words equal a 64-bit word), we subtract 2 from the AH header length.
- **Reserved** A 16-bit field reserved for future use.
- **Security Parameter Index (SPI)** A 32-bit value used in conjunction with a destination address and IPSec protocol (in this case, AH) to uniquely identify the security association relating to a particular packet. The SA defines the IPSec protocols and modes, cryptographic transforms, keying information, and key lifetimes associated with an IPSec session.
- **Sequence Number** A 32-bit field used by the anti-replay service to detect replay attacks. The field contains a monotonically increasing counter value that is initialized to 0 at SA creation. The first packet transmitted using a given SA has a sequence number of 1. The IPSec anti-replay service uses this value to identify retransmitted packets.

NOTE:
When anti-replay protection is enabled (it is enabled by default), the sender's and receiver's sequence number counters must be reset prior to transmitting 2^{32} *bytes of data protected with the same SA. This will prevent sequence numbers from overlapping.*

- **Authentication Data** A variable-length field containing the *integrity check value (ICV)* for the packet. The ICV is the result of the keyed MAC or HMAC computation; it must be a multiple of 32 bits in length.

AH Modes of Operation

In transport mode, AH authenticates the AH header, upper-layer protocol information, and parts of the IP header. The following diagram shows an IP packet protected with AH in transport mode:

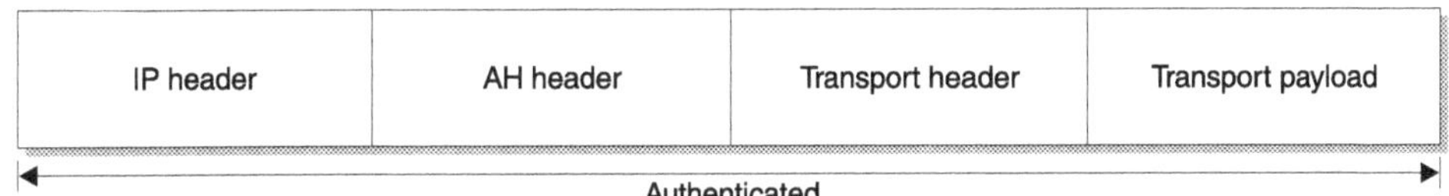

In tunnel mode, AH completely authenticates the AH header, the upper-layer protocol information, and the inner IP header, and authenticates the immutable fields within the outer IP header. Here you see an IP packet protected with AH in tunnel mode:

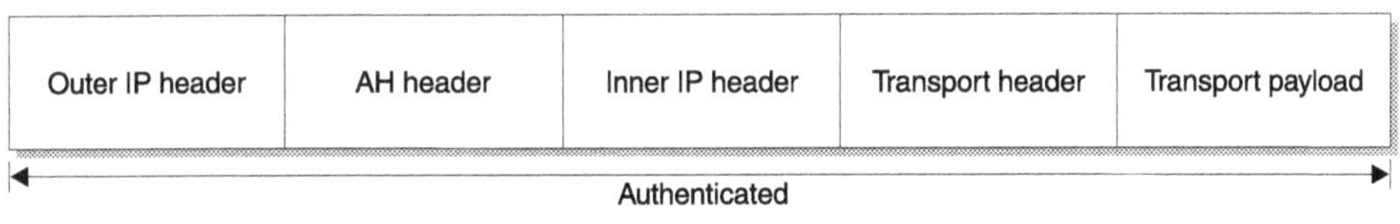

NOTE:
In practice, AH is rarely used in tunnel mode, because it duplicates much of the functionality of transport mode AH (Doraswamy and Harkins, 1999).

AH Processing

Let's look at how the AH protocol processes inbound and outbound IP packets. Prior to transmitting data, two communicating peers must first negotiate a pair of SAs or an SA bundle for the connection; they do this by using IKE or another key-management protocol. During SA establishment, the sequence number counter is set to 0, and an entry for the SA is inserted into the corresponding SAD. IPSec also creates an SPI for referencing the SA in the recipient's SAD. Once the sender and recipient have completed these steps, they can begin processing inbound and outbound IP packets.

Outbound Processing Outbound AH processing follows these steps:

1. The IPSec implementation inspects all outbound packets from the IP stack, and matches each packet to the policy information contained in the SPD. This is done using selectors (for instance, the source or destination address, name, protocol, and upper-layer ports). The matching entry in the outbound SPD points to an SA or an SA bundle within the SAD.
2. From the SA, the sender determines the appropriate authentication algorithms and keying information for use in securing the packet.

3. The sender increments the sequence number counter by 1 before creating the AH header and fills the remaining AH header fields with the appropriate values. All mutable IP header fields and the Authentication Data field are initialized to 0 before the ICV is calculated.
4. The sender calculates the ICV using the authentication algorithm specified by the SA. The keyed MAC or HMAC generates the ICV from the IP header, AH header, and upper-layer protocol headers and data. The resulting value is placed in the Authentication Data field.
5. Finally, the sender transmits the packet to its destination.

Inbound Processing Inbound AH processing follows these steps:

1. If the packet has been fragmented (that is, broken into smaller units in order to cross a network segment that has a smaller MTU than the source network), the recipient waits until all fragments have been received. It then reassembles the packet.
2. The recipient extracts the SPI, destination address, and IPSec protocol (AH, for example) from the IP and AH headers, and uses the values to locate a matching SA within the local SAD. IPSec drops the packet if no corresponding entry exists for the SA in the SAD.
3. The recipient uses the recovered SA to process the inbound packet. Processing begins with determining whether the selectors in the IP header match those specified in the SA—and thus determining whether the SA is appropriate for processing the packet.
4. The selectors identified in the packet are used to locate a matching policy entry in the inbound SPD, which is then compared to the SA from step 3. The packet is dropped if the policies defined by the SA and SPD do not match.
5. If the recipient has enabled anti-replay protection, it checks the sequence number to determine whether the packet has been replayed. If the check fails, IPSec discards the packet.
6. The recipient saves the ICV value contained within the packet and fills the Authentication Data field and all mutable fields in the IP header with 0's. The recipient then calculates the ICV using the algorithm specified in the SA, and compares it with the saved value. If the values do not match, the packet is discarded.

Encapsulating Security Payload

Like the AH protocol, the Encapsulating Security Payload (ESP) includes data origin authentication, connectionless integrity, and anti-replay protection. In addition to these services, ESP also supports data privacy and limited traffic flow confidentiality. As with AH, anti-replay protection is optional, however, the sender must include a valid sequence number within every ESP header. RFC 2406 (Kent and Atkinson, 1998c) defines the ESP protocol. The authentication and encryption algorithms that must be supported by any RFC 2406–compliant IPSec implementation are

- DES in CBC mode (although 3DES is now the standard, since the 56-bit key length used by DES is considered weak)
- HMAC using MD5
- HMAC using SHA-1
- The NULL Authentication algorithm
- The NULL Encryption algorithm

ESP Header and Packet Format

ESP headers and trailers encapsulate data from upper-layer protocols. In transport mode, the ESP header, illustrated in Figure 6-9, is inserted immediately following the IP header and preceding the upper-layer protocol payload. The ESP trailer and authentication data follows the upper-layer protocol information. In tunnel mode, the ESP header is inserted between the outer and inner IP headers. Similar to AH, the header is inserted following any IPv6 extension headers.

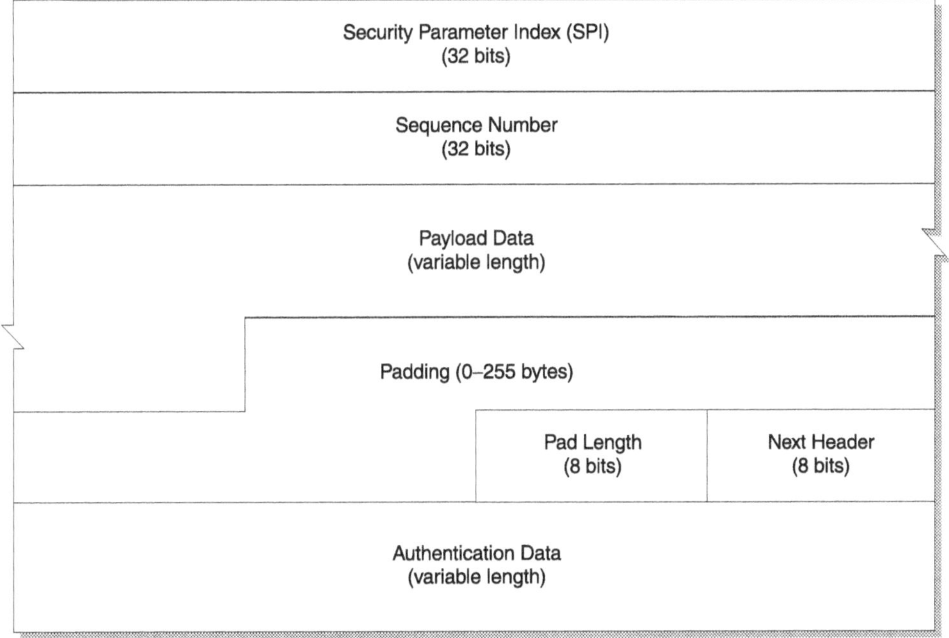

Figure 6-9 An ESP header and trailer

An ESP header and trailer consist of the following fields:

- **Security Parameter Index (SPI)** Similar to the SPI used with AH, this is a 32-bit value used in conjunction with a destination address and an IPSec protocol (in this case, ESP) to identify uniquely the security association relating to a particular packet. The SA defines the IPSec protocols and modes, cryptographic transforms, keying information, and key lifetimes associated with an IPSec session.
- **Sequence Number** A 32-bit field used by the anti-replay service to detect replay attacks. The field contains a monotonically increasing counter value that is initialized to 0 at SA creation. The first packet transmitted that uses a given SA has a sequence number of 1. The IPSec anti-replay service uses this value to identify retransmitted packets.
- **Payload Data** A variable-length field containing the upper-layer protocol headers and data to which ESP cryptographic transforms are applied. (An example might be an encrypted TCP or UDP segment.) If the cipher used to encrypt the payload requires an initialization vector (IV)—as most block ciphers do in certain modes—this field also contains the IV. RFCs 1829, 2405, and 2451 describe the use of various CBC mode ciphers with ESP. These RFCs outline the length, structure, and location of the IV within the Payload Data field.
- **Padding** A variable-length field (0–255 bytes) containing one of the following:
  - The padding required by the encryption algorithm to make the plaintext a multiple of the block size. (See Chapter 2 for a description of padding.)
  - Arbitrary data required to make the resulting ciphertext terminate on a 4-byte boundary.
  - Arbitrary data used to conceal the true length of the payload.
- **Padding Length** An 8-bit field representing the number of padding bytes in the Padding field.
- **Next Header** An 8-bit field indicating the data type (in other words, the upper-layer protocol information) contained in the Payload Data field. This value corresponds to one of the assigned IP protocol numbers contained in RFC 1700.
- **Authenticated Data** A variable-length field containing the *integrity check value (ICV)* for the packet. The ICV is the result of the keyed MAC or HMAC computation over the ESP packet not including the authentication data. The length of this field depends on the authentication algorithm used. For ESP, this field is optional, and is included when authentication services have been enabled for a given SA.

ESP Modes of Operation

In transport mode, ESP authenticates the ESP header, upper-layer protocol information, and ESP trailer. It also encrypts the encapsulated upper-layer protocol header and payload. The following diagram shows an IP packet protected with ESP in transport mode:

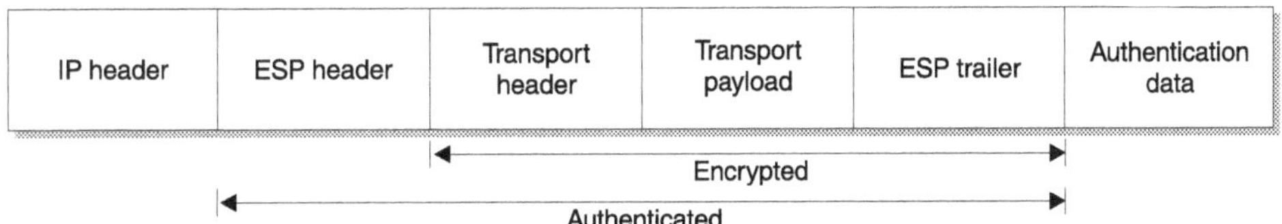

For tunnel mode operation, ESP authentication and encryption extend to include the inner IP header. Here you see an IP packet protected with ESP in tunnel mode:

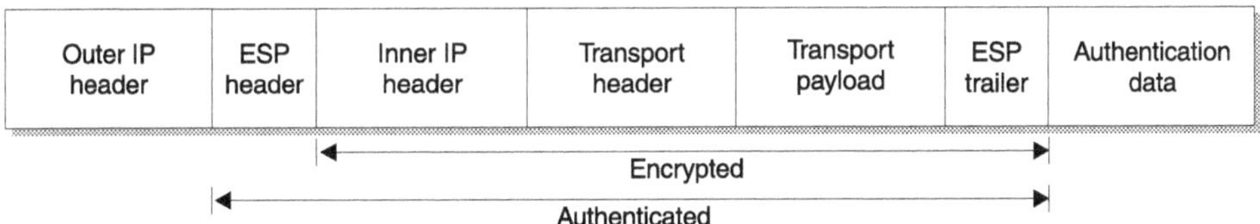

ESP Processing

Many of the processing steps for ESP match those used by AH; however, we must also account for encryption. Prior to transmitting data, two communicating peers must negotiate a pair of SAs, or an SA bundle, by using IKE or another key management protocol. They must then insert the SA into the SAD, set their corresponding sequence number counters to 0, and create an SPI for referencing the SA in the SAD. Once these operations are complete, processing may begin.

Outbound Processing Outbound ESP processing follows these steps:

1. The IPSec implementation inspects all outbound packets from the IP stack and matches each packet to the policy information contained in the SPD. This is done with selectors (for instance, source/destination address, name, protocol, and upper-layer ports). The matching entry in the outbound SPD points to an SA or an SA bundle within the SAD.
2. From the SA, the sender determines the appropriate cryptographic transforms and keying information to be used for securing the packet.
3. The sender increments the sequence number counter by 1 before creating the ESP header and fills the remaining ESP header fields with the appropriate values. Prior to encryption, the payload data field contains the upper-layer protocol data or an entire IP packet, depending on whether transport or tunnel mode has been selected.

4. If the policy settings contained within the SPD specify the use of confidentiality services, IPSec uses the cryptographic transforms referenced in the SA to encrypt the Payload Data, Padding, Padding Length, and Next Header fields.
5. If the policy settings contained within the SPD specify the use of authentication services, IPSec calculates the ICV by using the integrity mechanism specified by the SA. The resulting value is placed in the Authentication Data field in the ESP trailer.
6. The sender forwards the packet to its destination.

NOTE:
For outbound packets, encryption is always applied before authentication.

Inbound Processing Inbound ESP processing follows these steps:

1. If the packet has been fragmented, the recipient waits until all fragments have been received. It then reassembles the packet.
2. The recipient extracts the SPI, destination address, and IPSec protocol (ESP, for instance) from the IP and ESP headers, and uses the values to locate a matching SA within the local SAD. IPSec drops the packet if no corresponding entry exists for the SA in the SAD.
3. The recipient uses the recovered SA for processing the inbound packet. Processing begins with determining whether the selectors in the IP header match those specified in the SA—and thus determining whether the SA is appropriate for processing the packet.
4. The selectors identified in the packet are used to locate a matching policy entry in the inbound SPD, which is then compared to the SA from step 3. The packet is dropped if the policies defined by the SA and SPD do not match.
5. If the recipient has enabled anti-replay protection, it checks the sequence number to determine whether the packet has been replayed. If the check fails, IPSec discards the packet.
6. If inbound policy settings require authentication services, the recipient calculates the ICV using the integrity algorithm specified in the SA. It then compares the result with the value stored in the Authentication Data field. If the values do not match, the packet is discarded.
7. If specified by inbound policy settings, the final step in processing packets that are protected with ESP is decryption. The encrypted data—consisting of the Payload Data, Padding, Padding Length, and Next Header fields—is decrypted via the cipher and keying information specified by the SA.

NOTE:
Properly authenticating a packet prior to decryption will reduce the effectiveness of many types of DoS attacks. Only packets from trusted sources that have not been tampered with will be decrypted and passed to upper-layer protocols for processing.

If no failures occur during processing, IPSec passes the IP packet to the upper-layer protocol specified in the Next Header field of the ESP header. Refer to RFC 2401 (Kent and Atkinson, 1998a) for further details on inbound and outbound processing for the AH and ESP protocols.

Internet Key Exchange

As previously mentioned, before two parties can begin secure communication using IPSec, they must agree on at least one pair of SAs. How do two peers who have never communicated before negotiate SAs? The answer to this question lies in the Internet Key Exchange (IKE) protocol, whose primary responsibilities include the exchange of keying material related to IPSec transforms and the dynamic creation of SAs. IKE is actually a general-purpose key exchange protocol; and although it can be used by any protocol for key exchange and SA negotiation, it is most often associated with IPSec. RFC 2409 (Harkins and Carrel, 1998) discusses IKE operation as it pertains to IPSec. The majority of IKE functionality is based on three other protocols: the *Internet Security Associations and Key Management Protocol (ISAKMP)*, *Oakley*, and *SKEME*.

ISAKMP

ISAKMP defines procedures and message types used by communicating peers for authentication, transmission of information relating to a key exchange, and negotiation of security services through the establishment of SAs. However, ISAKMP does not define a specific key-exchange mechanism, nor does it describe the details of establishing IPSec SAs. IKE handles these operations in the context of IPSec. RFC 2408 (Maughan, Schertler, Schneider, and Turner, 1998) defines the ISAKMP specification.

ISAKMP Messages An ISAKMP message consists of a fixed-length ISAKMP header followed by an arbitrary number of ISAKMP payloads. The ISAKMP header, illustrated in Figure 6-10, consists of nine fields:

- **Initiator Cookie** An 8-byte field containing a value generated by the initiator of an ISAKMP exchange that uniquely identifies messages originating from the initiator. To prevent DoS attacks, the responder discards messages that contain cookies not previously received from the initiator.
- **Responder Cookie** An 8-byte field containing a unique value generated by the responder of an ISAKMP exchange; this value allows the initiator to quickly identify messages from the responder.

- **Next Payload** An 8-bit field identifying the first payload contained within the ISAKMP message. The available payload types are listed in Table 6-6, and are discussed later in this section.
- **Major Version** A 4-bit field indicating the major version number of the ISAKMP protocol currently in use.
- **Minor Version** A 4-bit field indicating the minor version number of the ISAKMP protocol.
- **Exchange Type** An 8-bit field identifying the ISAKMP exchange type for the given message. The available exchange types are listed in Table 6-7.
- **Flags** An 8-bit field indicating the options set for the current ISAKMP exchange. The first three bits—the Encryption, Commit, and Authentication Only bits—are used today. (See RFC 2408 for a description of the bits in the Flags field.)
- **Message ID** A 4-byte field containing a unique identifier for a Phase 2 ISAKMP exchange. This value is randomly generated by the Phase 2 initiator. We discuss the various phases of ISAKMP in the upcoming section entitled "ISAKMP Phases and Exchanges."
- **Message Length** A 4-byte field containing the length (in bytes) of the entire ISAKMP message, including the ISAKMP header.

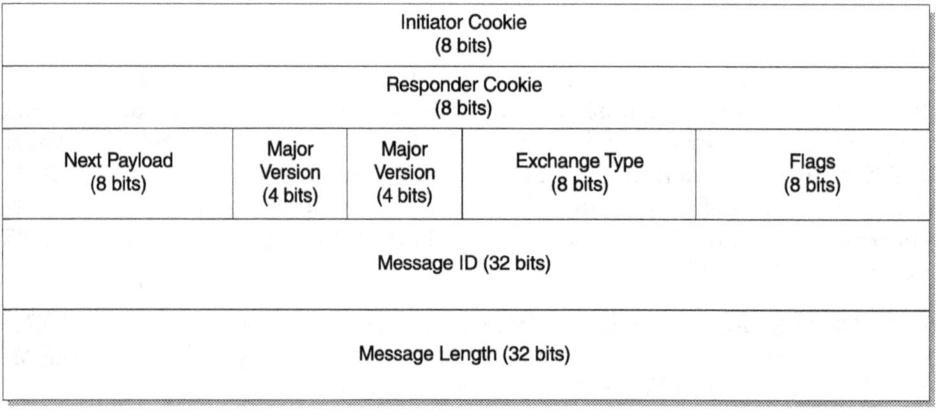

Figure 6-10 An ISAKMP header

Next Payload Type	Value
None	0
Security Association	1
Proposal	2
Transform	3
Key Exchange	4
Identification	5
Certificate	6
Certificate Request	7
Hash	8
Signature	9
Nonce	10
Notification	11
Delete	12
Vendor ID	13
RESERVED	14–127
Private Use	128–255

Table 6-6 Assigned Values for ISAKMP Next Payload Types

Exchange Type	Value
None	0
Base	1
Identity protection	2
Authentication only	3
Aggressive	4
Informational	5
Future use	6–31
DOI specific use	32–239
Private use	240–255

Table 6-7 Assigned Values for ISAKMP Exchange Types

NOTE:
A Domain of Interpretation (DOI) defines payload formats, exchange types, processing guidelines, and naming conventions specific to protocol usage in a particular environment. RFC 2407: The Internet IP Security Domain of Interpretation for ISAKMP *(Piper, 1998) describes the details associated with the creation of IPSec SAs using ISAKMP.*

ISAKMP defines a number of payload types for exchanging authentication data and keying information. Two or more ISAKMP payloads may be chained to form a single ISAKMP message. ISAKMP defines 13 specific payload types whose details can be found in RFC 2408:

- **Security Association** Defines attributes related to an ISAKMP or IPSec SA, such as the DOI under which the SA is negotiated and the required secrecy and integrity levels for the SA.
- **Proposal** Allows the initiator of an ISAKMP exchange to request a set of security services and mechanisms for the SA being established. The Proposal payload is used in conjunction with the Transform payload to indicate specific protocols (ISAKMP, AH, and ESP, for example) and transforms.
- **Transform** Defines the security mechanisms (in other words, the cryptographic transforms) used to protect the protocol referenced in the proposal payload.
- **Key Exchange** Contains information necessary for key exchange or generation, such as an RSA public key or a Diffie-Hellman public value.
- **Identification** Allows communicating parties to exchange information pertaining to their identity. Such identifying information includes fully qualified domain name or username strings, IP addresses or address ranges, and X.500 distinguished names.
- **Certificate** Provides communicating peers with a means to exchange digital certificates and other certificate-related material (for example, PKCS #7-formatted X.509 certificates, PGP certificates, Kerberos tokens, CRLs, and so on).
- **Certificate Request** Allows entities communicating via ISAKMP to request digital certificates.
- **Hash** Contains the output of a message digest operation, which is frequently used to ensure the integrity of data contained within an ISAKMP message.
- **Signature** Contains a digital signature for use in ensuring the integrity of an ISAKMP message; can also be used for non-repudiation.

- **Nonce** Contains a randomly generated value used in the detection of replay attacks.
- **Notification** Used to transmit error conditions and other informational data between ISAKMP peers. (RFC2408 contains a complete list of notification message types.)
- **Delete** Used to inform an ISAKMP peer that an SA has been deleted (that is, removed from the SAD) and is no longer valid.
- **Vendor ID** Contains vendor-defined constants.

Each of these payloads shares a generic header:

Next Payload (8 bit)	Reserved (8 bit)	Payload Length (2 byte)

The generic header is comprised of the following fields:

- **Next Payload** An 8-bit field identifying the next payload type contained in the message. This field contains a value of 0 if the payload is the last in the message.
- **Reserved** An 8-bit field that is currently unused by ISAKMP. Its value should be set to 0.
- **Payload Length** A 2-byte field indicating the length (in bytes) of the payload (including the payload header).

ISAKMP Phases and Exchanges ISAKMP defines two phases of operation for negotiating security services between hosts: *Phase 1* and *Phase 2*. During Phase 1, communicating peers establish ISAKMP SAs. Phase 1 SAs identify the security services for protecting future exchanges, and are later used to secure Phase 2 negotiations. Phase 2 exchanges result in the creation of SAs for security protocols other than ISAKMP. In the case of IPSec, Phase 2 negotiation results in the establishment of AH and ESP SAs. Many Phase 2 SAs can be established using the same Phase 1 SA.

ISAKMP includes five exchange types for communicating authentication data, keying information, and SAs:

- **Base** This type of exchange transmits authentication data and key exchange information together to reduce the number of messages required for SA negotiation. It does not protect the identity of the communicating peers, since encryption keys have not been exchanged. A base exchange consists of four messages.

- **Identity Protection** This exchange protects the identity of the ISAKMP peers by separating the keying information and the authentication data. The peers first exchange keying information, which they then use to encrypt their identities. This type of exchange requires two more messages than the Base exchange.
- **Authentication Only** In the event that encryption is not necessary or is already being provided, this exchange allows two peers to communicate only authentication data. An Authentication Only exchange consists of three ISAKMP messages.
- **Aggressive** This exchange allows keying information, authentication data, and an SA to be combined within a single message to further reduce the number of roundtrips required for SA negotiation. No identity protection is provided. An Aggressive exchange consists of three messages.
- **Informational** This exchange provides a mechanism for one ISAKMP peer to communicate SA management information to another. It consists of only a single message.

IKE Phases and Exchanges

As previously mentioned, IKE defines the actual key-exchange mechanism used by IPSec. A complete IKE exchange results in the creation of an SA that defines the security services associated with a particular connection (for example, encryption transform or keying material). IKE includes two phases of operation, conveniently named after the phases of ISAKMP. During a Phase 1 exchange, IKE negotiates ISAKMP SAs, which it uses to secure future Phase 2 exchanges. Phase 2 negotiations result in the creation of IPSec SAs. These SAs describe the security services used by the AH and ESP protocols. IKE defines all attributes related to Phase 1 SA negotiation; the DOI for IPSec defines the attributes associated with Phase 2 SA establishment.

IKE defines five exchange types used to negotiate IPSec SAs and communicate information between IKE peers:

- **Main Mode** A Phase 1 exchange very similar to the identity protection exchange of ISAKMP. A Main Mode exchange separates key information and authentication data to protect the identity of the communicating peers. The peers exchange keying material first, and then use the keys to encrypt their identities. This type of exchange requires six messages.
- **Aggressive Mode** A Phase 1 exchange modeled after the Aggressive Mode exchange in ISAKMP. This type of exchange does not protect identity information, but requires fewer messages (three in all) than does a Main Mode exchange.
- **Quick Mode** A Phase 2 exchange for negotiating SAs for a security protocol other than IKE (in this case, IPSec). A Quick Mode exchange

requires three messages, all of which are secured by means of the security services negotiated by a Phase 1 exchange.

- **New Group Mode** Allows two IKE peers to negotiate a new Diffie-Hellman group for key agreement. IKE defines five groups for use with the Diffie-Hellman public-key algorithm.
- **Informational Mode** Used to exchange status information and error codes between communicating peers.

NOTE:
We will revisit IKE during our discussion of PacketCable in Chapter 10. For a detailed discussion of ISAKMP and IKE exchanges, or for more information on IPSec, refer to Doraswamy and Harkins, 1999 and Davis, 2001.

SSL and TLS

SSL, or the *Secure Sockets Layer*, sits just above TCP, and provides authentication, encryption, and message integrity services. The original intent for SSL was to secure connections between web servers and web browsers, which remains the most common use for SSL today. While frequently associated with web-based transactions, SSL is not limited to securing the Hypertext Transfer Protocol (HTTP). Any upper-layer protocol or application that relies on TCP can employ the security services provided by SSL, as illustrated in Figure 6-11. In addition to securing web-based traffic, common uses for SSL include securing the following:

- Vendor-proprietary communication protocols
- Connections between back-end servers within enterprise and B2B environments
- Connections between network devices, such as provisioning equipment and routers, and the remote management consoles used to administer these devices

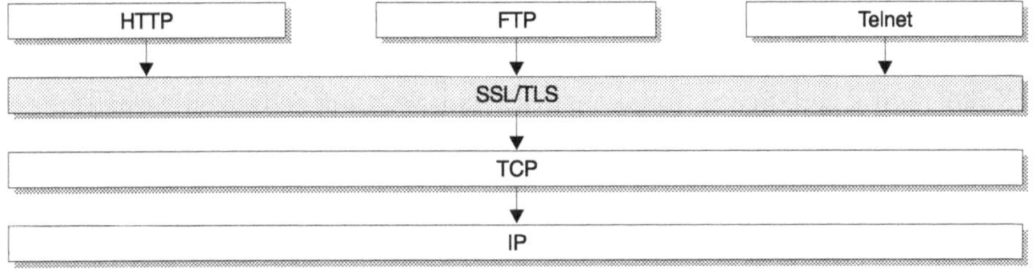

Figure 6-11 SSL can be used to secure many upper-layer protocols and applications, including HTTP, FTP, and Telnet.

A Brief History of SSL

SSL was originally developed by Netscape Communications in anticipation of the need for securing World Wide Web transactions. Netscape completed design of SSLv1 in 1994. However, this version was never publicly released. Later that year, Netscape finished development of the SSLv2 specification, which shipped with its Navigator web browser. Up until then, Netscape had developed and maintained SSL almost exclusively, with little input from other industry vendors.

In 1995, Microsoft published a competing specification known as Private Communications Technology (PCT). PCT was backward compatible with SSLv2, but provided a number of security enhancements. Many of these enhancements were incorporated in SSLv3.

As SSL became the de facto standard for securing Web traffic, the need for public input and review became apparent. In 1995, with help from the Web community, Netscape released SSLv3, which corrected many of the security shortcomings present in SSLv2. SSL remained the responsibility of Netscape until 1996, when the specification was transitioned to the IETF. The TLS Working Group within the IETF has since adopted a replacement standard for SSL known as *Transport Layer Security (TLS)*, which is outlined in RFC 2246. The TLS Working Group has defined the following RFCs in conjunction with TLS:

- RFC 2246: The TLS Protocol Version 1.0
- RFC 2712: Addition of Kerberos Cipher Suites to Transport Layer Security (TLS)
- RFC 2817: Upgrading to TLS Within HTTP/1.1
- RFC 2818: HTTP Over TLS

SSLv3 and TLSv1 are nearly identical, and you may often hear TLS referred to as SSLv3.1. To date, TLS is widely supported by many SSL implementations; however, SSLv3 remains the de facto standard. Because of their similarity, any reference to SSL in this book also applies to TLS as well, unless otherwise noted.

SSL in Detail

SSL provides a number of security services, including traffic encryption, client-side and server-side authentication, and message integrity; it supports the use of X.509 digital certificates for authentication. SSL defines specific roles for each host participating in an SSL session. The behavior of each host varies considerably during connection negotiation according to whether a host is acting as the client or the server. The client is responsible for initiating a connection, while the server listens on a predetermined port awaiting a connection.

SSL Protocol Architecture

SSL uses TCP to ensure reliable, end-to-end (connection-oriented) communication. Internally, SSL is a multilayer protocol consisting of four separate components, as illustrated in Figure 6-12. The *SSL Record Protocol* formats and encapsulates messages from higher-layer protocols before passing them on to the TCP layer. It also provides encryption and authentication services. The three higher-layer components defined by SSL are the *Handshake, Alert,* and *ChangeCipherSpec* protocols. Each of these protocols has its own associated message types, and manages some aspect of an SSL connection.

Session and Connection States

SSL and TLS are stateful protocols, which means that both the client and server maintain state information pertaining to each SSL session and connection. *Connections* are transient, and multiple connections can be associated with a single *session*. However, a single connection must be associated with only one session. A session state consists of a Session Identifier, a Peer Certificate, a Master Secret, a Cipher Suite, a Compression Method, and the "Is Resumable" flag, as detailed in Table 6-8 (Freier, Karlton, and Kocher, 1996). All of these parameters can be considered global values; they are persistent parameters that apply to all connections associated with a given session until that session expires. For example, all connections within the same session use a common Master Secret to generate connection-specific encryption and MAC keys.

In addition to session state parameters, the client and server maintain a random value, encryption and MAC keys, an initialization vector, and a sequence number for each connection, as outlined in Table 6-9 (Freier, Karlton, and Kocher, 1996). These parameters are destroyed once the client and server tear down the connection.

A new session is generated each time a client and server exchange keying information. This occurs during the handshake, and it is the responsibility of the Handshake protocol to synchronize state transitions between a client and server. A session marked as *resumable* can be reinstated under a new connection via a reference to an existing session identifier. Sessions eventually expire

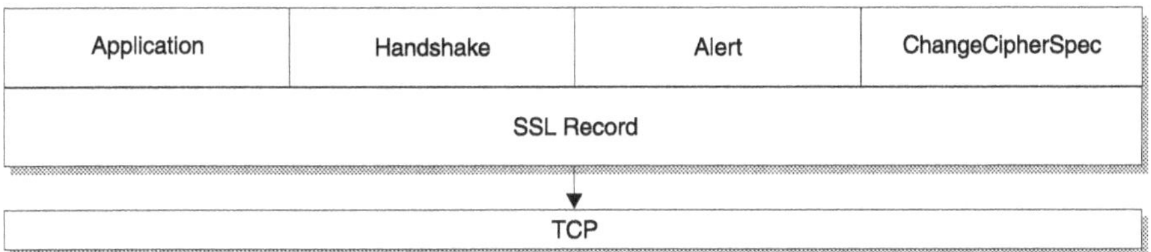

Figure 6-12 The SSL protocol architecture

Element	Description
Session Identifier	An arbitrary byte sequence chosen by the server to identify an active or resumable session state.
Peer Certificate	X509.v3 certificate of the peer. This element of the state can be *null.*
Compression Method	The algorithm used to compress data prior to encryption.
Cipher Spec	A parameter that specifies the bulk data encryption algorithm (such as null, DES, and so on) and a MAC algorithm (such as MD5 or SHA-1), and related attributes.
Master Secret	A 48-byte secret shared between the client and server, which is combined with a random value to generate connection-specific encryption and MAC keys.
Is Resumable	A flag indicating whether the session can be used to initiate new connections.

Table 6-8 Session State Elements

Element	Description
Server and Client Random	Byte sequences that are chosen by the server and client for each connection.
Server Write MAC Secret	The secret used in MAC operations on data written by the server.
Client Write MAC Secret	The secret used in MAC operations on data written by the client.
Server Write Key	The bulk cipher key for data encrypted by the server and decrypted by the client.
Client Write Key	The bulk cipher key for data encrypted by the client and decrypted by the server.
Initialization Vector	When a block cipher in CBC mode is used, an initialization vector (IV) is maintained for each key. This field is first initialized by the SSL handshake protocol. Thereafter, the final ciphertext block from each record is preserved for use with the following record.
Sequence Numbers	Each party maintains separate sequence numbers for transmitted and received messages for each connection. When a party sends or receives a *change cipher spec* message, the appropriate sequence number is set to 0.

Table 6-9 Connection State Elements

to ensure that a previously authenticated client in possession of an expired certificate cannot reauthenticate to a server by using an existing session identifier.

SSL Record Protocol

The SSL Record Protocol provides confidentiality and message integrity. It is responsible for encapsulating data from higher-layer protocols into a format understood by an SSL peer. The four higher-layer protocols and their protocol type values are listed in Table 6-10.

Figure 6-13 demonstrates the formatting used by the Record layer. Each record consists of the following fields:

- **Protocol Type** A 1-byte field indicating the higher-layer protocol encapsulated within the Record layer (see Table 6-10).
- **Version** A 2-byte field indicating the version of SSL/TLS in use. The higher-order byte contains the major version number, while the lower-order byte contains the minor version number. (The version number for TLS is 3.1.)
- **Length** A 2-byte field containing the length of the application data in bytes. The length of the data should not exceed 16,384 bytes.
- **Application Data** An *n*-byte field containing application data. SSL treats this data as transparent to the Record layer.
- **Message Authentication Code (MAC)** An optional field consisting of a 16-byte value for MD5 or a 20-byte value for SHA-1. (For the details of SSL MAC calculation, refer to Freier, Karlton, and Kocher, 1996 and Dierks and Allen, 1999.)

NOTE:

SSL encrypts both the Application Data and the MAC fields. When using block ciphers for encryption, the application data must fill an entire block. In most cases, it is necessary to add padding so that the application data is a multiple of the block size for the cipher.

Table 6-10 Higher-Layer Protocol Types

Type	Protocol
20	ChangeCipherSpec
21	Alert
22	Handshake
23	Application Data

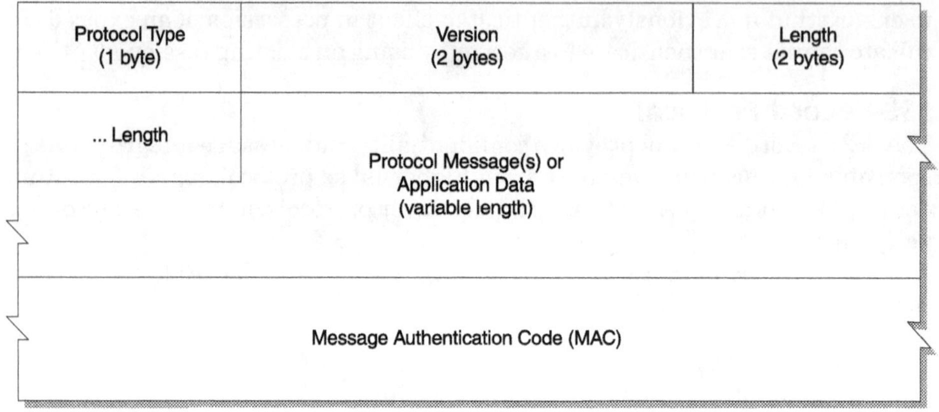

Figure 6-13 SSL Record layer encapsulation

Record Layer Processing During encapsulation, the Record layer can perform a number of operations on upper-layer protocol data before passing the data to the TCP layer. These operations include fragmentation, compression, and encryption. The Record protocol operates as follows:

1. Application data is fragmented into manageable blocks.
2. Compression is applied to the resulting fragments.
3. A MAC is computed and appended to the end of the compressed data.
4. The compressed data, plus the MAC, are encrypted.
5. An SSL record header is appended to beginning of the encrypted data.

Upon receipt, the data is decrypted, verified, decompressed, and reassembled before being passed to the upper-layer application. Compression is optional, and defaults to *null* for SSLv3 and TLS, which do not specify compression algorithms.

SSL Handshake Protocol

The Handshake protocol performs many crucial tasks related to session establishment. It manages session and connection states, authenticates communicating parties, negotiates cryptographic algorithms, and exchanges keying information. Figure 6-14 illustrates the format of a Handshake protocol message.

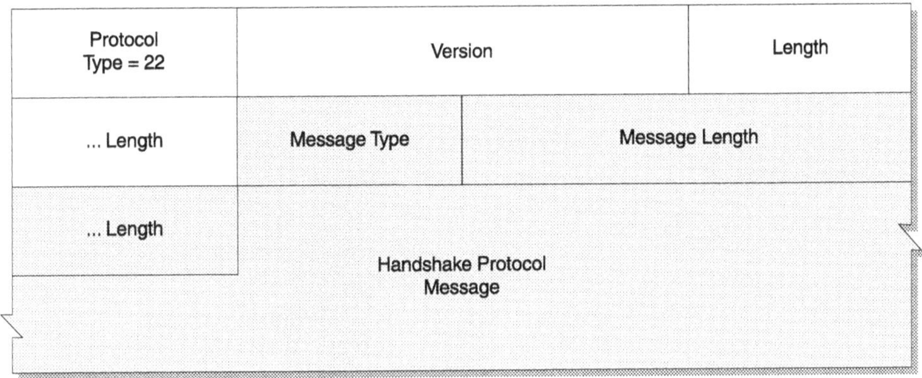

Figure 6-14 Handshake protocol message format

A single Record layer message can encapsulate multiple handshake messages. Each handshake message contains three fields:

- **Type** A 1-byte field containing one of the message types listed in Table 6-11.
- **Length** A 3-byte field containing the length of the message in bytes.
- **Content** An n-byte field ($n \geq 1$) containing handshake message parameters (see Table 6-12).

Type	Message	Description
0	*hello_request*	Request made by a server for a client to begin negotiation by sending a *client_hello* message.
1	*client_hello*	First message a client sends to a server when requesting a secure connection.
2	*server_hello*	Server's response to a *client_hello* message, provided that the *client_hello* does not result in a failure alert.
11	*certificate*	Contains a server certificate that the client will use to authenticate the server.
12	*server_key_exchange*	Used by a server to supply a client with keying material required for future secure communication

Table 6-11 Handshake Protocol Message Types

Type	Message	Description
13	*certificate_request*	Allows a server (non-anonymous) to request a certificate from a client.
14	*server_hello_done*	Indicates the end of the *server_hello* sequence.
15	*certificate_verify*	Provides explicit verification of a client certificate via a digital signature.
16	*client_key_exchange*	Client supplies server with keying material required for future secure communication.
20	*finished*	Sent immediately following a *change_cipher_spec* message to verify successful key exchange and authentication.

Table 6-11 Handshake Protocol Message Types *(continued)*

State Processing During normal operation (that is, post-handshake), both the client and server maintain a current read state for receiving data and a write state for sending data. Additionally, pending read and write states are created by the Handshake protocol. Once the client and server agree on a common cipher suite and swap *change_cipher_spec* messages, the pending read and write states are copied to the current read and write states. At this point, both parties can begin transmitting data using the newly negotiated keying information and encryption algorithms.

The SSL Handshake During an SSL handshake (see Figure 6-15), the client and server exchange authentication and keying information, and they choose a

Message Type	Parameters
hello_request	No parameter (*null*)
client_hello	Protocol Version, Random, Session ID, Cipher Suites, Compression Methods
server_hello	Protocol Version, Random, Session ID, Cipher Suite, Compression Methods
certificate	X.509 Version 3 Certificate Chain
server_key_exchange	Key-exchange parameters (RSA, DH, Fortezza), Signature
certificate_request	Certificate Type, List of Certificate Authorities
server_hello_done	No parameter (*null*)
certificate_verify	Signature
client_key_exchange	Keying material (RSA, DH, Fortezza); with RSA, the only parameter is the encrypted pre-master secret
finished	Hash Value

Table 6-12 Handshake Protocol Message Parameters

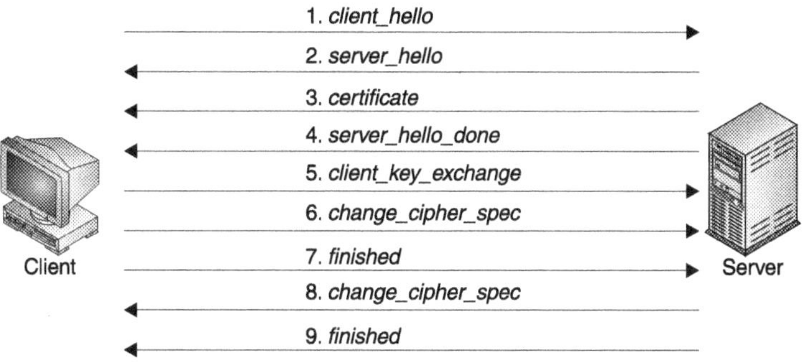

Figure 6-15 A typical SSL handshake consists of nine steps.

cipher suite—a combination of a symmetric cipher, public-key algorithm, and message digest. The Handshake protocol authenticates the client and server. A typical SSL handshake consists of the following steps:

1. The client sends a *client_hello* message presenting the SSL options it supports.
2. The server selects the SSL options, such as the SSL version and cipher suite, that it wishes to use during the session, and responds with a *server_hello* message.
3. The server sends a *certificate* message containing its public-key certificate.
4. The server transmits a *server_hello_done* message indicating to the client that it has completed this portion of the negotiation.
5. The client generates and encrypts a random value using the public key contained within the server's certificate and sends the value to the server in a *client_key_exchange* message. The random value will later be used to derive a shared session key for use by a symmetric stream or block cipher to encrypt traffic sent between the client and server.
6. The client sends a *change_cipher_spec* message to indicate that it is ready to begin using the negotiated security services.
7. The client transmits a *finished* message letting the server know that it has completed its portion of the handshake and is ready to begin transmitting data. The server uses this message to test the newly negotiated security services.
8. The server responds with a *change_cipher_spec* message to activate the negotiated security services.
9. Finally, the server sends a *finished* message of its own, which the client uses to test the negotiated security services.

NOTE:
Since multiple handshake messages can be combined within a single Record layer message, the entire SSL handshake requires only four TCP segments: one for step 1, one for steps 2–4, one for steps 5–7, and one for steps 8 and 9.

In the preceding example, the client authenticates the server based on a valid digital certificate, and on the fact that the server can decrypt the session key that has been encrypted with its public key. By doing so, the server proves that it possesses the corresponding private key. Server-side authentication can be separated from the private-key decryption, as illustrated in Figure 6-16.

1. The client sends a *client_hello* message presenting the SSL options that it supports.
2. The server selects the SSL options, such as the SSL version and cipher suite, that it wishes to use during the session, and responds with a *server_hello* message.
3. The server sends a *certificate* message containing its public-key certificate.
4. The server digitally signs the public key it wishes for the client to use when encrypting the session key, and sends the signed key to the client within a *server_key_exchange* message. This message type can only be used when the certificate message from step 3 does not contain enough information for the client to exchange the session key.
5. The server then transmits a *server_hello_done* message indicating to the client that it has completed this portion of the negotiation.
6. The client generates and encrypts a random value (later used to derive a shared session key) using the public key specified within the server's *server_key_exchange* message, and it sends the value to the server in a *client_key_exchange* message.

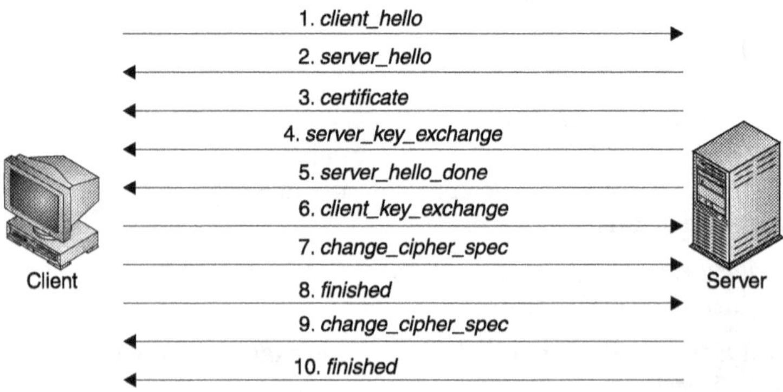

Figure 6-16 Server-side authentication and private-key decryption can be separated by means of the server_key_exchange message.

7. The client sends a *change_cipher_spec* message to indicate that it is ready to begin using the negotiated security services.
8. The client then transmits a *finished* message letting the server know that it has completed its portion of the handshake and is ready to begin transmitting data. The server uses this message to test the newly negotiated security services.
9. The server responds with a *change_cipher_spec* message to activate the negotiated security services.
10. Finally, the server sends a *finished* message of its own, which the client uses to test the negotiated security services.

The reason for separating authentication and encryption are twofold. Many countries (including the United States) restrict the export of strong encryption products to other countries. To get around this legal restriction, companies are required to ship export-grade products containing cryptographic algorithms that only support shortened key lengths for encryption. Many of these countries do not place similar limitations on the size of keys used for signing. To allow the longest key length possible for signing, while maintaining export restrictions on encryption keys, the functions of authentication and encryption must be separated. Second, some public-key algorithms, such as DSA, can only be used for signature generation and not encryption. Therefore, the key provided in the server's certificate cannot be used to encrypt the keying information exchanged in step 6 of the previous example (Thomas, 2000).

The two previous handshake scenarios provide no client-side authentication. SSL does support client-side authentication, as illustrated in Figure 6-17. The

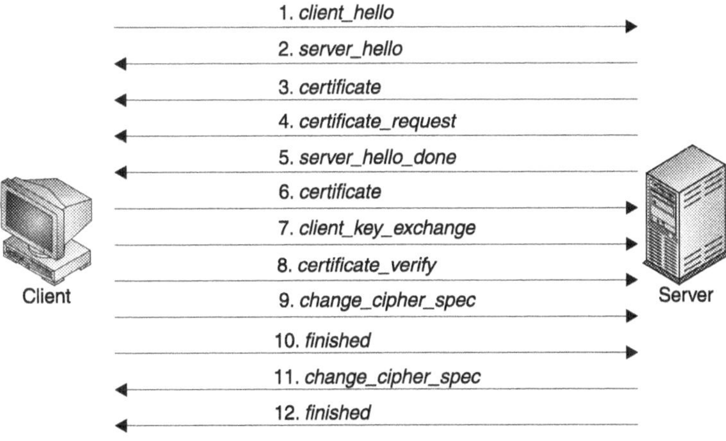

Figure 6-17 An SSL handshake with client-side authentication

process is very similar to that used for server-side authentication, but employs three more messages than a typical SSL handshake:

1. The client sends a *client_hello* message presenting the SSL options it supports.
2. The server selects the SSL options, such as the SSL version and cipher suite, that it wishes to use during the session, and responds with a *server_hello* message.
3. The server sends a *certificate* message containing its public-key certificate.
4. The server issues a *certificate_request* message asking the client for its public-key certificate. This indicates that the server wishes to authenticate the client.
5. The server then transmits a *server_hello_done* message indicating to the client that it has completed this portion of the negotiation.
6. The client sends a *certificate* message containing its public-key certificate.
7. The client generates and encrypts a random value (later used to derive a shared session key) using the public key contained within the server's certificate, and sends the session key in a *client_key_exchange* message.
8. The client digitally signs information relating to the current session using its private key and transmits the signed data to the server in a *certificate_verify* message; the private key corresponds to the public key in the certificate sent to the server in step 6.
9. The client sends a *change_cipher_spec* message to indicate that it is ready to begin using the negotiated security services.
10. The client then transmits a *finished* message letting the server know that it has completed its portion of the handshake and is ready to begin transmitting data. The server uses this message to test the newly negotiated security services.
11. The server responds with a *change_cipher_spec* message to activate the negotiated security services.
12. Finally, the server sends a *finished* message of its own, which the client uses to test the negotiated security services.

Remember that public-key cryptographic operations (especially those using a private key) are computationally expensive. For an SSL server responding to hundreds or thousands of incoming requests in a short period, encryption or decryption using a private key introduces considerable processing overhead. Each time the server generates a signature or decrypts a session key from a client, it uses its private key. By resuming an SSL session, a process known as *session reuse,* the server avoids having to generate a new signature or decrypt a session key, which reduces processing load. This equates to fewer Web servers needed to respond to incoming requests. Figure 6-18 demonstrates session reuse.

TIP:
Session reuse saves valuable processing resources by limiting the number of private-key encryption and decryptions an SSL server must perform.

Resuming an SSL session requires these steps:

1. The client sends a *client_hello* message containing the session ID of a previous SSL session.
2. The server replies with a *server_hello* message indicating that it accepts the session ID.
3. The server sends a *change_cipher_spec* message to activate the security services negotiated during the cached session.
4. The server then sends a *finished* message, allowing client to test the negotiated security services.
5. The client sends a *change_cipher_spec* message to indicate that it is ready to begin using the negotiated security services.
6. Finally, the client transmits a *finished* message letting the server test the newly negotiated security services.

ChangeCipherSpec Protocol

The ChangeCipherSpec protocol consists of a single message type, *change_cipher_spec,* which always has a value of 1. The purpose of this protocol is to signal changes in the ciphering schema used during transmission. Reception of a ChangeCipherSpec message indicates that subsequent records will be secured using the most recently negotiated encryption and hashing algorithms and keying information. This message also results in a state change where the receiver copies the pending read state into the current read state. Only one *change_cipher_spec* message can be contained within any given record. Figure 6-19 illustrates the format of a ChangeCipherSpec protocol message.

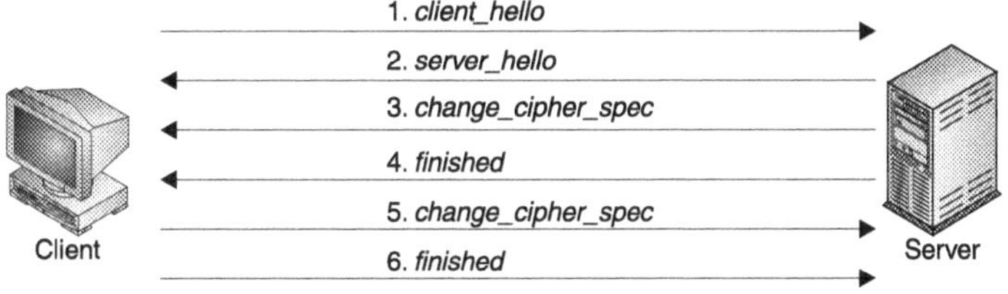

Figure 6-18 Resuming an SSL session requires fewer steps than the initial SSL handshake and avoids repeating costly cryptographic operations.

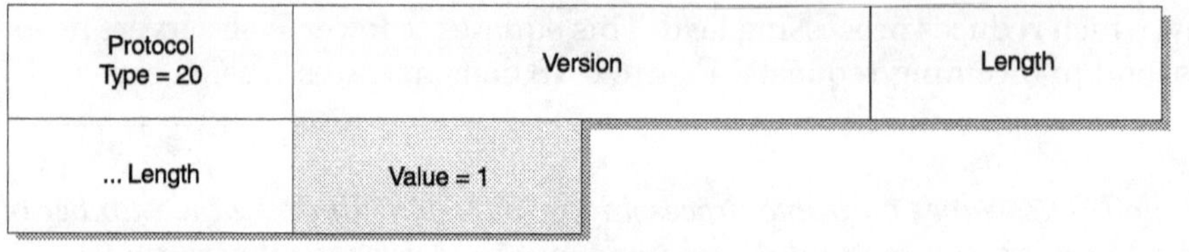

Figure 6-19 ChangeCipherSpec protocol message format

Alert Protocol

The Alert protocol generates a message when either communicating party encounters an error or wishes to close a session gracefully. Each alert message provides a description along with an associated severity level (see Figure 6-20), and is composed of two fields:

- **Level** A 1-byte field indicating the severity level of the alert; a value of 1 indicates a warning, and a value of 2 indicates a failure.
- **Description** A 1-byte field containing a description of the alert. Table 6-13 includes alert messages for both SSL 3.0 and TLS 1.0. Table 6-14 contains alerts only present in TLS 1.0.

Fatal messages result in immediate termination of the session. Warnings are more forgiving: A system receiving a warning may choose whether to continue communication.

SSL Cipher Suites

SSL supports a wide selection of cryptographic algorithms for authentication, encryption, and message integrity. The collection of a public-key algorithm,

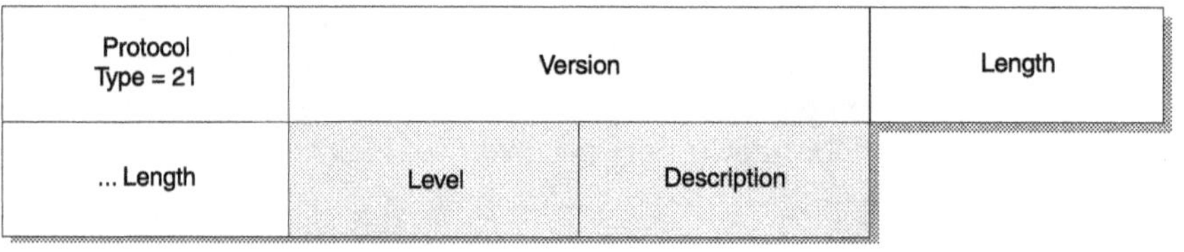

Figure 6-20 Alert protocol message format

Type	Message	Level	Description
0	*close_notify*	Either	Sender does not wish to transmit any more messages on the current connection.
10	*unexpected_message*	Fatal	Sender indicates that it received an unacceptable message.
20	*bad_record_mac*	Fatal	Sender indicates that it received a record for which the MAC verification failed.
30	*decompression_failure*	Fatal	Sender indicates that the decompression algorithm received malformed input.
40	*handshake_failure*	Fatal	Sender is unable to negotiate an acceptable set of security parameters (cipher suite).
41	*no_certificate*[2]	Either	Sender indicates that it does not have a suitable certificate satisfying the server's certification request; in this case, the sender is always a client.
42	*bad_certificate*	Either	Corrupt certificate received (for example, the signature did not verify).
43	*unsupported_certificate*	Either	Certificate is of a type that is not supported.
44	*certificate_revoked*	Either	Certificate has been revoked.
45	*certificate_expired*	Either	Certificate has expired.
46	*certificate_unknown*	Either	Certificate processing failed due to an unspecified error.
47	*illegal_parameter*	Fatal	Handshake message contained a field with a value that was illegal or inconsistent with other fields.

[2] TLS 1.0 does not specify the *no_certificate* alert message.

Table 6-13 SSL and TLS Alert Protocol Message Types

Type	Message	Level	Description
21	*decryption_failed*	Fatal	Sender could not decrypt a received message (for example, if the message was not the correct length or padding was not correct).
22	*record_overflow*	Fatal	Sender received an encrypted record with a length greater than 2^{14} + 2048 bytes, or a record that decrypted to a length greater than 2^{14} + 1024 bytes.

Table 6-14 TLS-Only Alert Protocol Message Types

Type	Message	Level	Description
48	*unknown_ca*	Fatal	Sender indicates that it cannot accept a certificate because it cannot locate the certificate of a matching trusted CA.
49	*access_denied*	Fatal	Sender received a valid certificate, but refuses to proceed with connection due to access control settings.
50	*decode_error*	Fatal	Sender could not decode a message due to an invalid field value or message length.
51	*decrypt_error*	Either	Sender is unable to verify a signature, decrypt keying information, or validate the MAC attached to a message.
60	*export_restriction*	Fatal	Sender detected an attempt to negotiate security parameters that do not meet export restrictions.
70	*protocol_version*	Fatal	Sender (always a server) indicates that the protocol version the client is trying to negotiate is recognized, but not supported.
71	*insufficient_security*	Fatal	Sender (always a server) indicates that it requires a more secure set of ciphers than the client requested.
80	*internal_error*	Fatal	An error, not caused by the peer (for example, memory allocation failure) has occurred preventing the connection from continuing.
90	*user_cancelled*	Fatal	Sender indicates that the current handshake is being cancelled for a reason other than a protocol failure.
100	*no_renegotiation*	Warning	Sent following a *hello_request* or a *client_hello* message to indicate that the sender does not wish to proceed with connection negotiation.

Table 6-14 TLS-Only Alert Protocol Message Types *(continued)*

symmetric cipher, and message digest is known as a *cipher suite*. Table 6-15 lists all the available SSL cipher suites supported by SSLv3. TLSv1 supports all the same cipher suites, except for those in the Fortezza suite. *Fortezza* is a hardware cryptographic token designed by the NSA (National Security Agency). It uses DSA and SHA-1 for digital signature generation, but includes its own key-agreement and encryption algorithms; for encryption, it uses *Skipjack*, and its key agreement algorithm is the *Key Exchange Algorithm (KEA)*. (KEA is a variant of Diffie-Hellman.) The motivation behind the development of Fortezza was

Cipher Suite	Key Exchange	Cipher	Hash	Export
SSL_NULL_WITH_NULL_NULL	NULL	NULL	NULL	Yes
SSL_RSA_WITH_NULL_MD5	RSA	NULL	MD5	Yes
SSL_RSA_WITH_NULL_SHA	RSA	NULL	SHA	Yes
SSL_RSA_EXPORT_WITH_RC4_40_MD5	RSA_EXPORT	RC4_40	MD5	Yes
SSL_RSA_WITH_RC4_128_MD5	RSA	RC4_128	MD5	No
SSL_RSA_WITH_RC4_128_SHA	RSA	RC4_128	SHA	No
SSL_RSA_EXPORT_WITH_RC2_CBC_40_MD5	RSA_EXPORT	RC2_CBC_40	MD5	Yes
SSL_RSA_WITH_IDEA_CBC_SHA	RSA	IDEA_CBC	SHA	No
SSL_RSA_EXPORT_WITH_DES40_CBC_SHA	RSA_EXPORT	DES40_CBC	SHA	Yes
SSL_RSA_WITH_DES_CBC_SHA	RSA	DES_CBC	SHA	No
SSL_RSA_WITH_3DES_EDE_CBC_SHA	RSA	3DES_EDE_CBC	SHA	No
SSL_DH_DSS_EXPORT_WITH_DES40_CBC_SHA	DH_DSS_EXPORT	DES40_CBC	SHA	Yes
SSL_DH_DSS_WITH_DES_CBC_SHA	DH_DSS	DES_CBC	SHA	No
SSL_DH_DSS_WITH_3DES_EDE_CBC_SHA	DH_DSS	3DES_EDE_CBC	SHA	No
SSL_DH_RSA_EXPORT_WITH_DES40_CBC_SHA	DH_RSA_EXPORT	DES40_CBC	SHA	Yes
SSL_DH_RSA_WITH_DES_CBC_SHA	DH_RSA	DES_CBC	SHA	No
SSL_DH_RSA_WITH_3DES_EDE_CBC_SHA	DHE_DSS_EXPORT	3DES_EDE_CBC	SHA	No
SSL_DHE_DSS_EXPORT_WITH_DES40_CBC_SHA	DHE_DSS_EXPORT	DES40_CBC	SHA	Yes

Table 6-15 Available SSL Cipher Suites

Cipher Suite	Key Exchange	Cipher	Hash	Export
SSL_DHE_DSS_WITH_DES_CBC_SHA	DHE_DSS	DES_CBC	SHA	No
SSL_DHE_DSS_WITH_3DES_EDE_CBC_SHA	DHE_DSS	3DES_EDE_CBC	SHA	No
SSL_DHE_RSA_EXPORT_WITH_DES40_CBC_SHA	DHE_RSA_EXPORT	DES40_CBC	SHA	Yes
SSL_DHE_RSA_WITH_DES_CBC_SHA	DHE_RSA	DES_CBC	SHA	No
SSL_DHE_RSA_WITH_3DES_EDE_CBC_SHA	DHE_RSA	3DES_EDE_CBC	SHA	No
SSL_DH_anon_EXPORT_WITH_RC4_40_MD5	DH_anon_EXPORT	RC4_40	MD5	Yes
SSL_DH_anon_WITH_RC4_128_MD5	DH_anon	RC4_128	MD5	No
SSL_DH_anon_EXPORT_WITH_DES40_CBC_SHA	DH_anon_EXPORT	DES40_CBC	SHA	Yes
SSL_DH_anon_WITH_DES_CBC_SHA	DH_anon	DES_CBC	SHA	No
SSL_DH_anon_WITH_3DES_EDE_CBC_SHA	DH_anon	3DES_EDE_CBC	SHA	No
SSL_FORTEZZA_DMS_WITH_NULL_SHA	FORTEZZA_DMS	NULL	SHA	No
SSL_FORTEZZA_DMS_WITH_FORTEZZA_CBC_SHA	FORTEZZA_DMS	FORTEZZA_CBC	SHA	No
SSL_FORTEZZA_DMS_WITH_RC4_128_SHA	FORTEZZA_DMS	RC4_128	SHA	No

Table 6-15 Available SSL Cipher Suites *(continued)*

to provide strong encryption, while allowing the NSA to eavesdrop on communication. The NSA accomplished this task by archiving a copy of the key contained within every Fortezza device (a process known as *key escrow*). When the NSA wished to eavesdrop, it simply decrypted the message by recovering the key original used to encrypt it.

Algorithm	Description	Key Size Limit
DHE_DSS	Ephemeral Diffie-Hellman with DSS signatures	None
DHE_DSS_EXPORT	Ephemeral Diffie-Hellman with DSS signatures	DH: 512 bits
DHE_RSA	Ephemeral Diffie-Hellman with RSA signatures	None
DHE_RSA_EXPORT	Ephemeral Diffie-Hellman with RSA signatures	DH: 512 bits RSA: none
DH_anon	Anonymous Diffie-Hellman, no signatures	None
DH_anon_EXPORT	Anonymous Diffie-Hellman, no signatures	DH: 512 bits
DH_DSS	Diffie-Hellman with DSS-based certificates	None
DH_DSS_EXPORT	Diffie-Hellman with DSS-based certificates	DH: 512 bits
DH_RSA	Diffie-Hellman with RSA-based certificates	None
DH_RSA_EXPORT	Diffie-Hellman with RSA-based certificates	DH: 512 bits RSA: none
FORTEZZA_DMS	Key Exchange Algorithm (KEA)	1024
NULL	No key exchange	N/A
RSA	RSA key exchange	None
RSA_EXPORT	RSA key exchange	RSA: 512 bits

Table 6-16 SSL Key-Exchange Algorithms

Table 6-16 contains a list of SSL key exchange algorithms. Tables 6-17 and 6-18 list all available encryption and message digest algorithms, respectively.

Cipher	Type	Key Material (in bytes)	Key Size (in bits)	IV Size (in bytes)	Block Size (in bytes)
NULL	Stream	0	0	0	N/A
FORTEZZA_CBC	Block	10	80	20	8

Table 6-17 SSL Encryption Algorithms

Cipher	Type	Key Material (in bytes)	Key Size (in bits)	IV Size (in bytes)	Block Size (in bytes)
IDEA_CBC	Block	16	128	8	8
RC2_CBC_40	Block	5	40	8	8
RC4_40	Stream	5	40	0	N/A
RC4_128	Stream	16	128	0	N/A
DES40_CBC	Block	5	40	8	8
DES_CBC	Block	8	56	8	8
3DES_EDE_CBC	Block	24	168	8	8

Table 6-17 SSL Encryption Algorithms *(continued)*

Application Layer—Kerberos

Developed in the mid 1980s at the Massachusetts Institute of Technology (MIT), *Kerberos* is a network authentication protocol allowing users and services to strongly authenticate one another via a trusted third party. The name Kerberos comes from Greek mythology—it's the name of the three-headed dog that guarded the entrance to Hades. Kerberos was designed in part to address key management issues in the absence of public-key cryptography (which had not yet become mainstream). It uses symmetric encryption techniques to distribute shared secret keys, or *session keys,* to users and services over insecure networks. The use of cryptography thwarts packet-sniffing attacks by concealing the session keys during distribution. Possession of a session key indicates successful authentication to the Kerberos system. As we will soon see, this is a bit of an oversimplification, but it suffices to illustrate the functionality of Kerberos at a high level. RFC 1510 describes Kerberos Version 5 (Neuman, 1993). This RFC is currently undergoing revision as an IETF draft (Ts'o, 2001).

Before we discuss the operational details of Kerberos, let's familiarize ourselves with some important terminology. Kerberos uses the term *realm* to describe an administrative domain consisting of a *key distribution center (KDC)* and all resources protected by that KDC. A KDC is responsible for responding to end-user authentication requests for access to protected services (Telnet, FTP,

Table 6-18 SSL Message Digest Algorithms

Hashing Algorithm	Hash Size (in bytes)	Padding Size (in bytes)
NULL	0	0
MD5	16	48
SHA	20	40

SMTP, and so on) within its realm. The KDC maintains a list of user/service IDs and corresponding long-term encryption keys associated with the realm to which the KDC belongs. In the following section, we describe how the KDC uses these IDs and long-term encryption keys for authentication.

A *principal* is an entity within a Kerberos system, such as a user, service, or authentication server, and Kerberos identifies each entity by a *principal name*. For users, a principal name can be thought of as a realm user ID that, when combined with a password, can be used to authenticate to a Kerberos realm.

KDCs are logically divided into two separate components: an *authentication server (AS)* and a *ticket granting server (TGS)*. The AS handles the initial authentication steps, authenticating the user to the TGS. The TGS performs subsequent authentication to protected services. As we will soon see, the functional distinction between the AS and TGS prevents users from having to enter their passwords every time they request access to a service.

Kerberos Authentication

Kerberos authenticates users to services, and may optionally authenticate services to users (if requested by the user). Both the user and service must register keys with the KDC to take advantage of the authentication services provided by Kerberos. Kerberos authentication follows these steps:

1. A user sends an authentication request to the KDC asking for access to a protected service running on a host within the realm to which the KDC belongs. For this example, the user is requesting access to an FTP server (see Figure 6-21). The user supplies a principal name and a password. The Kerberos client (usually part of the client application) generates a cryptographic key by passing the password through a message digest. This key is later used to decrypt responses from the KDC.
2. The KDC indexes into the Kerberos database to obtain two long-term encryption keys—one for the user and one for the service—using the principal name supplied by the user and that of the protected service. The KDC then encrypts the session key along with the name of the FTP server using a symmetric cipher and the long-term encryption key belonging to the user (Message 1).

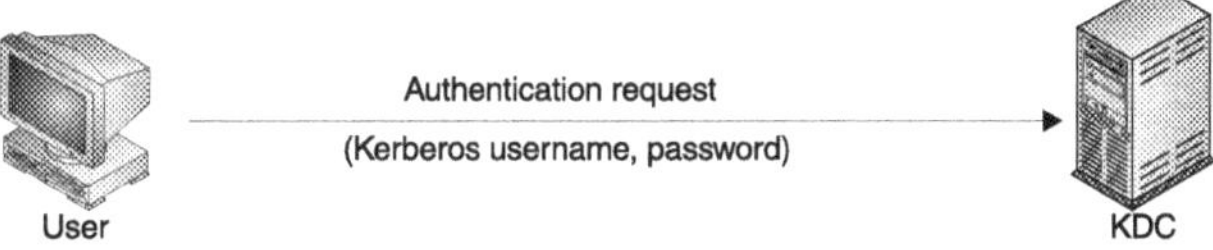

Figure 6-21 When authenticating via Kerberos, a user obtains a session key from a key distribution center (KDC).

3. The KDC encrypts a second copy of the session key along with the user's name using the long-term encryption key belonging to the FTP server (Message 2). The combination of the encrypted session key and user name is known as a *ticket*.
4. The KDC returns both Message 1 and Message 2 to the user (see Figure 6-22).
5. The user decrypts Message 1 using the key generated from his or her password to recover the session key and name of the FTP server. The key generated from the user's password and the key stored in the Kerberos database are the same symmetric key.
6. Without the long-term encryption key belonging to the FTP server, the user cannot decrypt the ticket. Instead, the user encrypts the current system time with the session key to create an *authenticator* (Message 3).
7. The user then sends both Message 2 and Message 3 to the FTP server (see Figure 6-23).
8. The FTP server decrypts the ticket (Message 2) using the long-term encryption key—which it shares with the KDC—to obtain the session key. The FTP server then decrypts the authenticator (Message 3) using the session key to obtain the timestamp. If the authenticator has not been previously used and the timestamp verifies correctly, the user successfully authenticates to the FTP server.
9. (Optional) If the user requests *mutual authentication,* the FTP server encrypts the timestamp retrieved from step 8, along with the FTP server's host name, using the session key. It then returns the message to the user for verification.

There is actually an intermediate step that may occur between steps 1 and 2. Before a user requests a ticket for authenticating directly to a service, he may first request a *ticket-granting ticket (TGT)* from the AS. The TGT and a

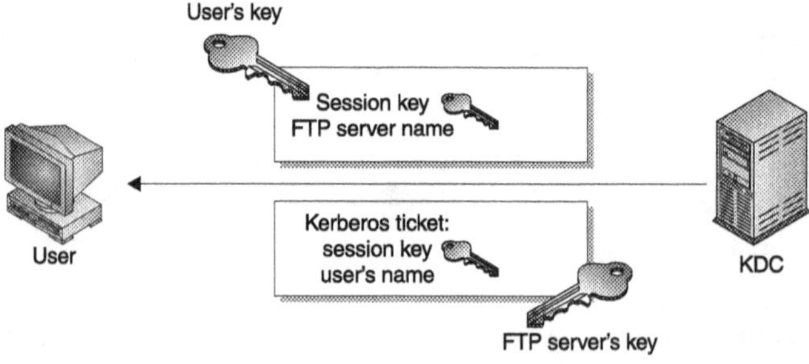

Figure 6-22 Following an authentication request, a KDC supplies a user with an encrypted session key and a Kerberos ticket.

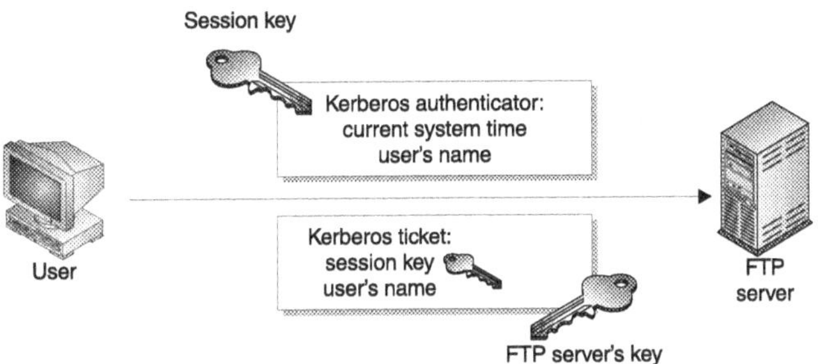

Figure 6-23 After decrypting the session key, the user sends the Kerberos ticket and authenticator to the FTP server. The FTP server uses the authenticator to verify the identity of the user.

long-term session key may then be used to authenticate to the TGS. The TGS is responsible for issuing shorter-lived application server tickets. Any time the user needs to request a new ticket, he first contacts the TGS using the established TGT and session key. This prevents the user from having to enter a password for each ticket request.

NOTE:

Including the name of the FTP server in the encrypted response in Step 2 allows the user (in other words, the, Kerberos client application) to verify that it has decrypted the response successfully, and that the response actually came from the KDC (Tung, 1999).

The timestamp supplied in an authenticator allows for detection of replay attacks. The host running the service also maintains a list of recently received authenticators for the same purpose.

Cross-Realm Authentication

In most situations, a single Kerberos realm encompasses an entire organization. Nevertheless, there may be cases when multiple realms are necessary. For example, a very large corporation might create separate realms for geographically disperse regions or to mimic an administrative hierarchy already present in the company. In addition, a single realm may not scale well for a very large organization, resulting in sluggish authentication. When using separate realms, each realm must contain a KDC and its own list of principals (for instance, users and services).

How does a user from one realm authenticate to a service belonging to another realm? The user cannot obtain a session key and ticket from its own KDC—the session key and ticket will not be valid in another realm. Instead, Kerberos includes mechanisms for authenticating users across realm boundaries through an existing relationship between the KDCs, a process known as *cross-realm authentication*. Authentication across realm boundaries adds additional steps to the Kerberos authentication process presented earlier in this chapter.

Cross-realm authentication requires that the administrators of the two realms exchange secret keys. (Separate keys can be used for authentication in each direction.) This exchange registers the TGS of each realm as a principal in the other. To authenticate to a service in the remote realm, a user first contacts the AS within its realm (the local AS) to receive a session key/TGT for the local TGS. The local TGS issues a ticket for accessing the TGS in the remote realm. The user then contacts the remote TGS directly to authenticate to a service in the remote realm. The remote TGS recognizes that the ticket issued by the TGS in the user's realm does not belong to its own realm. Before issuing a session key and ticket to the user, the remote TGS uses the prenegotiated secret key(s) to decrypt the ticket issued by the user's TGS.

Kerberos V5 supports an additional cross-realm authentication technique permitting users to authenticate via a hierarchical authentication path. A user first contacts the KDC in the local realm, which issues the user a ticket for another realm. The user then contacts the remote realm, which issues the user a ticket for a third realm. This process continues until the user reaches the realm containing the desired service. For this process to work, each intermediate KDC must verify that it trusts all realms in the authentication path up to that point. At each hop, the intermediate KDCs modify the ticket to reflect the current authentication path.

Public-Key Authentication with Kerberos

We mentioned previously that Kerberos uses symmetric cryptography for key exchange and authentication. However, two IETF drafts augment the functionality of RFC 1510 to include public-key authentication. *PKINIT* (Neuman et al., 2001) addresses initial public-key authentication within a single realm. *PKCROSS* (Hur et al., 2001) describes public-key cryptography for use in cross-realm authentication. In Chapter 10, we will see how these two specifications are used within PacketCable security.

For additional information on Kerberos functionality and administering Kerberos realms, refer to (Tung, 1999), (Tung, 1996), and (Hornstein, 2000).

References

Davis, Carlton. 2001. *IPSec: Securing VPNs*. Berkeley, California: Osborne/McGraw-Hill.

Dierks, T., and C. Allen. 1999. *RFC 2246: The TLS Protocol Version 1.0*. Internet Engineering Task Force.

Doraswamy, Naganand, Dan Harkins. 1999. *IPSec: The New Security Standard for the Internet, Intranets, and Virtual Private Networks*. New Jersey: Prentice-Hall PTR.

Freier, Alan, Philip Karlton, and Paul Kocher. 1996. *The SSL Protocol Version 3.0*. The Internet Engineering Task Force.

Harkins, D., D. Carrel. November, 1998. *RFC 2409: The Internet Key Exchange (IKE)*. Internet Engineering Task Force.

Hornstein, Ken. 18 August, 2000. *Kerberos FAQ, v2.0*. Naval Research Laboratories. 16 January, 2002 (http://www.faqs.org/faqs/kerberos-faq/general/preamble.html).

Hur, Matthew, Brian Tung, Tatyana Ryutov, Clifford Neuman, Gene Tsudik, Ari Medvinsky, Bill Sommerfeld. 8 November, 2001. *Public Key Cryptography for Cross-Realm Authentication in Kerberos*. Internet Engineering Task Force.

Kent, S., R. Atkinson. November, 1998. *RFC 2401: Security Architecture for the Internet Protocol*. Internet Engineering Task Force.

—November, 1998. *RFC 2402: IP Authentication Header (AH)*. Internet Engineering Task Force.

—November, 1998. *RFC 2406: IP Encapsulating Security Payload (ESP)*. Internet Engineering Task Force.

Maughan, D., M. Schertler, M. Schneider, J. Turner. November, 1998. *RFC 2408: Internet Security Association and Key Management Protocol (ISAKMP)*. Internet Engineering Task Force.

Neuman, Clifford, J. Kolh. September, 1993. *RFC 1510: The Kerberos Network Authentication Service (V5)*. Internet Engineering Task Force.

Neuman, Clifford, Brian Tung, Mathew Hur, Ari Medvinsky, Sasha Medvinsky, John Wray, Jonathan Trostle. 28 November, 2001. *Public Key Cryptography for Initial Authentication in Kerberos*. Internet Engineering Task Force.

Piper, D. November, 1998. *RFC 2407: The Internet IP Security Domain of Interpretation for ISAKMP*. Internet Engineering Task Force.

Reynolds, J., J. Postel. October, 1994. *RFC 1700: Assigned Numbers*. Internet Engineering Task Force.

Thomas, Stephen. 2000. *SSL and TLS Essentials: Securing the Web*. New York: John Wiley & Sons, Inc.

Ts'o, Theodore, Clifford Neuman, John Kolh, Ken Raeburn, Tom Yu. 20 November, 2001. *The Kerberos Network Authentication Service (V5)*. Internet Engineering Task Force.

Tung, Brian. 1999. *Kerberos: A Network Authentication System*. Massachusetts: Addison-Wesley.

Tung, Brian. 19 December, 1996. *The Moron's Guide to Kerberos, Version 1.2.2*. Information Sciences Institute. 16 January, 2002 (http://www.isi.edu/~brian/security/kerberos.html).

Chapter 7

Placing Security Services and Mechanisms

In Chapter 6, we introduced a number of popular network security protocols. We described how each protocol operates within a different layer of the OSI or TCP/IP network reference models, and as such, each offers different capabilities. In this chapter, we continue this discussion by explaining when and where security protocols are most applicable. Our discussion continues with a description of the various techniques used for implementing security protocols as part of a protocol stack, as a bump in the stack, and as a bump in the wire. We conclude the chapter with a discussion on the use of host-based security versus security gateways, and we discuss when each method is most appropriate.

Binding Security Services and Mechanisms to Data

The first thing for you to consider when placing security services and mechanisms is whether practical requirements exist for binding security to application data (in other words, data-oriented security). For instance, a legal obligation for non-repudiation on all financial transactions electronically exchanged between two companies may exist. The most appropriate security mechanism for implementing non-repudiation is a digital signature, and a sample transaction may proceed as follows:

1. Company A transmits an electronic request to Company B to purchase 100,000 widgets. Before sending the request, Company A digitally signs the request using its RSA private key and binds the signature to the data by creating a PKCS #7 signed data message. See Chapter 2 for a discussion of the Public-Key Cryptography Standards (PKCS).
2. Upon receipt of Company A's request, Company B verifies the signature contained within the PKCS #7 message to ensure that the request is legitimate.
3. If the signature verifies correctly, Company B processes the request, charges Company A's account, and arranges for the shipment of 10,000 widgets.
4. Company B transmits a response back to Company A indicating that it has received and processed the order. The response is a nested PKCS #7 signed data message that contains the original signed request from Company A, an order confirmation from Company B, and a digital signature generated over the contents of the message using Company B's RSA private key.
5. Both companies store the response message so that neither party can later deny involvement in the transaction.

Regardless of how the transaction is transmitted, processed, or stored, the signatures must remain with the application data at all times. This allows either party to verify the transaction at a later date. In this case, your choices for placement of security mechanisms are limited—you must implement security at the Application layer (within the application responsible for generating the data).

If, on the other hand, there is no need to bind security mechanisms to the data, your placement choices become far more plentiful. Many applications require only that data be secured while in transit between two communicating parties; they are not concerned with how the data is handled before it is transmitted and after it is received. If session-based security (security that exists only during a communication session) is appropriate for your application, you must select a suitable network layer at which to implement the various security services. You must also determine whether security mechanisms can (and should) be placed within end-user hosts or security gateways. The majority of this chapter addresses the placement of security services and mechanisms when using session-based security.

NOTE:
In this chapter, we generically use the term "network layer" to indicate any layer within the OSI and TCP/IP network reference models. When we are referring specifically to Layer 3 of the OSI Reference Model or the Network layer of the TCP/IP Reference Model, we will use "Network layer" with a capital N.

Which Network Layer?

The placement of a security protocol within a protocol stack greatly affects the type and extent of services it can offer. Your choice of existing security protocols or the layer at which you decide to create a new protocol determines how transparent its operation is to an application. This decision often dictates how difficult the protocol will be to integrate into a new or existing network environment. It also determines whether the protocol can be used only to protect traffic across a single hop in a network or whether it can secure data all the way from its source to its destination regardless of the number of hops. In addition to the network layer at which a security protocol operates, its behavior may vary depending on whether it is part of the protocol stack of an operating system or exists outside of the stack. In this section, we distinguish between the two cases when the location—inside or outside of a protocol stack—affects the operation of the security protocol or the level of effort required to integrate the protocol.

Application Transparency

Transparency refers to the level of interaction between an application and a security protocol and whether an application must be modified to take advantage of the services offered by the protocol. In general, security becomes more transparent as you move down the protocol stack. From the point of view of an application, security implemented at or below the Transport layer requires little or no modification to the application when the security services are included as part of a protocol stack. (Note that the operating system of a host manages the protocol stack.) To illustrate, consider that a single Transport layer security protocol implemented within a protocol stack can transparently secure multiple applications that communicate using the same Transport layer protocol (TCP or UDP). A security protocol operating within the Network layer can transparently secure multiple applications that employ different Transport layer communication protocols. IPSec, for example, can protect both TCP and UDP traffic without altering the applications or the Transport layer protocols. Taking this concept one step further, security protocols running below the Network layer can secure any number of applications, all communicating over different Transport and Network layer communication protocols. Figure 7-1 demonstrates the relative position of many common security protocols within the TCP/IP Reference Model.

To demonstrate the concept of transparency, let's look at an example using IPSec. In this example, two applications each use a different Transport layer protocol (see Figure 7-2). Both applications require security, but neither application includes support for any security services. For reliability and error recovery, Application 1 communicates over TCP; Application 2 employs UDP for its reduced overhead. Both applications process user data and hand the data over to their respective Transport layer protocols for transmission over the network. TCP and UDP both encapsulate the user data within headers and pass the data to the Network layer. IPSec then intercepts and processes the TCP and UDP data to determine what security services it should apply to each Transport layer segment datagram based on the security policy that has been configured for the host. Following IPSec processing, IP adds its own protocol headers and hands the resulting packets to the Link layer. Neither the applications nor the Transport layer protocols need be modified, nor do they need to be aware of IPSec, to take advantage of the security services offered by IPSec.

Up to this point, we have assumed that all security services operating at or below the Transport layer are provided by a protocol stack. However, when you're implementing security outside a protocol stack, the level of transparency varies greatly depending on the security protocol and the security services required. For instance, most Secure Sockets Layer (SSL) and Transport Layer Security (TLS) implementations operate on the border of the Application and Transport layers (see Figure 7-1). This allows SSL/TLS functionality to be added to an application without altering the protocol stack. However, SSL/TLS both require that an application be modified slightly to take advantage of the security offered. The modifications usually entail replacing all TCP socket calls within the application with calls to an SSL/TLS *software development kit (SDK),* as illustrated in Figure 7-3. The SDK performs all cryptographic operations and maintains all session-state information for each SSL/TLS session initiated by the application.

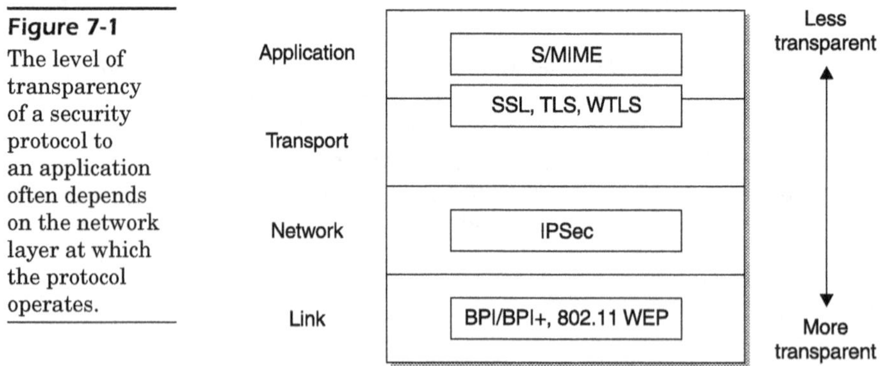

Figure 7-1 The level of transparency of a security protocol to an application often depends on the network layer at which the protocol operates.

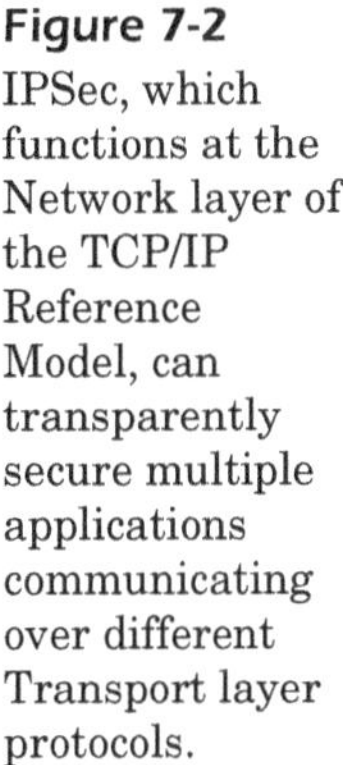

Figure 7-2
IPSec, which functions at the Network layer of the TCP/IP Reference Model, can transparently secure multiple applications communicating over different Transport layer protocols.

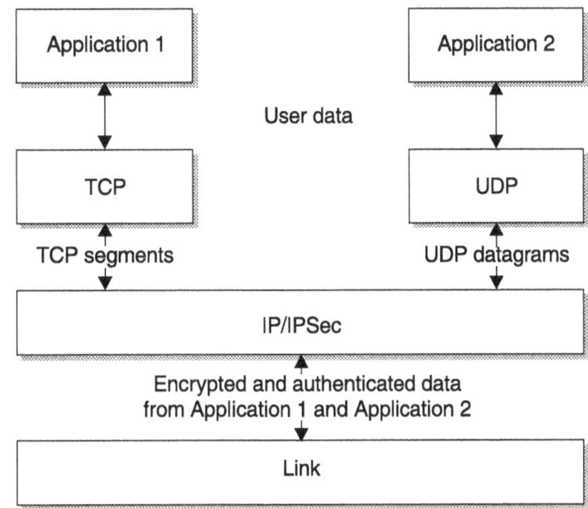

NOTE:
Some popular SSL / TLS SDKs are BSAFE SSL-C (C version) and SSL-J (Java version) from RSA Security, Inc.; Sun Microsystems' Java Secure Socket Extension (JSSE); and OpenSSL. These SDKs all provide an application programming interface (API) *that a software application can call to access the functionality of the SDK. An API defines the format of all function calls to the SDK, the arguments each function expects, and the values returned by the function upon completion of an operation. To date, the author knows of no SSL / TLS APIs implemented directly within a protocol stack. However, this does not mean that it cannot be done.*

Figure 7-3
Most SSL and TLS implementatios operate on the border of the Application and Transport layers to prevent a protocol stack from having to be modified to take advantage of security implemented within the Transport layer.

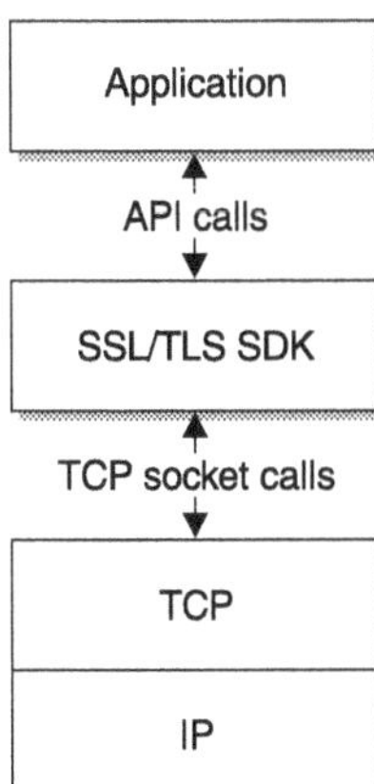

Table 7-1 compares and contrasts the level of transparency offered by many popular security protocols.

Ease of Integration

The level of difficulty associated with integrating a security protocol often correlates to the layer at which the protocol operates. Protocols running at the Application layer or on the border of the Application and Transport layers, such as S/MIME and most (if not all) SSL and TLS implementations, require that an application be modified to take advantage of the security services provided by

TCP/IP Layer	Level of Transparency	Example Protocol
Application	• Can secure traffic to and from a single application • Must be implemented completely within the application • Requires each application be modified to become security enabled	Secure/Multipurpose Internet Mail Extensions (S/MIME)
Transport	• Can secure traffic to and from multiple applications all using the same Transport layer protocol (e.g., TCP, UDP, and so on) • Can be implemented within an application or protocol stack • May not require direct modification to an application, but usually does when implemented on the border of the Application and Transport layers	SSL, TLS, and Wireless Transport Layer Security (WTLS)
Network	• Can secure network traffic to and from multiple applications using any Transport layer communication protocol compatible with the Network layer protocol in use (for example, IP requires TCP or UDP for Transport layer communication) • Can be implemented within a protocol stack or using a shim layer (see "Ease of Integration" for details) • Usually *does not* require direct modification to an application	IPSec
Link	• Can secure network traffic to and from multiple applications using any combination of Transport and Network layer communication protocols • Usually implemented within a protocol stack at the device driver level (see "Ease of Integration" for details) • Rarely (if ever) requires direct modification to an application	DOCSIS BPI/BPI+ and 802.11 Wired Equivalent Privacy (WEP)

Table 7-1 The Level of Transparency of Various Security Protocols to Applications

the protocol. While these modifications are relatively simple for an experienced programmer with access to an application's source code and the right development tools, all applications must be enabled separately. As mentioned earlier in the section "Application Transparency," a number of SSL/TLS SDKs are available for adding encryption, integrity, and authentication services to an application. When using these SDKs, there is no need for you to modify the protocol stack, as the SDKs operate on the edge of the Transport layer. As a result, Application layer protocols are often the least difficult and invasive to implement when securing a single application or small number of applications.

NOTE:
For applications running on devices with limited storage, it is important that you keep the footprint of security code as small as possible. When multiple applications use different protocol implementations (that is, different SDKs), they waste precious storage space. In constrained environments, it is best to have a single shared API available for use by all applications.

Integrating Network layer (Layer 3) security protocols, such as IPSec, often involves modifications to an existing protocol stack. These modifications require direct access to the source code for the stack, as well as expertise in protocol stack development. An alternative approach that does not involve altering a protocol stack entails the use of a shim layer that splices into the protocol stack between the Network and Link layers. IPSec SDKs, such as RSA Security's BSAFE IPSEC-C and IPSec Express from SSH, implement IPSec using a shim layer (often termed a *packet interceptor*). We discuss both techniques later in the section "Security Protocol Implementation."

Link layer security protocols are often implemented within the *device driver* of a network interface card (NIC). The device driver controls the operation of the NIC and is responsible for performing any Link layer processing, such as adding protocol headers and trailers, before the NIC converts upper-layer data to a physical signal. Although Link layer protocols are transparent to the Network layer and do not require that you make modifications to an existing TCP/IP stack, they do require that you have some low-level knowledge of device-driver development. As you can see, the further down the protocol stack you move, the more specific the knowledge required to integrate a security protocol into a system.

NOTE:
The layer at which a protocol operates is not the only factor that determines the level of difficulty associated with implementing the protocol. A protocol like IPSec, which incorporates packet filtering and policy management, is generally more complex than SSL or S/MIME in terms of functionality; as a result, it will most likely require more effort to implement.

Configuration and Policy Settings

As you move down the protocol stack, configuration and policy settings become more general, meaning that they apply to traffic from more upper-layer communication protocols and applications. For example, SSL configuration settings, such as authentication methods and cipher suites, apply to a single communication session within one application. IPSec *configuration* settings, on the other hand, can apply to all upper-layer protocols and applications at the same time. This flexibility is due in part to the connectionless nature of IP. Connectionless protocols treat each data packet as an autonomous unit. This allows the security protocol to apply security to all packets sent to or from a host without having to consider the session or application to which the packet belongs. In addition, IPSec policy settings can apply to multiple upper-layer protocols and applications. Since IPSec has more visibility into the nature of a packet than upper-layer protocols (more protocol headers from which to obtain information), it can provide a detailed set of packet-filtering options; IPSec can filter packets based on their source and destination addresses, source and destination ports, Transport layer protocols, and user or system names.

Extent of Coverage

Some security protocols offer end-to-end protection, while others do not. Security protocols operating below the Network layer lack the logical addressing mechanisms necessary for end-to-end security in routed environments. At best, Link layer security protocols can protect only a single hop in the path between the source and destination hosts. Due to this limitation, Link layer protocols are often referred to as *point-to-point*. Those security protocols running at or above the Network layer—namely, IPSec, SSL/TLS, and S/MIME—can take advantage of the logical addressing provided by the Internet Protocol. Therefore, they *can* secure traffic that spans multiple network hops.

NOTE:

The flat (hardware) addressing scheme used by the Link layer limits security implemented within this layer to point-to-point connections. Security mechanisms must be implemented at or above the Network layer to provide true end-to-end protection in routed environments.

Referring back to our discussion in the section "Binding Security Services and Mechanisms to Data," what if end-to-end security also implies that we must protect stored data? This requirement is common in store-and-forward applications, such as e-mail (see Figure 7-4). Most network security protocols protect data *only* during transmission. Link, Network, and Transport layer security protocols decrypt all incoming data and remove any integrity and authentication information

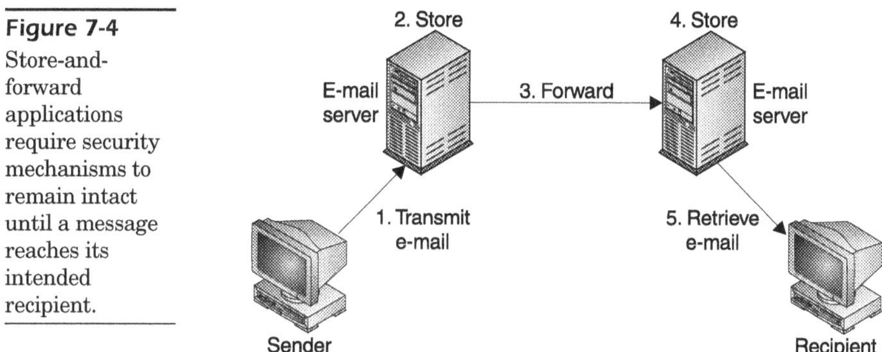

Figure 7-4 Store-and-forward applications require security mechanisms to remain intact until a message reaches its intended recipient.

prior to passing data up to the calling application. The lack of persistent security mechanisms prevents these protocols from securing stored data.

In the case of securing stored data, we are obliged to implement an Application layer solution. We can bind security services to data only by implementing security within the application responsible for generating and processing the data. S/MIME, for example, applies all security mechanisms to user data before passing the data to the protocol stack. When the receiving host processes the incoming data, its protocol stack strips all Link, Network, and Transport layer header information but leaves the S/MIME-related security mechanisms intact. The application decides whether to decrypt and verify the data immediately or place the data into storage. This makes S/MIME suitable for providing end-to-end security for e-mail and other store-and-forward applications where data may sit on a server indefinitely before being retrieved by the intended recipient. Figure 7-5 illustrates the extent of coverage offered by many popular security protocols.

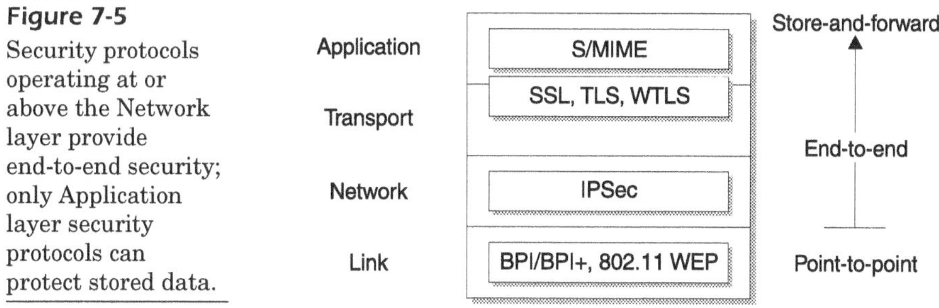

Figure 7-5 Security protocols operating at or above the Network layer provide end-to-end security; only Application layer security protocols can protect stored data.

S/MIME

In 1995, RSA Data Security, Inc., in collaboration with a number of major e-mail client vendors, developed *Secure Multipurpose Internet Mail Extensions (S/MIME)* for securing MIME messaging systems. The MIME standard defines a protocol for extending the functionality of text-based ASCII messages commonly used by e-mail systems. Among its most important features is an extensible set of formats for nontext (that is, multimedia) and multiple-part message bodies. The associated security protocol—S/MIME—was originally intended to add security to electronic mail applications, but S/MIME works equally well when applied to any communication system transporting MIME data.

S/MIME includes mechanisms for confidentiality, message-origin authentication and integrity, and nonrepudiation. RFC 2633 (Ramsdell, 1999) defines S/MIME v3, the most current version of the standard. According to RFC 2633, S/MIME messages are a collection of MIME message bodies and Cryptographic Message Syntax (CMS) objects. Since CMS is based on PKCS #7, you may sometimes hear S/MIME described as PKCS #7 with MIME headers. CMS (Housley, 1999) supports many content types; however, S/MIME v3 specifies the use of only three types—**Data**, **SignedData**, and **EnvelopedData**—for representing signed and encrypted messages. For example, an e-mail signed using S/MIME would consist of a SignedData message containing both a digital signature and a Data message component. The Data message contains the data over which the signature has been generated (that is, the original text or multimedia content of the e-mail). The **EnvelopedData** message type applies to data encrypted using the digital enveloping technique discussed in Chapter 2.

NOTE:

In 1996, Security Dynamics Technologies, Inc. acquired RSA Data Security, Inc. Security Dynamics later changed its name to RSA Security, Inc.

Performance

Largely, the performance of a security protocol relates little to the network layer at which it operates. Protocol performance depends on a number of factors including, but not limited to, the following:

- The complexity of the protocol (that is, overhead associated with establishing and maintaining a secure connection)
- The type and frequency of cryptographic operations
- The extent to which cryptographic algorithms have been optimized, and whether operations occur in hardware or software

- The network bandwidth consumed by protocol headers
- The level of policy management and processing granularity offered by the protocol (for example, IPSec provides many policy configuration options for filtering upper-layer traffic, whereas SSL incorporates no policy configuration other than choosing a cipher suite and turning on authentication.)
- Whether the protocol supports session caching to facilitate connection reestablishment
- The priority granted to the protocol by the operating system over other processes running on a host

We discuss the performance of cryptographic algorithms further in Chapter 8.

Comparing Existing Security Protocols

Each security protocol discussed in this chapter offers its own advantages and disadvantages. We have already discussed how some protocols are capable of providing end-to-end security, while others are not. We have also discussed how some protocols integrate more easily into certain network environments than others. Table 7-2 identifies many of the advantages and disadvantages associated with the more popular security protocols in use today.

Protocol	Advantages	Disadvantages
S/MIME	• Appropriate for store-and-forward applications and applications requiring end-to-end security • Does not require low-level modification to a protocol stack	• Each application must be enabled separately. • Requires access to application source code. • All configuration settings apply only to a single application.
SSL and TLS	• Can provide end-to-end security • Simple to add functionality to an application when using a third-party SDK • Can transparently secure multiple applications that use the same Transport layer communication protocol when implemented within a protocol stack	• Each application must be enabled separately when implemented outside a protocol stack. • Frequently requires access to application source code when implemented outside a protocol stack. • All configuration settings apply only to a single application.

Table 7-2 Advantages and Disadvantages of Several Popular Security Protocols

Protocol	Advantages	Disadvantages
IPSec	• Can provide end-to-end security • Can transparently secure multiple applications employing multiple Transport layer protocols • Detailed policy configuration based on addresses, ports, upper-layer protocols, and user and host names • A single policy setting can apply to multiple applications	• Overly-complex for simple applications. • True integration with a protocol stack requires access to source code and knowledge of protocol stack development.
BPI/BPI+ and 802.11 WEP	• Offers the greatest level of transparency; can transparently secure multiple applications that employ a variety of Transport and Network layer communication protocols	• Limited to point-to-point connections. • May require detailed knowledge of device-driver development.

Table 7-2 Advantages and Disadvantages of Several Popular Security Protocols *(continued)*

Security Protocol Implementation

Three primary methods exist for incorporating security protocols into any network environment: security protocols may be implemented as native code, a bump in the stack, or a bump in the wire. All three techniques offer advantages and disadvantages, and some are better suited to certain environments than others.

The most efficient method of implementing a security protocol is natively as part of the protocol stack. Unless the protocol stack ships with built-in support for the required security services, native implementation requires source-code-level modifications to the stack. Native code is controlled by the operating system of a host, and it operates within the network layer responsible for providing security. This format is common among routers, dedicated firewall and virtual private network (VPN) equipment, and other hardware security devices, since these devices require the high-performance processing offered by native implementations.

NOTE:
Because version 6 of the Internet Protocol designates the use of IPSec, some hosts running IPv6 may already include support for the IPSec Authentication Header (AH) and Encapsulating Security Payload (ESP) protocols.

When the network protocol stack of a host cannot be altered to support a security protocol natively, a *bump-in-the-stack (BITS)* implementation may be appropriate. For instance, security personnel may not have access to the source code for a protocol stack, or they may not wish to introduce complications or vulnerabilities by modifying the stack to support a new security protocol. The BITS approach permits security services to be inserted into an existing protocol stack using a shim layer. The shim layer splices into the stack just below the layer at which the security protocol operates. However, it does not modify the stack itself. Instead, the shim layer is responsible for intercepting incoming and outgoing protocol data units (PDUs) and applying the appropriate security services to the PDUs before passing them down to the next layer in the protocol stack. The BITS technique, illustrated in Figure 7-6, pertains mostly to securing hosts.

Many software-based IPSec clients, such as those from Cisco, Nortel, SafeNet, and Aventail, implement IPSec as a BITS. RSA Security's IPSEC-C SDK, which allows software developers to add IPSec functionality to any application, uses a BITS approach to implement IPSec. IPSEC-C consists of a number of modules for implementing the various components of IPSec, including modules for Internet Key Exchange (IKE), packet processing, policy management, certificate management, and cryptographic operations. The packet processing engine for IPSEC-C includes a *packet interceptor* (see Figure 7-7), which sits just below the Network layer of an IP stack. To the Network layer, the interceptor looks and acts like a network adapter driver. It captures packets exchanged between the Network and Link layers and hands the packets over to the packet processing engine. The packet processing engine applies the appropriate cryptographic transforms and passes the data back over to the packet interceptor. The interceptor then reinserts the "secured" packets into the appropriate level of the protocol stack (that is, the Link layer for outbound traffic and the Network layer for inbound traffic).

As an alternative to native code or BITS, the *bump-in-the-wire (BITW)* approach allows encryption to be transparently dropped into a network without making modifications to existing applications, hosts, or routers. BITW

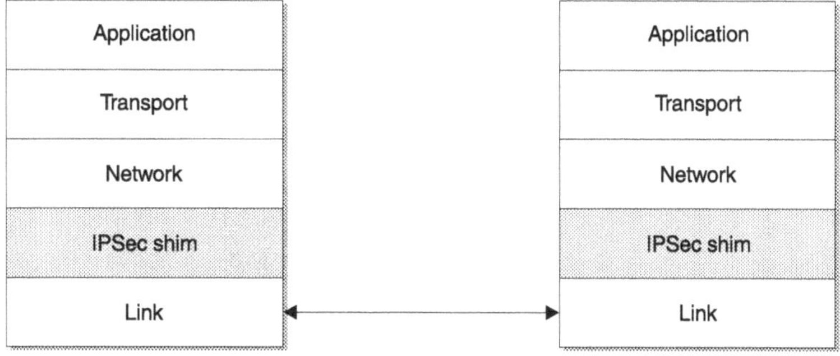

Figure 7-6 The BITS method allows security services to be inserted into an existing protocol stack without modifying the stack itself—in this example, the shim layer is for IPSec.

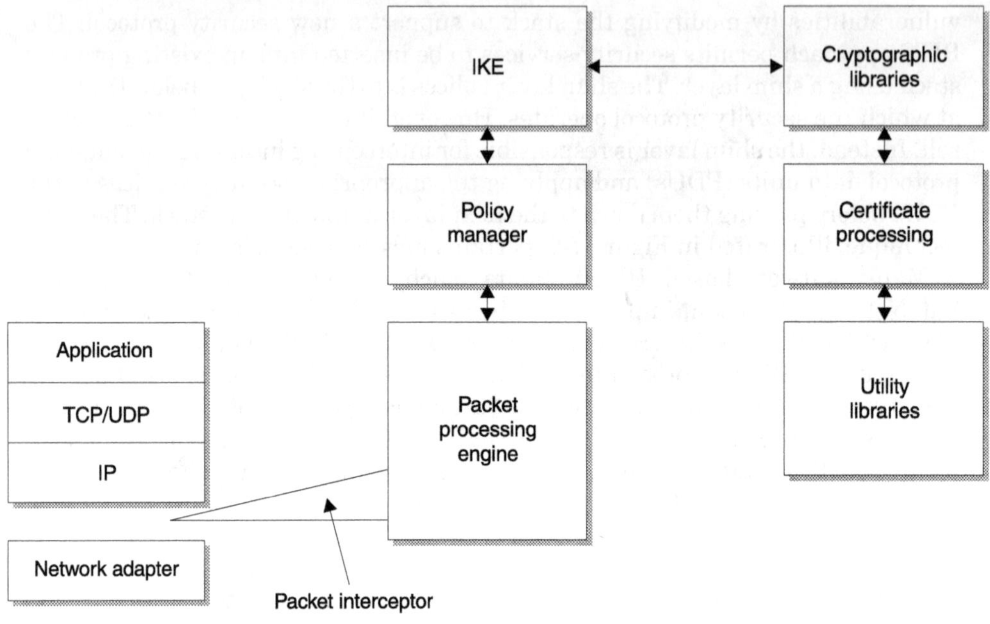

Figure 7-7 The IPSEC-C software development toolkit from RSA Security, Inc. consists of six separate modules. The packet interceptor operates as a BITS of the host operating system's protocol stack (RSA Security, 2001).

implementations use external cryptographic hardware devices or dedicated hosts acting as security gateways to encrypt all traffic to and from a host or network (see Figure 7-8). The most common hardware-based BITW implementations use high-performance, point-to-point line encryption devices to secure communication between two persistent endpoints. This approach is frequently employed by the military for its tamper resistance. Outboard BITW devices often have their own IP addresses; and to other machines on the network, they appear as addressable hosts. When used to secure the communication of a single

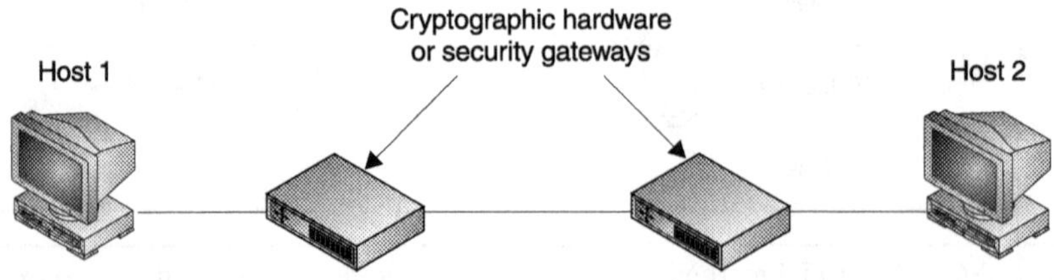

Figure 7-8 The BITW approach allows security services to be transparently added to a host or network. The security services are implemented within a hardware device or dedicated security gateway.

host, a BITW device usually appears as a BITS to the sending host. When used in conjunction with a router, a BITW device operates as a separate security gateway. The former approach allows a BITW device supporting IPSec to operate in both transport and tunnel mode, while the latter approach limits the functionality of the BITW device to tunnel mode only.

Keep in mind the effects of tunnel mode operation when implementing IPSec using the BITW method. If the BITW device operates as a security gateway in tunnel mode, it adds an additional 20 bytes of header data to each packet. This increases the header-to-payload ratio, and it may adversely affect network performance (especially in communication systems transmitting large quantities of small packets).

TIP:

On an Ethernet network, a BITW device or another security gateway may be forced to fragment a packet when adding the additional IP header required for tunnel mode operation. This occurs when the total size of the packet (outer IP header + inner IP header + encapsulated data) exceeds 1500 bytes. Fragmentation not only incurs additional processing time but also consumes network bandwidth by increasing header-to-payload ratios. Decreasing the maximum IP packet size for all other hosts on the network to less than 1480 bytes may help to avoid this problem. Although a host may still have to fragment large IP packets, the resulting fragments will not undergo additional fragmentation on the security gateway.

Host-Based Security vs. Security Gateways

In addition to choosing the appropriate network layer at which to place a security protocol, you must also decide whether to implement security services within security gateways or within each individual host. A *security gateway* is a dedicated host or hardware device responsible for securing network traffic to and from a network or other hosts and devices. The use of security gateways provides a number of benefits, and a number of drawbacks, over host-based security when it comes to coverage, transparency, implementation, configuration, and maintenance. Your use of host-based security versus security gateways will likely depend on the quantity and type of applications you are securing, the level of effort and cost associated with securing crucial applications, and whether the communication and security endpoints coincide. Keep in mind that the best solution may employ a combination of host-based security and gateways.

NOTE:

In this section, the term "host" encompasses both general-purpose computers and end-user devices, such as modems and set-top boxes.

Extent of Coverage

The first major characteristic distinguishing host-based security from an approach employing security gateways is the extent of coverage: only host-based security offers true end-to-end protection (see Figure 7-9). For two communicating parties to guarantee end-to-end security, they must communicate using devices that support the necessary security services; they cannot rely on third-party hosts or devices to implement these services. All privacy, integrity, authentication, and nonrepudiation mechanisms should be applied by the source host prior to transmitting data packets or messages over a network and should not be removed until the data reaches its final destination. This prevents intermediate devices in the path between the source and destination from decrypting and modifying the data or altering its source.

NOTE:

When using symmetric cryptography, end-to-end security implies that only the sender and recipient share the information (such as encryption and authentication keys) necessary to decrypt and verify the authenticity of data exchanged between them. In the case of public-key cryptography, end-to-end security dictates that key exchanges occur only between end-user devices, and that any digital signature generation takes place directly on the source host.

Although it provides end-to-end protection, host-based security cannot protect against traffic flow analysis. Going back to our discussion of IPSec in Chapter 6, traffic flow confidentiality necessitates the use of security gateways to conceal the source and destination addresses contained within protocol headers. In the case of tunnel-mode IPSec, a security gateway encrypts all IP packets (IP header and en-

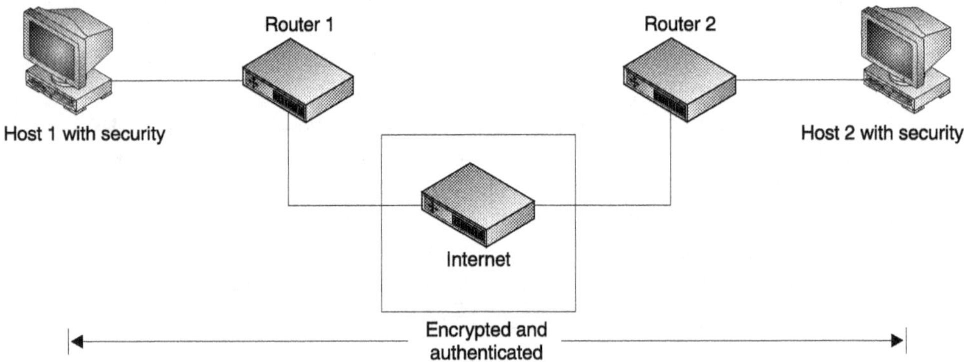

Figure 7-9 Only host-based security offers true end-to-end protection of data.

capsulated upper-layer data) leaving a network and encapsulates each encrypted packet within a new IP header. Since the header contains the source address of the gateway, the true source of the packet is disguised. If the destination address of the outer IP header is another security gateway, the true destination address is also masked. This is something that host-based security cannot accomplish.

While security gateways can protect against traffic-flow analysis, they cannot provide end-to-end security in the absence of host-based security. At best, they can protect only a subset of the path between the source and destination. For instance, a VPN typically protects data flowing between two or more remote locations over a public network, as illustrated in Figure 7-10. As data leaves the internal network of Site A, a VPN router applies encryption and authentication algorithms to all packets and forwards them to their final destination in Site B. Upon reaching the destination network, another VPN router strips the encryption and authentication information and passes the packets onto the internal network at Site B. However, neither VPN router protects network traffic after it enters the internal network at either site. Without the appropriate perimeter defenses (that is, a strong firewall and intrusion detection system) or physical security mechanisms (such as security guards and routine physical inspections), an attacker may be able to install a packet sniffer just inside the gateway and capture all traffic to and from the VPN connection. When using security gateways, there will always be a point in the path between the sender and recipient where the data is unprotected (unless the gateways are used in conjunction with host-based security). An attacker who has managed to compromise a security gateway may even be able to monitor traffic from the gateway itself.

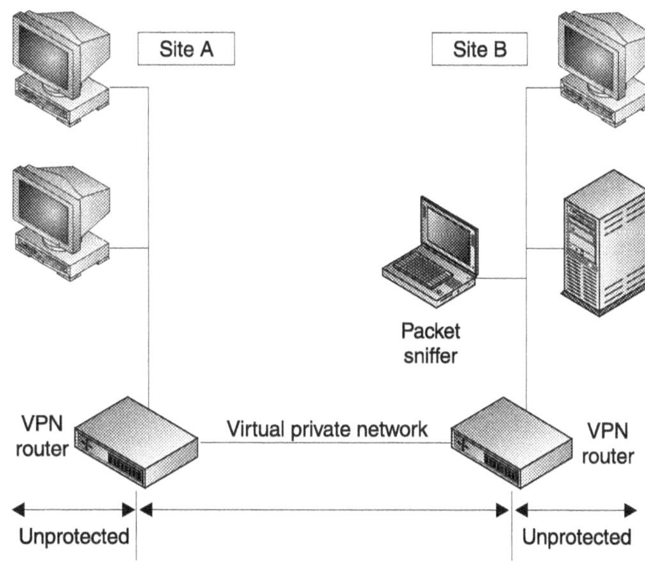

Figure 7-10 When using security gateways, it may be possible for an attacker to capture network traffic as it passes from a gateway to an internal network.

The WAP Gap

The *Wireless Application Protocol (WAP) Gap* is one of the most publicized, and probably one of the most exaggerated, security design flaws in recent times. As we will soon discuss, the problem with WAP stems from the use of a security gateway to simulate end-to-end security.

Before discussing the security aspects of WAP, let's first introduce the protocol itself. To facilitate the delivery of services and information to mobile wireless devices (such as mobile phones, pagers, and two-way radios), the WAP Forum developed the Wireless Application Protocol. WAP defines a communication protocol and an application environment aimed at devices with limited processing power and available memory, and narrowband network connections (typically 19.2 KBs or less). One of the first services supported by WAP was the delivery of Web-based content to *micro-browsers* embedded within mobile devices. Instead of using the Hyptertext Markup Language (HTML), WAP formats Web pages according to the *Wireless Markup Language (WML)* specification. When formatting Web pages, WML takes into account the small screen area and computational resources available on most handheld mobile devices. Since most micro-browsers support only WML, the WAP infrastructure relies on gateways to strip all HTML formatting and replace it with WML.

Along with WML, WAP incorporates a Transport layer security protocol known as *Wireless Transport Layer Security (WTLS)*. WTLS, which is based on the TLS protocol, provides encryption, integrity, and authentication mechanisms similar in functionality to TLS, but it eases some of the computational burden associated with establishing a secure session on devices with limited processing power and low-bandwidth connections. WTLS also incorporates support for dynamic key refreshing and datagram delivery using unreliable transport protocols (WAP Forum, 2001).

The problem with WAP does not lie in the WTLS protocol, but in how WAP attempts to simulate the end-to-end security offered by TLS in the wired world. WTLS and TLS are not compatible, so a mobile device cannot open a secure connection to an SSL/TLS-enabled web server using WTLS. Instead, the mobile device must first contact a WAP gateway. The gateway acts as a proxy by establishing a secure connection to the web server and relaying data from the mobile device to the web server (see Figure 7-11). The basic process is as follows:

1. A mobile device establishes a secure WTLS session with the WAP gateway.
2. The WAP gateway determines the host to which the mobile device wants to connect and establishes an SSL/TLS session with the destination web server.
3. The gateway decrypts data from the mobile device and stores the data within the gateway's volatile memory.
4. The gateway then re-encrypts and transmits the data under the newly established SSL/TLS session.

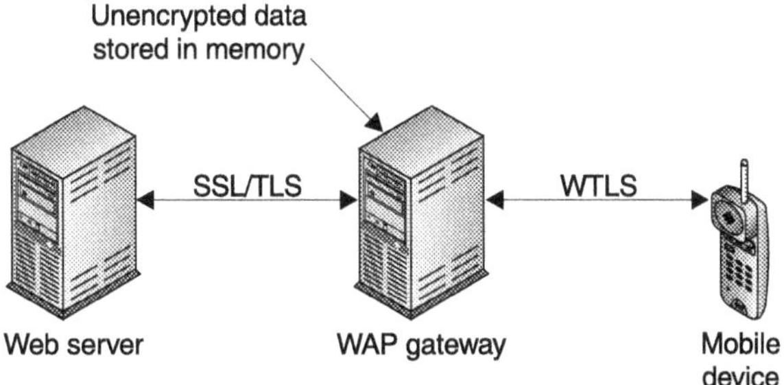

Figure 7-11 A WAP gateway must decrypt and store sensitive information within memory as it proxies WTLS and SSL/TLS sessions between Web servers and mobile devices, resulting in the infamous WAP Gap. An attacker who has compromised the gateway may view or capture the unencrypted data as it traverses the WAP Gap.

Do you see a problem with this approach? The WAP gateway must place unencrypted data into memory between steps 3 and 4—hence the term "WAP Gap." The fear of many IT security professionals—especially in the financial industry—is that sensitive data could be captured by performing a dump of the gateway's memory in the event that the gateway is compromised. No matter how small the possibility of an attacker compromising a WAP gateway, many people felt that the risk might be too high for use in the communication of large financial transactions or highly sensitive data. Keep in mind that although the end-user device implements the appropriate security services, it breaks the cardinal rule of end-to-end security by relying on an intermediate device to convert from WTLS to SSL/TLS. For more information on WAP and WTLS, visit http://www.wapforum.org.

NOTE:

The iMode phones produced by NTT DoCoMo, a large Japanese mobile phone manufacturer, natively support SSL/TLS. Because they do not require a security gateway, they provide true end-to-end security.

Implementation, Configuration, and Maintenance

While security gateways do not provide end-to-end security, they do offer many advantages over host-based security in terms of implementation, configuration, and maintenance. One important advantage is transparency. The use of security gateways does not require modifications to end-user hosts, devices, or applications. When implementing security within a router, a dedicated VPN device, or another security gateway, the end-user hosts and devices need not be aware of the security provided by the gateway. They simply transmit data packets to the intended recipient, and it is the responsibility of the gateway to apply the necessary cryptographic transforms (such as encryption and authentication algorithms).

Use of security gateways also centralizes policy administration. For instance, if a corporate IT department were to decide that DES encryption is too weak for remote users coming into the network, they may decide to change the security policy to employ 3DES or RC5. Instead of individually reconfiguring each host to adhere to the policy change, an administrator would need to update only a single gateway to reflect the change. Granted, each remote user coming into the internal network from outside must still reconfigure his VPN client software; however, no changes are necessary to hosts on the internal network.

In addition to simplified installation and configuration, security gateways reduce maintenance costs. For instance, each time a new vulnerability is discovered in an operating system or protocol stack, every host running that version of the software must be patched. Automated tools exist for deploying the patches, but they do not always work flawlessly. It is then the responsibility of human administrators to troubleshoot and complete failed upgrades. When a vulnerability is discovered in a security gateway, only the gateway need be patched.

CAUTION:

Although upgrading, patching, and maintaining a single gateway requires less time and effort than a host-based solution, security gateways frequently provide a single point of failure in a network. A misconfigured security gateway (or one containing a vulnerability) can compromise or halt the operation of an entire network.

Securing Traffic Between a Large Number of Hosts or Applications

Security gateways are frequently used when there is a need to encrypt traffic between a large number of hosts or applications residing within a small number of networks. If, in the VPN example discussed earlier in this chapter, numerous hosts at Site A exchange information with many hosts at Site B, the use of security gateways makes sense. Not only can the security services provided by security gateways protect traffic originating from multiple hosts, but the services can be inserted transparently. A security gateway is also a sensible solution when securing several disparate applications, because security gateways offer a simpler and often less costly solution than enabling security separately within each application. However, if only a single pair or a handful of hosts need to exchange traffic securely, it may be difficult to justify the expense of a dedicated VPN. A better solution might involve implementing a host-based solution. If two hosts exchange information within the context of a single application, SSL provides the simplest solution. Conversely, if the hosts must exchange data belonging to multiple applications, an IPSec solution might make more sense. Remember that IPSec allows you to secure multiple upper-layer applications transparently without directly modifying the applications.

Table 7-3 presents a number of scenarios and possible solutions for protecting communication between two networks.

Distinct Traffic Flows

When implementing security within a host, it is relatively easy for you to protect individual traffic flows. Even within the same application layer protocol (HTTP, FTP, Telnet, and so on), a host can choose to secure a session when exchanging highly sensitive information and can bypass security altogether for less confidential data. Security gateways do not offer such flexibility. Depending on the policy settings configured on the gateway, all traffic to and from the same source and destination addresses or traffic using the same upper-layer protocols must share similar security settings. A gateway cannot distinguish between individual HTTP sessions, for instance. The ability to choose when to apply security to a communication session helps avoid the overhead associated with a security protocol when security is not necessary.

User Contexts

Routers and other security gateways often do not maintain user contexts for each connection they proxy (Doraswamy and Harkins, 1999). (A *user context* includes identity information necessary for authentication during session establishment.) For a large number of sessions, this requires considerable memory. However, a host can easily maintain a context for each session it initiates. This is useful for environments in which a host must repeatedly authenticate to establish multiple secure connections. Without maintaining user contexts, a security gateway has no way of authenticating a session for a host. For this reason, a security gateway is rarely used in lieu of host-based security to establish a connection directly with a destination host. Instead, the source and destination hosts secure communication using a host-based security solution.

Table 7-3 Scenarios and Possible Solutions for Securing Traffic Exchanges Between Remote Networks

Scenario	Possible Solution
Large number of hosts with a large number of applications	Security gateways (a dedicated VPN, for example)
Large number of hosts with a small number of applications (that is, no more than two or three applications)	Security gateways or a host-based solution using SSL
Small number of hosts with a large number of applications	Host-based solution using IPSec
Small number of hosts with a small number of applications	Host-based solution using SSL

Coordination with Existing Security Policy

The security policy of many organizations prohibits the establishment of arbitrary VPN tunnels between hosts on their internal network and hosts external to the network. For example, most organizations configure their firewalls to pass Web-based traffic over ports 80 and 443 and FTP traffic over port 21, but they may be hesitant to open additional ports. This often limits the effectiveness of host-based security protocols. Unless the appropriate ports have been opened on the firewall the secure traffic will not be allowed to pass. Some applications overcome this restriction by using HTTP over port 80 as a transport protocol. The firewall allows the data to pass because, to the firewall, the data looks as though it was generated by a Web browser or Web server. The same is true with traffic sent over port 443 that is protected with SSL and TLS. Most firewalls are configured to allow SSL/TLS-protected traffic to pass unhindered, as long as it uses the correct port(s). Most corporate IT departments prefer the use of a dedicated VPN solution over host-based security when it comes to protecting traffic flowing into and out of their networks.

Table 7-4 compares and contrasts host-based security and security gateways.

Characteristics	Host-Based Security	Security Gateway
Extent of coverage	Provides end-to-end security between source and destination hosts.	Can protect only a subset of the path between the source and destination hosts.
Operational modes available to host or gateway	Transport and tunnel modes.	Tunnel mode only.
Effort associated with integration, configuration, and maintenance	• Must implement security within each individual host. • Must configure and maintain all hosts separately.	• Implementation is transparent to end-user hosts. • Provides centralized policy management. • Only the gateway must be maintained (application of security patches, bug fixes, and upgrades).
Support for distinct traffic flows	• Can secure individual traffic flows. • Security settings may vary from one session to the next even within the same application.	All traffic to/from a particular host or traffic using the same upper-layer protocols must share similar security.
Maintenance of user contexts	Host can maintain user contexts to facilitate authentication during session establishment.	Gateways generally do not maintain a user context for each session.

Table 7-4 A Comparison of Host-Based Security and Security Gateways

A Final Word on Encryption and Protocol Headers

The use of transport mode (end-to-end) security services implies that the critical header fields—source and destination addresses—necessary for routing a data packet throughout a network cannot be encrypted by the source host. If encrypted, intermediate routing devices would have no way of determining the information needed to route the packets to their destinations. This prevents transport mode encryption services from being used to protect against traffic-flow analysis.

On the other hand, tunnel-mode encryption may conceal quality-of-service (QoS) information carried within encapsulated protocol headers making all data packets appear as though they have the same priority. The *type of service* field within IPv4 headers and the *traffic class* field within IPv6 headers are used by routers in differentiated services (DiffServ) environments to categorize IP packets. Routers give those packets belonging to high-priority applications preferential treatment over best-effort traffic. However, tunnel-mode encryption, which conceals the original IP header of all packets from the source host or network, may affect how routing devices treat the packets. As a result, the QoS afforded to a packet may not meet prescribed service-level agreements. For network applications relying on a minimum level of end-to-end QoS, it is best to use transport mode encryption.

References

Doraswamy, Naganand, and Dan Harkins. 1999. *IPSec: The New Security Standard for the Internet, Intranets, and Virtual Private Networks*. New Jersey: Prentice-Hall PTR.

Housley, Russ. June, 1999. *RFC 2630: Cryptographic Message Syntax*. Internet Engineering Task Force (IETF).

Ramsdell, B. June, 1999. *RFC 2633: S/MIME Version 3 Message Specification*. Internet Engineering Task Force (IETF).

RSA Security. June, 2001. *RSA BSAFE IPSEC-C: Security Protocol Components for C - Quick Start Guide v1.0*. RSA Security, Inc.

WAP Forum. April, 2001. *Wireless Transport Layer Security Version 06-Apr-2001*. WAP Forum.

Chapter 8

Security Side Effects

Up to this point, we have discussed various security services and cryptographic mechanisms, and have acquainted ourselves with many popular security protocols. However, we have not addressed the effects of security on network and device performance, cost, and manageability. Security is not free, nor is it completely transparent. Seamless integration into a broadband network environment requires effort on the part of designers, implementers, and administrators. It is our duty as security architects to moderate the adverse effects of security so that our solutions are easy for users and providers to accept, implement, and maintain.

This chapter is intended as a practical tool for gauging the impact of cryptographic security mechanisms and protocols on broadband service operation. Here, we will address network performance concerns relating to quality of service (QoS), and introduce the operating constraints present in embedded devices. We'll then discuss in detail the performance characteristics of many common cryptographic algorithms, and review some guidelines for selecting the appropriate algorithms for your environment. We'll follow up with some tips on high-performance security protocol design and usage, and finish the chapter by discussing some important security management considerations.

Network Performance and QoS

To facilitate the successful and efficient operation of network services, a security infrastructure must not squander network resources or hinder the delivery of application content. Real-time multimedia services, such as audio- and video-on-demand, voice telephony, and video conferencing, call for creative security solutions that account for the specific QoS characteristics of each network application. In Chapter 4, we introduced a number of metrics for measuring QoS, such as bandwidth, latency, jitter, packet loss, and availability. While security provides obvious benefits in terms of availability by reducing the likelihood of denial of service (DoS) attacks, it adversely impacts most other aspects of QoS:

- *Security consumes bandwidth and reduces throughput.* Service providers rarely over-provision their networks. Over-provisioning is costly and inefficient, because it does not account for usage patterns. An over-provisioned network always offers surplus bandwidth (and other network resources), regardless of when and where the bandwidth is needed. A more common approach to improving overall QoS is to manage access to a limited set of resources by assigning each application (or traffic flow) a particular level of QoS. Those applications with higher priorities will be given preferential treatment over low-priority applications when competing for shared network resources. While this method of provisioning offers a more efficient solution to resource management than does over-provisioning, it requires creative security solutions that minimize wasted bandwidth and maximize throughput on network devices. As we will discuss in the upcoming section "Cryptography and Performance," security mechanisms—cryptographic mechanisms in particular—often consume valuable bandwidth when used improperly, and can severely limit the rate at which network devices transmit data.

- *Security introduces latency and jitter.* The initialization of security mechanisms may result in unacceptable wait times during device provisioning or when an application is loaded. The effects are compounded when the process involves numerous security protocol exchanges over a busy network or multiple public-key cryptographic operations. In addition to provisioning delays and increased load times, the overhead associated with reestablishing a secure communication session or renegotiating security parameters (key refreshes and renewals, for example) may introduce latency in the delivery of application content.

  Real-time multimedia applications are extremely sensitive to jitter (variations in inter-packet arrival or playback times). Since security mechanisms (especially those based on cryptography) add to signal processing times on end devices, they may contribute to jitter. Jitter also results from packet processing on routers and security gateways.

- *Security may contribute to packet loss.* Packet loss occurs when a network device, such as a router, cannot process incoming data packets as fast as it receives them. Most real-time multimedia applications require a packet-loss ratio of well under five percent. Retransmission is not an option for these applications because it introduces delay and jitter (in most cases, lost packets are ignored). Cryptographic operations and the application of security policy in network devices slow the rate at which these devices can process incoming data; this reduction in speed can lead to packet loss.

In Chapter 4, we listed the bandwidth, latency, jitter, and packet-loss requirements for a number of real-time network applications. Keep those values in mind when selecting cryptographic security mechanisms.

Embedded Device Constraints

As if maintaining network performance and QoS weren't difficult enough, the embedded world tosses in a few more challenges. Consumer devices, such as mobile phones, modems, and set-top boxes, typically have limited processing power, RAM, and storage. Limited resources restrict the types of security mechanism these devices are capable of supporting. To demonstrate, Table 8-1 lists the resources available for some popular embedded consumer devices.

NOTE:
Aside from smart cards, most of the processing power and memory of the devices lists in Table 8-1 are dedicated to the devices' primary purpose, not security. For instance, a gaming console exhausts much of its processing resources on maintaining game states, displaying video, and playing audio. This leaves few resources for cryptographic and other security-related operations.

Device	Processor	RAM	Flash/Storage
Smart card	1–5 MHz	512 bytes	2–16 KB
PDAs	16–33 MHz	96–256 KB	2–8 MB
Cable modems and set-top boxes	80–100 MHz	8–16 MB RAM	2–4 MB
Gaming consoles	33–300 MHz	2–32 MB	512+ KB

Table 8-1 Typical Embedded Device Resources

Efficient security design reduces the costs associated with manufacturing embedded devices. Fast security mechanisms reduce the processing power required to implement security, which makes the use of less expensive processors possible. Efficient software-based approaches may mean the difference between implementing security solely within software or having to incorporate costly hardware-based solutions. (We'll discuss cryptographic hardware later, in "Dedicated Cryptographic Hardware.") Compact security code further diminishes production costs by reducing the amount of flash memory required to store software images. Even with a difference of only pennies per device, the cost savings quickly add up during mass production of consumer products.

High-end network equipment, such as routers and dedicated VPN hardware, are not as constrained as the devices listed in Table 8-1. However, this does not mean that high-performance security isn't a necessity for these devices as well. While routers and VPN hardware generally contain faster processors and more RAM and flash memory, they must process much more data far more quickly than must typical end user devices. Consequently, high-performance security is just as important in service provider equipment as it is in home consumer products.

Cryptography and Performance

Once you have determined the security services appropriate for protecting your network, you must select cryptographic mechanisms to implement those services. Part of the decision-making process will involve choosing the necessary cryptographic algorithms. Your selections should be based on a number of factors, including the overall functionality, level of security, and operational characteristics (performance, code size, and memory usage) of each algorithm. Since most cryptographic operations involve complex mathematical calculations that consume valuable processing time and resources, the selection of high-performance cryptographic algorithms is critical to the successful operation of broadband network services. This is especially true in the case of real-time multimedia applications. However, it is not always a matter of selecting the fastest algorithm, since some situations call for functionality that can only be provided by a particular type or set of algorithms.

This section addresses many of the key decision-making considerations that go into selecting symmetric, public-key, and message digest algorithms. It also presents actual performance figures for various cryptographic algorithms in terms of raw encryption and decryption rates, key and parameter generation times, and signature generation and verification times.

CAUTION:
In this section, we present a number of performance metrics for common cryptographic algorithms. These figures are intended as rough estimates and are somewhat conservative. It's always a good idea to test cryptographic implementations on your specific processor and operating system platform, and you should not base your design decisions solely on the material presented here. (Note that all performance figures in this section are for memory-to-memory operations.)

General Considerations for Choosing Cryptographic Algorithms

All cryptographic algorithms share a common set of characteristics that influence the selection process. When choosing cryptographic algorithms to implement a security mechanism, you should consider the following:

- **Known weaknesses** Some algorithms include design flaws or shortcomings that limit their effectiveness. Two examples help to illustrate this point:
    - DES is no longer considered secure for *long-term* encryption of *highly sensitive* data due to its use of relatively short 56-bit keys.
    - Although MD4 is one of the fastest message digests available, cryptanalysts have discovered a number of exploitable vulnerabilities in its design.
- **Data processing rates** Processing rates directly affect the throughput of network devices and can introduce signal latency and jitter. Some encryption algorithms and message digests may be too slow for a particular application. For instance, the encryption and decryption rates for public-key algorithms are generally one to two orders of magnitude less than symmetric ciphers. This severely limits their effectiveness for bulk data encryption or the encryption of streaming multimedia. Even among public-key algorithms, the times required for individual private-key and public-key operations (signature generation and verification, for instance) tend to vary considerably. When encrypting and authenticating a media stream, the processing rate of ciphers and message digests should be greater than the throughput required for transmission of the media stream. Later in this section, we'll list the data processing rates of symmetric and public-key algorithms.
- **Key generation and setup times** Key-generation and key-setup times contribute to delay during session establishment, and to jitter during key refreshes and renewals. Longer keys usually result in more lengthy key-generation and key-setup times, although the setup times

for some algorithms, such as RC6, remain constant regardless of the key length chosen. In general, key-setup times are much less for symmetric ciphers than for public-key algorithms. Keep in mind that the key-generation times for some of the public-key algorithms can become quite lengthy with larger key sizes.

TIP:
To avoid introducing signal latency and jitter during key setup, you may wish to perform the setup operations ahead of time, so that the keys are available before they are needed.

- **Code footprint** Code size varies considerably depending on the type of algorithm and the specific vendor implementation chosen, and affects the amount of flash memory required to store a software image. Generally, there are tradeoffs between algorithm performance and code size. With programming techniques, such as unrolling *for* loops, using inline code instead of function calls, and increasing the size of precomputed lookup tables for cryptographic operations, performance can be improved at the cost of additional code size. (See Schneier and Whiting, 1997 and Pfitzmann and Aßmann, 1993 for more information on these optimization techniques.) Some cryptographic toolkits, such as RSA Security's line of micro edition products (BSAFE Crypto-C ME, Cert-C ME, and SSL-C ME), provide multiple versions of many popular algorithms; some versions are optimized for performance, while others are optimized for code size.
- **RAM consumption** Like code footprint, RAM consumption varies somewhat from algorithm to algorithm. In addition to the memory required for encryption and decryption operations, the key context for some algorithms can be relatively large (1–2KB or more). This may be an issue for memory-constrained devices when maintaining multiple key contexts.

NOTE:
A key context consists of the information and state variables required for the derivation of encryption keys.

- **Processor-specific optimizations** Some algorithms are designed to perform well when running on particular types of processors. For example, Rijndael and RC6 offer similar functionality, but they were designed with different processing environments in mind. The designers of Rijndael were more concerned with operation on 8-bit processors, such as those found in smart cards. The creators of RC6, on the other hand, focused on 32-bit and 64-bit processors, having anticipated the importance of these processors in future computing environments (Robshaw, 2001).

Furthermore, many cryptographic operations (especially those for public-key cryptosystems) involve the manipulation of very large integer values. For this reason, processors offering high-performance integer calculations will often outperform those that focus heavily on floating point operations. (Integer operations involve whole numbers, whereas floating point calculations involve fractions.)

NOTE:
Both Rijndael and RC6 were finalists for the Advanced Encryption Standard (AES). While both algorithms met the strict AES selection criteria, the AES selection committee eventually chose Rijndael as the standard.

Symmetric Ciphers

Symmetric ciphers are typically used for traffic and bulk data encryption, and can be used in conjunction with public-key algorithms for high-performance key exchange. They offer much better performance than public-key algorithms in terms of raw encryption and decryption rates. In addition to the general characteristics mentioned earlier in this section, the following considerations apply directly to selecting symmetric encryption algorithms:

Which type of symmetric cipher—block or stream—is more appropriate for a given scenario? In general, stream ciphers are very fast, but they do not scale well when encrypting a sizeable number of messages or data packets individually. Since you should never reuse keying material with a stream cipher, each message or packet must use a completely different encryption key. Stream ciphers are much better suited for session-based traffic encryption where a single key (and keystream) may be used to encrypt an entire communication session from beginning to end as one contiguous stream of data. Block ciphers, on the other hand, do not share the same key reuse restrictions as stream ciphers. When the situation calls for encrypting the contents of multiple messages, or the payloads of a large number of data packets separately using the same key, then a block cipher is probably the answer.

Stream ciphers are also notably difficult to synchronize in network environments that are prone to packet loss and that do not retransmit lost packets (for example, networks that use connectionless and unreliable Transport layer protocols, such as UDP). Stream ciphers can easily desynchronize when encrypting an entire communication session as one contiguous media stream. Loss of a single packet causes the keystreams of the sender and recipient to desynchronize by an amount equal to the length of the encrypted data contained within the packet. Once the keystreams desynchronize, all future decryption operations fail until the keystreams are resynchronized (and accomplishing this can be a very difficult task). This limits the effectiveness of stream ciphers for encrypting QoS-assured, real-time traffic, because resynchronization negatively impacts content delivery.

Block ciphers are much easier to keep synchronized because the same key can be used to individually encrypt each packet. Since only the IV is needed for synchronization, and it is usually contained within the packet, block ciphers rarely desynchronize.

NOTE:

The only time when block ciphers might desynchronize is during key-renewal operations when two communicating parties begin using newly negotiated encryption keys at different times.

What are the effects of key size and number of encryption rounds on algorithm performance? For those algorithms supporting multiple key sizes, you can always improve security by selecting larger keys. Some algorithms, such as RC5 and RC6, even allow you to adjust the number of encryption rounds to improve security. The added security, however, comes with a price. The use of larger keys and more encryption rounds generally slows an algorithm down, so you must strike a balance between QoS and the level of security by choosing appropriate values for those parameters.

NOTE:

The RC4 stream cipher allows you to increase the key size without affecting the throughput (raw encryption and decryption rates) of the algorithm. While larger keys may not affect performance, they do result in longer RC4 key setup times.

Which block cipher mode offers the best performance? Depending on the processing environment and cryptographic algorithm chosen, each block cipher mode offers a different level of performance. The various modes in order of decreasing performance on a single processor system are ECB, CBC, CFB, and OFB.

Moreover, some block cipher modes allow parallelization of encryption and decryption operations in multiprocessor environments. ECB mode supports parallelization during both encryption and decryption, which means that multiple blocks of data can be encrypted or decrypted simultaneously. CBC and CFB modes support parallelization during decryption only, and OFB mode supports neither parallelized encryption nor parallelized decryption. Keep in mind that, in addition to performance, each block cipher mode offers different security characteristics in terms of concealing plaintext patterns and the level of error propagation.

NOTE:

As mentioned in Chapter 2, the throughput of n-*bit CFB and OFB mode ciphers is reduced by a factor of* m / n *compared to CBC mode, where* m *is the block size of the cipher, and* n *is the number of bits encrypted per encryption cycle (Menezes, Oorschot, and Vanstone, 1996).*

What are the effects of padding and IVs on available network bandwidth? With block ciphers, the space consumed by padding and initialization vectors (IVs) may become significant when many small messages are encrypted individually. Padding is required when the plaintext data being encrypted is not a multiple of the block size of the block cipher. When 8-byte blocks are used, as much as 8 bytes of padding may be required prior to encryption. Each message should also contain a unique IV equal in length to the block size of the cipher, and this will consume additional bandwidth.

For example, imagine that you are encrypting 53-byte quantities using a cipher with a block size of 8 bytes. (Note that ATM uses 53-byte cells.) Each encryption operation would add 3 bytes of padding plus 8 bytes for the IV. After 100,000 such datasets were encrypted, 1,100,000 bytes of bandwidth (or 17.19 percent of each message) will have been consumed by padding and IVs. This wouldn't even account for the bandwidth required for protocol headers. What if you were encrypting 56-byte quantities using the same block size? In this case, the padding and IV would consume 16 bytes (or 23.19 percent) of each message, resulting in 1,600,000 bytes of wasted bandwidth.

NOTE:

*According to version 2 of the PKCS #5 standard (RSA Laboratories, 1999), padding is required regardless of whether the amount of data being encrypted is a multiple of the block size. PKCS #5 specifies that each encrypted message contain 8– (*M *mod 8) bytes of padding, where* M *is the length of the plaintext data. Therefore, the largest amount of padding occurs when the plaintext data is a multiple of the block size. Why is this so, you ask? All messages must follow an identical syntax; otherwise, the recipient will have no way of determining where the encrypted payload of a message ends and the padding begins. The sender, then, must always specify some value for padding. For a message whose length is a multiple of the block size, the extra padding information overflows into a new block. The sender must then add enough padding to complete the additional block (which happens to be 8 bytes in size).*

When there is a need to conserve as much bandwidth as possible, a stream cipher may be a better choice. Stream ciphers generally do not require padding, so their output is identical in length to their input. There also exist nonstandard block cipher modes of operation, such as *offset codebook (OCB),* whose input and output lengths are nearly identical.

TIP:

Some nonstandard block cipher modes provide benefits over those discussed in Chapter 2. For instance, counter (CTR) *mode conceals plaintext patterns while allowing for the parallelization of encryption and decryption operations (Lipmaa, Rogaway, and Wagner, 2001). OCB mode efficiently combines encryption and message authentication into a single operation, eliminating the need for separate encryption and MACing algorithms (Rogaway et al., 2001).*

- **How do the code sizes of the various symmetric ciphers compare?** Stream ciphers, such as RC4, are generally the simplest to implement, and result in the smallest amount of code. A reference application that includes 8–10KB of program overhead for performing RC4 encryption and decryption may run anywhere between 30KB and 55KB. The complexity of some block ciphers, such as DES and 3DES, increases their footprints somewhat when they are implemented in software. The difference in the code footprint for the various symmetric ciphers varies by 10–25KB depending on the algorithm and the particular implementation.

Tables 8-2 and 8-3 list the raw encryption and decryption rates and key setup times for many popular symmetric ciphers on various processor and operating system platforms. The ranges provided in the table account for the different block cipher modes.

Algorithm	Pentium II 350 MHz Running Windows 2000 (MB/sec)	SPARC Ultra 400 MHz Running Solaris 2.8 (MB/sec)	StrongARM 200 MHz Running Windows CE 3.0 (MB/sec)
DES	7–8	5.5–6.5	1.5–1.8
3DES	2.5–3.0	2–2.5	0.75–0.90
RC4	50–55	35–38	9–11
RC5 (16 rounds)	20–23	13–15	4–5
RC6[1] (20 rounds)	12–14	6–8	2.8–3.5
AES (128 bits)	13–15	9–10	2.5–2.8

[1] RC5 and RC6 using similar round counts offer comparable performance.

Table 8-2 Raw Encryption and Decryption Rates for Various Block Ciphers

Algorithm	Pentium II 350 MHz Running Windows 2000 (microseconds)	SPARC Ultra 400 MHz Running Solaris 2.8 (microseconds)
DES	7.4–7.5	7.45–7.60
3DES (EDE mode)	13–14	14–15
RC4	14–15	13–14
RC5 (16 rounds)	6.5–7.0	10–11
RC6 (20 rounds)	9–10	13–14
AES (128 bits)	8–9	9–10

Table 8-3 Key Setup Times for Various Symmetric Ciphers

Public-Key Algorithms

While symmetric ciphers provide better performance for bulk data encryption than public-key cryptosystems do, there are a number of tasks that cannot be accomplished using symmetric encryption alone. Authentication and nonrepudiation based on digital signatures, as well as most initial key-exchange operations, require the use of public-key encryption. Due to their relatively poor performance, public-key algorithms may have a greater impact on network and device operation than the symmetric algorithms used for traffic encryption. The following considerations apply to selecting public-key algorithms:

- **For those algorithms supporting digital signatures, how much time is required for signature generation and verification?** For some public-key algorithms, an operation involving the use of a private key requires far more time than an equivalent operation using a public key. This is true for the RSA algorithm. The opposite is true for DSA and algorithms based on DSA (Elliptic Curve DSA, for instance). As a result, RSA outperforms DSA-based algorithms for signature verifications, but does not perform as well for signature generation.

  When signature-generation times and encryption/decryption rates using *private* keys are more critical, you may wish to use a DSA or an Elliptic Curve DSA. If, on the other hand, signature verification times and encryption/decryption rates using *public* keys are crucial, then RSA may prove to be a wiser choice. The latter case is common on web-enabled wireless devices, where there is little need for client-side authentication during SSL session establishment. Instead, the wireless device will use the public key of the web server to encrypt a session secret, and may need to authenticate the server by verifying its digital signature. See Table 8-4 for a performance comparison of the RSA and DSA algorithms for signature generation and verification.

Algorithm	Sign (in milliseconds)	Verify (in milliseconds)
RSA (512-bit modulus)	4–4.5	0.45–0.52
RSA (768-bit modulus)	10–11	0.8–1.0
RSA (1024-bit modulus)	20–22	1.25–1.40
RSA (2048-bit modulus)	115–120	4–4.5
DSA (512-bit modulus)	4.5–5.0	8–9
DSA (768-bit modulus)	7.5–8.5	14–15
DSA (1024-bit modulus)	12–13	22–24
DSA (2048-bit modulus)	35–38	68–72

Table 8-4 RSA and DSA Signature Generation and Verification Times on an Intel Pentium II 350 MHz Running Windows 2000

- **What are the effects of modulus size on key and parameter generation?** Key-generation times increase very quickly as the size of the modulus increases. For some algorithms, the difference in key-generation times for 1024- and 2048-bit modulus may be several seconds to several minutes, depending on the available processing power. See Tables 8-5 and 8-6 for a comparison of key-generation and key-agreement times for different modulus sizes.

NOTE:
The terms modulus size *and* key size *are often used interchangeably in conjunction with public-key cryptosystems.*

Table 8-5 Key-Generation Times for RSA and DSA on an Intel Pentium II 350 MHz Running Windows 2000

Algorithm	Key-Generation Time (in seconds)
RSA (512-bit modulus)	0.10–0.12
RSA (768-bit modulus)	0.25–0.30
RSA (1024-bit modulus)	0.60–0.70
RSA (2048-bit modulus)	6.5–7.5
DSA (512-bit modulus)	2.0–2.5
DSA (768-bit modulus)	7.5–8.5
DSA (1024-bit modulus)	18–20
DSA (2048-bit modulus)	210–230

- **What are the effects of transmitting public keys and digital certificates on network bandwidth?** The key sizes for most public-key algorithms are far larger than those used by symmetric ciphers. Common key sizes range between 512 and 2048 bits. Typical RSA public-key certificates vary in size from several hundred bytes to a few kilobytes. Frequent exchange of public keys and certificates may consume valuable bandwidth on extremely low-bandwidth networks, such as those used by current- generation mobile phone systems.
- **How does code size for public-key algorithms compare to that of symmetric ciphers?** Public-key cryptosystems are more complex than symmetric encryption algorithms, and their code size reflects their complexity. A minimal application performing public-key operations (with 8–10KB of program overhead) may run 70–125KB in size, depending on the algorithm selected and the subset of functionality required (encryption, decryption, generation, and verification). This includes any message digests used during signature generation and verification.

Although we do not discuss them here, the elliptic-curve variants of DH and DSA are faster than the values shown for DSA in Tables 8-4 and 8-6. Elliptic-curve operations frequently take advantage of acceleration tables (precomputed tables of points on elliptic curves) for performing elliptic-curve calculations. These tables can improve verification and signing times by factors of 1.5 and 10 or more, respectively. Moreover, algorithms using elliptic curves are generally believed to be as secure as RSA, DH, and DSA using much smaller key sizes. The smaller key size is advantageous in bandwidth-constrained environments where it is easier to transmit a 160-bit key than a 1024-bit key.

Diffie-Hellman Modulus Size	Parameter Generation Time (in milliseconds)	Key Agreement Time (in milliseconds)
512 bits	4.0–4.5	9.5–10.2
768 bits	18–20	28–31
1024 bits	40–45	55–60

Table 8-6 Diffie-Hellman Performance on an Intel Pentium II 350 MHz Running Windows 2000

RSA Modulus Length	Encryption/Decryption Rate Using Public Key (in Kbps)	Encryption/Decryption Rate Using Private Key (Kbps)
512-bit	145–155	14–16
768-bit	125–132	8–9
1024-bit	105–110	5.5–6.0
2048-bit	60–64	1.9–2.2

Table 8-7 RSA Encryption and Decryption Rates on an Intel Pentium II 350 MHz Running Windows 2000

In addition to the considerations mentioned earlier in this section, you may find the following guidelines helpful when deploying public-key cryptosystems:

- **Avoid repeated use of public-key cryptographic operations.** Limit the use of public-key algorithms to key exchange, key agreement, and signature generation and verification. Never use public-key cryptography for bulk data encryption unless you are creating digital envelopes. The relatively slow performance of the RSA algorithm is evident in Table 8-7.
- **Consider multi-tier key-exchange mechanisms that employ a combination of public-key and symmetric encryption algorithms.** Public-key cryptography can be used initially to exchange a symmetric key encryption key (KEK). The KEK can then be used in conjunction with a symmetric encryption algorithm for regular key refreshes and renewals. This multi-tier approach eliminates the need for repeated public-key operations.

MultiPrime RSA MultiPrime RSA, patented by Compaq Computer Corporation, improves the performance of the standard RSA algorithm by increasing the number of prime factors used in public-key calculations. Increasing the number of prime factors reduces the effective length of each prime, resulting in less computational effort for operations involving a private key. MultiPrime RSA using three primes results in nearly a two-fold increase in the performance of RSA signature generation and private-key encryption and decryption rates with very little reduction in security. The technique employed by MultiPrime RSA is platform independent, so the two-fold performance increase applies to any processor and operating system combination.

CAUTION:

When using MultiPrime RSA, the number of prime factors is not limited to three. You may improve performance even further by increasing the number of primes. However, the resulting reduction in the size of the prime factors decreases the strength of the RSA algorithm. To maintain an adequate level of security, RSA Security, Inc. has chosen to implement MultiPrime RSA using three primes.

Message Digests and HMACs

One of the key characteristics of message digests is their speed. Message digests are designed to process large quantities of data very quickly. High performance is critical when generating digital signatures over large messages, and when authenticating network traffic using HMACs. In addition to raw digest performance, the following considerations apply specifically to selecting and using message digest algorithms:

What are the effects of HMACs on network bandwidth? HMACs consume considerable bandwidth when authenticating very small quantities. The output of an HMAC is a fixed size regardless of the length of its input. An HMAC based on the 160-bit variant of SHA-1, for instance, always produces a 160-bit (20-byte) output even when the input is a single byte or bit. Significant bandwidth is wasted unless the output length of the HMAC is relatively small in comparison to the quantity of data being authenticated.

A simple example may suffice to demonstrate this concept. When a 1500-byte packet is authenticated, the message authentication code (MAC) produced by HMAC SHA-1 accounts for only 1.32 percent of the bandwidth consumed by the packet. However, when a 128-byte packet is authenticated, the MAC accounts for almost 14 percent! (Note that MACs are usually appended to the end of a message.) On high-traffic networks, such wasted bandwidth can adversely affect network performance.

How do the code sizes for the various message digests compare? The code size for all message digests and HMACs are comparable. A minimal application (with 8–10KB of program overhead) for performing an MD5 or SHA-1 message digest or HMAC operation may run anywhere between 35–45KB.

Tables 8-8 and 8-9 summarize the performance of various message digest and HMAC algorithms.

Algorithm	Intel Pentium II 350 MHz Running Windows 2000 (in MB/sec)	Sun SPARC Ultra 400 MHz Running Solaris 2.8 (in MB/sec)	Strong ARM 200 MHz Running Windows CE 3.0 (in MB/sec)
MD2	0.8–1.0	1.0–1.2	0.38–0.45
MD5	45–50	24–28	5–6
SHA-1	24–28	14–17	3.0–3.5

Table 8-8 Digest Generation Times for Popular Message Digest Algorithms

Algorithm	Processing Rate (in MB/sec)
MD5	45–50
HMAC-MD5	20–24
SHA-1	24–28
HMAC-SHA-1	16–18

Table 8-9 A Comparison of Message Digest and HMAC Performance on an Intel Pentium II 350MHz Running Windows 2000

Dedicated Cryptographic Hardware

When software-based cryptographic solutions just don't offer enough performance, you can opt for dedicated cryptographic hardware. In practice, infrequent cryptographic operations, such as authentication and key exchanges, take place in software. Although they usually involve slow public-key operations, they tend to occur only during session establishment or at regular, but infrequent, intervals during normal operation. Traffic encryption algorithms, on the other hand, are implemented almost exclusively in hardware, especially within broadband network devices. Even though traffic encryption employs fast symmetric algorithms, it occurs very frequently. Without dedicated hardware, much of the processing resources on a broadband network device would be expended performing mundane cryptographic operations.

Although they are very fast, dedicated hardware solutions have drawbacks in terms of cost and portability. More often than not, dedicated cryptographic hardware costs more to develop and manufacture (or purchase) than software-based solutions. Aside from supporting and maintaining a code base, the costs associated with developing software cease once the software has been written (unless you are paying royalties on code licensed from a reputable security vendor). In contrast, hardware-based solutions have recurring expenses related to material and fabrication costs.

Dedicated hardware is also less portable than software. While a software-based solution can often be moved from one system to another without any physical modifications, hardware solutions are much more restrictive. Hardware accelerators in the form of computer interface cards, such as the nFast series of accelerators from nCipher and Rainbow Technologies' CryptoSwift Secure Server Accelerator, can be removed from one computer and inserted into another using nothing more than a screwdriver. However, the hardware accelerators included within many broadband consumer devices and network infrastructure equipment cannot be removed or upgraded. Although limited upgradability may be provided through firmware updates, one cannot remove and replace the circuitry responsible for performing cryptographic operations.

Encryption and Compression

On the surface, encryption and compression may not appear to have much in common other than their ability to convert data into an incomprehensible collection of 1's and 0's. However, as we move just beneath the surface, we find that they are quite complementary. Many encryption algorithms focus on concealing plaintext patterns that, when present, may aid in cryptanalytic attacks. Many encryption algorithms conceal plaintext patterns using a feedback mechanism, that creates a nonlinear relationship between the input and output of a block cipher. (We discussed cryptographic feedback in Chapter 2.) This nonlinear behavior helps to prevent input patterns from carrying over to the ciphertext output. However, some modes of operation—namely, Electronic Code Book (ECB) mode—do not incorporate feedback mechanisms. As a result, any pattern that exists in the input to an ECB mode block cipher results in a similar pattern in the output.

What if we want to conceal plaintext patterns, but our particular situation calls for ECB mode operation? For instance, we might want to simultaneously encrypt multiple plaintext blocks in parallel. Parallelization during encryption is supported by ECB mode, but is difficult to impossible for block cipher modes employing feedback mechanisms. This is where compression comes into play. To reduce the size of a file, a compression algorithm must decrease the level of redundancy in the data contained within the file. A compression algorithm accomplishes this by removing repetitive patterns of 1's and 0's. Before encrypting the data, we can effectively remove most of the plaintext patterns by running it through a compression algorithm.

NOTE:

Compression improves the effectiveness of any encryption algorithm, not just those operating in ECB mode.

In addition to concealing plaintext patterns, compressing data prior to encryption offers benefits in terms of performance. We mentioned earlier in this chapter that encryption involves complex mathematical operations that consume valuable processing resources. On small devices, where these resources are often very scarce, encryption can slow the device to a crawl. The larger the dataset, the longer the processor is tied up encrypting the data. By reducing the size of the original dataset, compression decreases the amount of computational effort required for encryption (if the compression algorithm far outperforms the algorithm used for encryption, and if that compression results in a notable reduction in the size of the data set).

The order in which you apply encryption and compression is very important: compression algorithms should always be applied first. The output of a robust encryption algorithm should be largely unpredictable and unbiased, which means that it should not contain patterns. Data that does not contain patterns cannot be compressed, so the effects of compression on encrypted data are minimal. In fact, if the encryption algorithm is doing its job correctly, the size of the compressed image should be almost identical to—or even slightly larger than—that of the uncompressed data.

Security Protocol Tuning

Like cryptography, security protocol overhead can adversely affect the delivery of real-time content by consuming bandwidth and introducing latency and jitter. When deploying an existing protocol or designing a custom protocol, you may find the following performance tips useful:

- **Limit the number of exchanges (handshaking messages) required to initiate a session** These exchanges are very dependent on network delivery and response times. In general, the greater the number of exchanges, the more latency that is introduced during session establishment.
- **Limit the number of public-key cryptographic operations during session establishment** For example, SSL includes support for multiple authentication modes. An SSL server can be authenticated in one of two ways:
  - Using two separate authentication and key exchange transactions
  - Using a single transaction that combines authentication and key exchange

The first approach requires that the server generate a digital signature to authenticate itself to the client, and then to perform a private-key decryption to obtain the master secret supplied by the client. (In Chapter 6, we discussed why this method might be necessary.) The second approach eliminates the need for generating a separate signature. Instead, the server proves possession of its private key by decrypting the master secret key and beginning encrypted communication.

- **If possible, design protocols to perform simultaneous client- and server-side (or peer-to-peer) processing during session establishment** For instance, when using a combination of digital signatures and certificates for authentication, one communicating party (Party A) might transmit a certificate to its peer (Party B) prior to generating and transmitting a signature. This allows Party B to begin validating the certificate and extracting the public key required to verify Party A's signature while Party A generates the signature. By the time Party B receives the message containing Party A's signature, it has already established trust in the public key contained within Party A's certificate.

- **Consider caching session state information** Caching allows sessions to be resumed quickly based on previously negotiated security parameters. This greatly reduces the time required for session reestablishment, and is supported by many common security protocols, such as SSL, TLS, and IPSec. The downside to session caching is that it can require a great deal of memory for storing session state variables. This is particularly true in client/server applications where a single server accepts requests from a large number of clients. To conserve memory, restrict session caching to clients that reconnect within 10 or 15 minutes.
- **Use the largest payload length possible while still maintaining prescribed bandwidth, delay, and jitter requirements** Larger payloads reduce the ratio of protocol headers, padding, IVs, and message authentication information to actual message content. This results in efficient use of network bandwidth. For SSL, this corresponds to using larger record sizes. (See Rescorla, 2001 for a discussion of the effects of record size on SSL/TLS performance.) It is important to note here that some real-time multimedia applications limit payload length to facilitate the timely delivery and processing of application content.

Additional Tips for Improving Security in Real-Time Multimedia Applications

As previously mentioned, real-time multimedia applications have the most restrictive QoS requirements of any network-based applications. Slight variations in QoS can result in unintelligible voice and video signals that hamper communication. The following performance tips apply to improving the delivery of multimedia content in real-time applications:

- **Use buffering to reduce the effects of jitter** Buffering helps to iron out signal variations resulting from erratic interpacket delivery times and cryptographic processing. Nevertheless, buffering introduces additional signal delay, and it should not be used to offset slow encryption algorithms. As mentioned earlier in this chapter, the throughput of an encryption algorithm should always be faster than the transmission rate required by the application.
- **Consider using out-of-band (asynchronous) key-management techniques** Key refreshes and renewals can increase signal delay and jitter when performed synchronous to media stream transport. This occurs when communicating devices halt transmission to generate and exchange information for deriving new encryption and authentication keys. By separating key management from media stream encryption and authentication, key updates can take place at any time without

affecting content delivery. Key updates should always occur before the old keying information expires, so that the communicating parties can immediately switch to the new keys upon expiration (or shortly before expiration) of the old ones.

Manageability

In the final section of this chapter, we'll address some of the fundamental challenges that you are likely to encounter when managing a security infrastructure. Since manageability is a difficult topic to tackle in one chapter (let alone a single section), we won't attempt to cover all aspects of security management here. Instead, we focus on identifying some of the more important management concerns, and we present a number of resources that address these topics in further detail. In the context of the material presented in this book, the top concerns include administration, key management, and certificate management.

Security administration encompasses many common tasks, including the configuration of user and host security policies, software installation and upgrades, security patches, and the periodic monitoring and auditing of security systems (for both intrusions and performance bottlenecks). The level of administrative effort required depends on many factors, including, but not limited to

- The number of users on the network
- The role each user plays within an organization (whether he or she is an employee, administrator, business partner, customer, subscriber, and so on)
- The number, type, and location of protected resources, and the duration for which these resources must remain secure
- The various devices, operating systems, applications, and security protocols deployed within the network, and how well these components interoperate
- The need to regularly upgrade components of the network (including hardware replacement and software upgrades and patches)
- The network layer at which security services and mechanisms are placed
- Whether security services are implemented on end-user hosts and devices or within security gateways

The way in which you choose to design a security infrastructure greatly affects the level of effort associated with administration. For example, choosing standards-based and interoperable components, considering the performance and scalability of security mechanisms during the design stages, minimizing the number of hosts and devices that must be individually configured and maintained, and incorporating automated, remote software upgrades all help to simplify administration. (Refer to the sections "Configuration and Policy Settings" and "Implementation, Configuration, and Maintenance" in Chapter 7 for more information on how the placement of security services and mechanisms affect administrative effort. See Chapter 9 for information on enabling secure, remote software upgrades.)

Key management can pose a significant task when protecting stored data, such as customer databases, billing records systems, and store-and-forward messaging applications. The most common approach for securing large quantities of data is to use symmetric encryption or a combination of a public-key and symmetric encryption. When choosing a system for managing encryption keys, you must consider the following issues:

- **How will you secure the keying material?** You can secure encryption keys through a number of techniques. Password-based encryption provides a simple, inexpensive solution. However, the use of weak passwords can compromise the security of an entire system. On the other hand, hardware-based solutions—such as smart cards and hardware storage modules (HSMs)—improve security considerably, but are more expensive. (We'll discuss HSMs briefly in Chapter 9.)
- **Who should have access to the keying material?** Should one individual have sole access to keying information? In general, the answer is "most definitely not." A more common approach is the use of an *m*-of-*n* password-sharing scheme that requires multiple individuals (*m* people out of a total of *n* people with passwords) to simultaneously supply their passwords to unlock the encryption keys. Access to keying information should always be provided to individuals based on their responsibilities within an organization, and only on a need-to-know basis.
- **How much data should be encrypted using the same key?** Although you can encrypt large amounts of data with a single key when using a block cipher, this is not advisable. It increases the chances of successful cryptanalysis, and does not adequately partition the data in the event of a key compromise (in other words, an attacker who has obtained the key may have access to all the data). By breaking a large data set up into many smaller sets, and encrypting each of the smaller sets using a different key, you limit the amount of data available to an attacker who has managed to compromise one or two encryption keys. However, you must now safely store and manage access to multiple encryption keys.
- **How often should you renew the encryption keys?** Frequent key renewals greatly reduce the effectiveness of brute-force key searches and cryptanalytic attacks. For session-based encryption, key renewal is a simple process. Once you generate a new key, you simply discard the old key and begin encrypting traffic with the new one. However, when you are renewing keys for stored data, the data must be decrypted with the old key and re-encrypted with the new key. This process may

require considerable time for very large data sets, especially if you've employed a partitioning scheme that uses numerous encryption keys.

Public-key cryptography eases many of the burdens associated with exchanging symmetric encryption keys, but it is susceptible to impersonation and man-in-the-middle attacks. Digital certificates, which bind keying material to an identity (a user or device), prevent these types of attacks when implemented properly. However, they introduce yet another layer of management complexity. Their functionality stems from the existence of public key infrastructures (PKIs), which can be very complex. A PKI defines the guidelines for certificate issuance, management, and revocation, and establishes trust relationships between the various components of the PKI. If you plan to use digital certificates, you will need to implement your own PKI. Here is a list of some of the questions you should ask when considering a PKI:

- **What type of trust hierarchy is appropriate for your environment?** The trust hierarchy describes the relationship between root certification authorities (CAs), subordinate CAs, cross-certified CAs, and end entities (the users and devices).
- **Should you operate your own CA(s), or should you outsource management of the CA(s) to a third-party PKI provider, such as Verisign?**
- **If you choose to manage your own CA, should you build or buy?** Building a CA is a monumental task. Unless you are a security vendor, your best bet is to purchase a CA from a reputable PKI vendor. Popular CA vendors include RSA Security, Entrust, and Baltimore Technologies.
- **How should you protect the private key for your root CA(s)?**
- **What certificate request and management protocols should you employ, and is there a need for these protocols to interoperate with other PKIs?**
- **What values should appear in the certificate fields, and what certificate extensions should you use?**
- **How often should the certificates be renewed (if they can be renewed)?**

We'll discuss PKI concepts, such as certificate life cycle management and trust hierarchies, in Appendix B. Chapter 9 provides a sample certificate hierarchy for the DOCSIS environment, and describes various scenarios for operating a CA.

NOTE:

An authoritative resource on PKI design, implementation, and management is PKI: Implementing and Managing E-Security *by Andrew Nash, William Duane, Celia Joseph, and Derek Brink (Nash et al., 2001).*

References

Lipmaa, Helger, Phillip Rogaway, and David Wagner. 2001. *Comments to NIST Concerning AES Modes of Operation.*

Menezes, A., P. van Oorschot, and S. Vanstone. 1996. *Handbook of Applied Cryptography*. Boca Raton: CRC Press.

Nash, Andrew, William Duane, Celia Joseph, and Derek Brink. 2001. *PKI: Implementing and Managing E-Security*. New York: Osborne/McGraw-Hill.

Pfitzmann, Andreas, and Ralf Aßmann. 1993. "More Efficient Software Implementations of (Generalized) DES." *Computer and Security*. Volume 12, number 5.

Rescorla, Eric. 2001. *SSL and TLS: Designing and Building Secure Systems*. Upper Saddle River, NJ: Addison-Wesley.

Robshaw, M. J. B. 9 January, 2001. *RC6 and the AES*. RSA Laboratories.

Rogaway, Phillip, Mihir Bellare, John Black, and Ted Krovetz. 19 September, 2001. *OCB: A Block-Cipher Mode of Operation for Efficient Authenticated Encryption.*

RSA Laboratories. 25 March, 1999. *PKCS #5 v2.0: Password-Based Cryptography Standard.*

Schneier, Bruce, and Doug Whiting. January, 1997. "Fast Software Encryption: Designing Encryption Algorithms for Optimal Software Speed on the Intel Pentium Processor." *Fast Software Encryption, Fourth International Workshop Proceedings*. New York: Springer-Verlag.

References

Part III

Case Studies

Chapter 9

Securing Broadband Internet Access: DOCSIS BPI+

DOCSIS, or *Data-Over-Cable Service Interface Specifications*, is a collection of functional specifications for ensuring the interoperability, quality-of-service, and security of data communication over cable networks. DOCSIS 1.0 supports the bidirectional transmission of best-effort IP data services, such as e-mail, Web browsing, online chat, and other common Internet access applications. DOCSIS 1.1 and 2.0 also incorporate support for constant bit rate (assured QoS) and multicast (one-to-many) group services over IP. These services include voice telephony, video conferencing, and other real-time multimedia applications.

Figure 9-1 shows a typical DOCSIS network. From right to left in the diagram, the network consists of *customer premise equipment (CPE)*—such as personal computers, a cable modem for every subscriber, the *hybrid fiber-coax (HFC)* cable network, a CMTS, and an interface to the cable provider's backend network. Each consumer attaches his or her CPE to a cable modem using an Ethernet connection. The cable modem attaches directly to the cable network via a coaxial cable. On the other side of the HFC network, the connection terminates at a *cable modem termination system (CMTS)*. The CMTS connects to the cable provider's distribution and backbone networks via a high-bandwidth WAN connection. The CMTS and cable modem convert Layer 2 traffic (Ethernet and ATM, for example) to *radio frequency (RF)* signals appropriate for transmission across the cable network, and vice versa. A single CMTS may service thousands of subscribers.

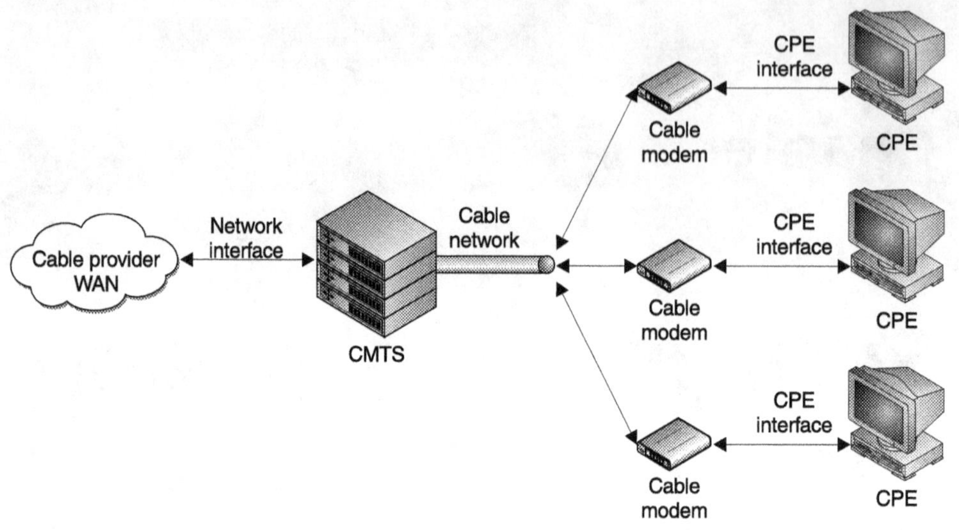

Figure 9-1 The DOCSIS network environment consists of customer premise equipment, cable modems, the cable network, a CMTS, and a network connection to the provider's back-end network.

DOCSIS 1.0, 1.1, and 2.0 include specifications for describing each interface present in Figure 9-1. These include

- Data interface specifications describing the connections (typically Ethernet connections) between the CPE and cable modem, and between the CMTS and the cable provider's backend network.
- An RF interface specification defining the cable modem-to-cable network interface and both the upstream and downstream CMTS-to-cable network interfaces.
- An *operations support system (OSS)* interface specification outlining the interfaces between each network element and the network management components (initialization and configuration, billing, and fault and performance management).
- A telephony return specification defining the interface between the cable modem and the telephony return path. In the event that the cable network does not support bidirectional communication, a cable modem may transmit upstream data using a dial-up connection.
- Security specification describing the security framework for enforcing data privacy and theft-of-service protection across the cable modem-to-CMTS segment of a DOCSIS network. DOCSIS 1.0 includes the *Baseline Privacy Interface Specification (BPI)*, and DOCSIS 1.1 and 2.0 incorporate the *Baseline Privacy Plus Interface Specification (BPI+)*.

In this chapter, we will discuss BPI+ (CableLabs, 2001a), the more comprehensive of the two security-related specifications. For more information on DOCSIS 1.0, 1.1, and 2.0, refer to the CableLabs Certified Cable Modem project website at http://www.cablemodem.com.

NOTE:
Before a cable modem or CMTS vendor can claim compliance with DOCSIS 1.0, 1.1, or 2.0, it must submit its cable modem or CMTS design to CableLabs for certification testing and approval. The certification process ensures interoperability among modems and CMTSs from different vendors.

An Overview of the Baseline Privacy Plus Interface

DOCSIS BPI+ (CableLabs, 2001a) addresses the need for data privacy and authentication on cable networks. Unlike a telephone network, where a dedicated wire runs from a central office directly to each customer's home, a cable network uses a single coaxial cable to distribute services to many subscribers. The broadcast nature of cable television has made its distribution much better suited to shared-media networks; and as we discussed in Chapter 5, these networks tend to be much more susceptible to eavesdropping attacks than dedicated lines. By attaching a packet-sniffing utility to a cable network, an eavesdropper might capture traffic to and from any other subscriber communicating over that segment of the cable network. On a typical cable network, an eavesdropper could monitor the communications of an entire neighborhood!

In addition to the damaged reputation and lost customers resulting from publicized eavesdropping attacks, theft of service poses another sizeable economic threat to service providers. Without appropriate levels of authentication, malicious users can potentially steal services by attaching a cloned cable modem to an HFC network. An attacker who can duplicate the identifying characteristics (for instance, the make, model, serial number, and MAC address) of a paying subscriber's device might create a clone. To the service provider, the cloned device would look identical to that of a paying subscriber, and allow the attacker to access service under another user's account. Weak authentication schemes, such as authentication based on a valid MAC address, afford little protection, since an attacker can determine a MAC addresses by querying the network interface of a cable modem. Even paying subscribers might abuse a service by altering his or her cable modem to request enhanced levels of QoS (for example, increased bandwidth, lower delay and jitter times, or reduced packet loss rates) for which they are not authorized. This consumes valuable network resources and, as a result, can deny service to subscribers who are paying premiums for guaranteed QoS.

BPI+ addresses both of these concerns by strongly authenticating each cable modem, associating the cable modem with a paying subscriber, and encrypting traffic flows between the cable modem and the CMTS. BPI+ consists of two primary components: an encapsulating encryption protocol and the *Baseline Privacy Key Management (BPKM)* protocol. The encapsulating encryption protocol offers privacy equal to or greater than that of dedicated line networks by encrypting the contents of MAC layer frames transmitted between a CMTS and all cable modems on the network segment serviced by the CMTS. The protocol defines the format of the encrypted frames and the cryptographic algorithms used to encrypt each frame.

Cable modems use BPKM to request authentication and keying information from a CMTS, and to periodically request traffic encryption keys. To facilitate strong authentication, every DOCSIS 1.1 cable modem must contain an RSA public/private key pair and an X.509 digital certificate, both of which the cable modem vendor must install during manufacturing. In the case of upgrading a DOCSIS 1.0 cable modem to DOCSIS 1.1, the modem must either contain a manufacturer-installed certificate, or be capable of accepting and loading a digital certificate sent to it by a service provider. Using BPKM, a cable modem and CMTS exchange a series of authorization messages allowing the CMTS to authenticate the cable modem and supply the modem with keying material and a list of services that the cable modem is allowed to access. It also enables the cable modem to periodically reauthorize and refresh encryption and authentication keys. A CMTS controls access to cable services by ensuring that keying material is distributed only to authenticated and authorized cable modems.

BPKM employs security associations to identify the services that a paying subscriber is authorized to access. A BPI+ *security association (SA)* describes the encryption keys, block cipher initialization vectors, and cipher suites (in other words, a combination of an encryption and an authentication algorithm) used by a cable modem and CMTS to protect a particular traffic flow. BPI+ defines three types of security associations for securing upstream (cable modem-to-CMTS) and downstream (CMTS-to-cable modem) traffic: primary, static, and dynamic. Following MAC registration, every cable modem registers a single, unique *primary SA* with the local CMTS. All upstream traffic must be encrypted under the cable modem's primary SA, and a CMTS can only use a primary SA for encrypting downstream traffic destined for a single cable modem.

Static and dynamic SAs, on the other hand, can only be used to encrypt downstream traffic sent from a CMTS to a cable modem. During authorization, the CMTS supplies each cable modem with a list of *static SAs*—one for every traffic flow associated with the services the subscriber is allowed to access. *Dynamic SAs* are created and destroyed on-the-fly by a CMTS. The lifetime of a dynamic SA coincides with the establishment and termination of the downstream traffic flow. Multiple cable modems can share a single static or dynamic SA; and in multicast environments, a CMTS will most frequently encrypt traffic under a shared SA. All SAs are identified by a 14-bit value known as a *security association identifier*, or *SAID*. Figure 9-2 demonstrates the usage of the various SA types. (We'll discuss BPKM further in the section entitled "Baseline Privacy Key Management Protocol," later in this chapter.)

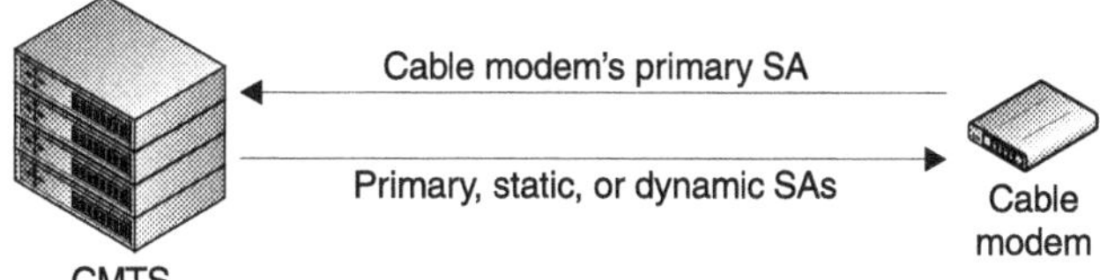

Figure 9-2 Primary, static, and dynamic SAs secure traffic between a cable modem and CMTS

DOCSIS MAC Layer Frame Formats

Both DOCSIS and BPI+ operate at the Data Link layer of the OSI Network Reference model (see Figure 9-3). As such, BPI+ provides point-to-point security between a cable modem and CMTS.

DOCSIS includes formats for both complete and fragmented MAC layer frames. A complete MAC frame contains a frame header, an extended header, and an encrypted payload, as shown here:

FC	ELEN	Length	Extended header	HCS	Packet PDU

The frame is comprised of the following fields:

- **Frame Control (FC)** An 8-bit field identifying the remaining contents of the MAC frame
- **Extended Header Length (ELEN)** An 8-bit field indicating the length (in bytes) of the extended header field
- **Length** A 16-bit field specifying the length (in bytes) of the MAC frame (the length is the sum of the number of bytes in the extended header and the number of bytes of data following the HCS field)
- **Extended Header** Described later in this section
- **Header Check Sequence (HCS)** A 2-byte field containing a checksum for ensuring the integrity of the frame header (including the FC, ELEN, Length, and extended header fields)
- **Packet PDU** Described later in this section

Application
Presentation
Session
Transport
Network
Data Link
Physical

Figure 9-3 DOCSIS and BPI+ operate at Layer 2 of the OSI Network Reference model.

The *Extended Header (EH)* field is 5–240 bytes in length and consists of various elements, including a Baseline Privacy EH (BP EH) element. The BPI+ extended header element, shown here, specifies parameters necessary for encrypting user data.

Type	Length	Key Seq.	Version	E	T	SID/SAID	Request/ Reserved	Additional EH elements

The BP EH element for complete frames contains the following fields:

- **Type** A 4-bit field indicating the type of the extended header. In the case of BPI+, this field contains the type identifier for a BPI+ extended header.
- **Length** A 4-bit field specifying the length of the extended header element (not including the type or length field).
- **Key Sequence Number** A 4-bit field used to uniquely identify an authorization key; at any given time, a cable modem and CMTS can maintain multiple authorization keys.
- **Version** A 4-bit field containing the version of the encapsulated protocol. For variable-length DOCSIS MAC headers, this value is set to 1.
- **Flags (E/T)** Two 1-bit fields for determining the encryption status of a PDU. When set, the ENABLE (E) bit indicates that encryption is enabled for the PDU; the TOGGLE (T) bit matches the least significant bit of the key sequence number.
- **SID/SAID** A 14-bit field containing the SID/SAID of the traffic stream to which the frame belongs. Cable modems and CMTSs use the SID/SAID to determine which security services to apply to the contents of a DOCSIS MAC layer frame.
- **Request/Reserved** An 8-bit field used by a cable modem to request additional upstream bandwidth from a CMTS. This field is set to zero in downstream frames.

The final component of a DOCSIS MAC layer frame, the *Packet PDU (protocol data unit),* is shown here:

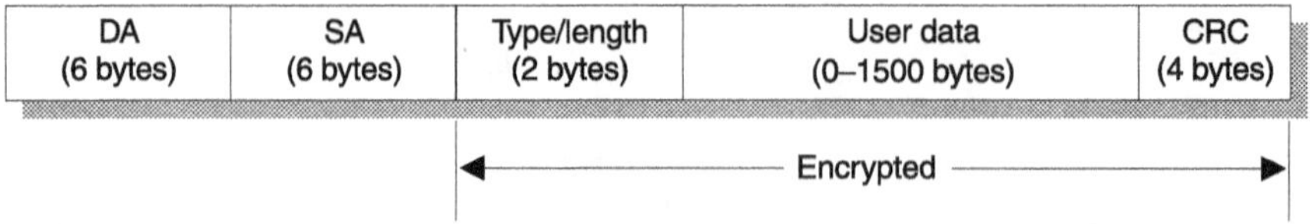

The Packet PDU is the portion of the DOCSIS MAC layer frame containing encrypted user data. Notice that only the type/length, user data, and CRC fields within the PDU are encrypted; BPI+ does not encrypt the source or destination hardware address of the frame. The Packet PDU contains the following fields:

- **DA** A 6-byte field containing the destination Ethernet address of the frame
- **SA** A 6-byte field identifying the source Ethernet address of the frame
- **Type/Len** A 2-byte field indicating the type and length of the data contained in the User Data field
- **User Data** A 0–1500 byte field containing upper-layer protocol data
- **CRC** A 4-byte field containing a checksum calculated over the PDU

In support of fragmentation, DOCSIS includes a modified BP EH, which includes information necessary for reconstructing a frame from fragments. A fragmented DOCSIS MAC layer frame takes this form:

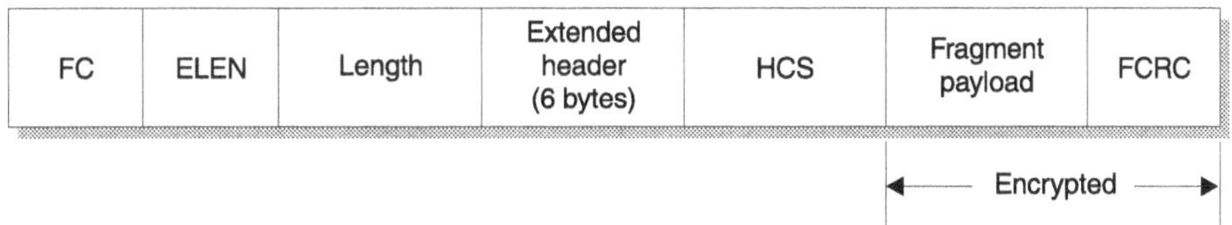

The entire fragment payload and CRC are encrypted under BPI+.

The BP EH for a fragmented frame, shown here, modifies the contents of its Extended Header element to include fragmentation control information:

Type	Length	Key Seq	Version	E	T	SID	Request	Fragment control

Notice that this is almost identical to the format of a complete frame, except that the final field has been replaced with the Fragmentation Control field. This field contains control information required to reconstruct the original frame from its fragments. BPI+ encrypts all fragments independently.

Baseline Privacy Key Management Protocol

BPKM is the BPI+ protocol responsible for cable modem authorization and key exchange. One can describe the operation of BPKM using two *state machines (SMs)*: one for authorization and the other for key exchange. The Authorization SM controls initial cable modem authentication and authorization, as well as periodic reauthorization. The TEK (Traffic Encryption Key) SM enables a cable modem and CMTS to synchronize keying material and to periodically exchange and refresh encryption and authentication keys. Both SMs consist of a number of operational states and events (for instance, an authorization request or reply, a timeout, or a key request or reply) that trigger transitions from one state to another.

Authorization State Machine

The BPKM Authorization state machine, used for cable modem authentication and authorization, consists of six states and eight events. Each of the eight events triggers a transition to a new state. All states, events, and associated messages and parameters are listed in Tables 9-1 through 9-4.

NOTE:
A cable modem is required to reauthorize before its current authorization expires, or when it receives a message from the CMTS indicating that its current authorization is no longer valid

---

State	Description
Authorized	The cable modem has received an *Authorization Reply* from the CMTS containing an authorization key and any static SAIDs.
Authorize Reject Wait	The cable modem has received an *Authorization Reject* message indicating a nonpermanent error code. Upon expiration of a timer, the cable modem transitions back to the Start state.
Authorize Wait	The cable modem has completed RF MAC registration with the CMTS, has sent both an *Authentication Information* and an *Authorization Request* message to the CMTS, and is awaiting an *Authorization Reply*.
Reauthorize Wait	The cable modem has requested reauthorization by sending another *Authorization Request* message to the CMTS, but has not yet received a response.
Silent	The cable modem has received an *Authorization Reject* message indicating a permanent error code, and does not attempt to reauthorize.
Start	The initial state of the Authorization SM.

Table 9-1 States of the BPKM Authorization State Machine

Message	Description
Authentication Information	Sent from a cable modem to a CMTS. Contains the X.509 certificate of the cable modem's manufacturer; this message can be ignored by the CMTS.
Authorization Invalid	Sent from a CMTS to a cable modem. Indicates that the modem's current authorization is invalid and that the modem should reauthorize.
Authorization Reject	Sent from a CMTS to a cable modem. Indicates an authorization failure.
Authorization Reply	Sent from a CMTS to a cable modem. Contains an authorization key encrypted using the cable modem's public key and a list of SAIDs that the cable modem is allowed to access.
Authorization Request	Sent from a cable modem to a CMTS. A request for an authorization key and a list of SAIDs.

Table 9-2 Authorization State Machine Messages

Event	Description
Authorization Grace Timeout	Indicates expiration of the *Authorization Grace timer*; following expiration of this timer, the cable modem must reauthorize.
Authorization Invalid	Sent from a cable modem to a CMTS. Generated when a *Key Reply, Key Reject,* or *TEK Invalid* message fails authentication (that is, when it contains an invalid MAC). Sent from a CMTS to a cable modem when the CMTS receives a *Key Request* message containing an invalid message authentication code (MAC).
Authorization Reject	Sent from a CMTS to a cable modem (following an *Authorization Request*) to indicate a nonfatal error code. The cable modem can attempt reauthorization following a timeout period.

Table 9-3 Authorization State Machine Events

Event	Description
Authorization Reply	Sent from a CMTS to a cable modem. Generated in response to an authorization request to supply a cable modem with an authorization key and a list of valid SAIDs.
Permanent Authorization Reject	Sent from a CMTS to a cable modem (following an *Authorization Request*) to indicate a fatal error condition (such as an unknown manufacturer, an invalid certificate signature, or an inability to negotiate a common set of security capabilities).
Provisioned	Generated following cable modem RF MAC registration with a CMTS; indicates that the cable modem can begin requesting keying material.
Reauthorize	Triggered when a previously authorized cable modem wishes to reauthorize (for example, when the Authorization Grace timer has expired or the CMTS requests that the cable modem reauthorize).
Timeout	Marks the expiration of a timer, and usually results in retransmission of a request by a cable modem.

Table 9-3 Authorization State Machine Events *(continued)*

NOTE:
Transmission of an Authorization Invalid *message indicates that the Authorization keys maintained by the CMTS and a cable modem are no longer synchronized.*

Cable modem authorization follows these steps, as illustrated in Figure 9-4:

1. During initialization, the cable modem sends an *Authentication Information* message to the CMTS. This message contains the X.509 certificate of the cable modem's manufacturer. If the CMTS wishes to obtain this information using an out-of-band method, it can ignore *Authentication Information* messages.
2. The cable modem transmits an *Authorization Request* message to the CMTS to obtain an authorization key and the SAIDs for any static security associations for the services that the subscriber can access. The *Authorization Request* includes the following.

Parameter	Description
Authorization Grace Timeout	The maximum time that can pass before a cable modem must attempt reauthorization.
Authorization Reject Wait Timeout	The period that an Authorization SM can sit in the Authorize Reject Wait state before returning to the Start state.
Authorize Wait Timeout	The period that must pass following transmission of an *Authorization Request* by a cable modem in the Authorize Wait state before the cable modem can send another *Authorization Request* message.
Reauthorize Wait Timeout	The period that must pass following transmission of an *Authorization Request* by a cable modem in the Reauthorize Wait state before the cable modem can attempt another reauthorization.

Table 9-4 Authorization State Machine Parameters

- The serial number and manufacturer of the cable modem.
- The MAC (or hardware) address of the cable modem.
- The RSA public key of the cable modem.
- An X.509 digital certificate issued to the cable modem by its manufacturer.
- The cable modem's security capabilities, which consist of a list of cipher suites supported by the cable modem. Each suite consists of an encryption and authentication algorithm. BPI+ current supports 40-bit and 56-bit DES for packet data encryption. The specification defines no authentication algorithms for traffic flows. The security capabilities selection feature allows for future migration to algorithms (such as AES) without modifying the existing BPI+ specification.
- The SAID of the primary SA that the cable modem wishes to register with the CMTS.

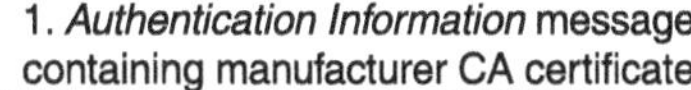
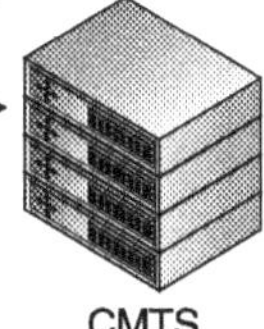
1. *Authentication Information* message containing manufacturer CA certificate
2. *Authorization Request* message containing cable modem certificate
Cable modem
3. Modem authentication and generation of authorization key by CMTS
4. *Authorization Reply* message containing RSA-encrypted authorization key
CMTS
5. Decryption of authorization key by cable modem using private RSA key

Figure 9-4 The cable modem authorization process includes certificate-based authentication and public-key cryptography.

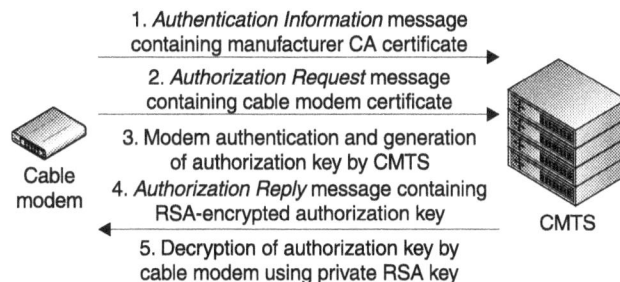

3. The CMTS authenticates the cable modem by verifying its digital certificate, and binds the cable modem to a paying subscriber. It then selects a cipher suite it shares in common with the cable modem, and generates/activates an *authorization key*.
4. The CMTS encrypts the authorization key using the RSA public key extracted from the cable modem's X.509 certificate, and transmits the encrypted authorization key to the cable modem in an *Authorization Reply* message. The *Authorization Reply* message includes
   - The encrypted authorization key.
   - A 4-bit key sequence number. A cable modem and CMTS must simultaneously maintain two authorization keys. The keys have varying validity periods to prevent loss of synchronization in the event that one of the keys expires. The sequence number allows the cable modem and CMTS to differentiate between successive authorization keys.
   - The key lifetime.
   - A list of the SAIDs for which the cable modem can obtain keying material. The list can contain a single primary and multiple static SAs, but cannot contain dynamic SAs.
5. The cable modem decrypts the authorization key using its RSA private key. Only the cable modem in possession of the private key corresponding to the public key in the X.509 certificate sent to the CMTS can decrypt the message. The cable modem and CMTS use the authorization key to derive 3DES key encryption keys (KEKs) and message authentication keys. The KEKs are later used to exchange DES traffic encryption keys (TEKs), which are used for bulk traffic encryption.

NOTE:
Cable modems and CMTSs must deter physical access to keying material stored in memory, and both devices must meet FIPS 140-1 Level 1 requirements for protection of keying material in multiple-chip, standalone modules. These requirements include the use of production-grade chips, metal or hard plastic enclosures, and a sealant for covering exposed circuitry (National Institute of Standards and Technology, 1982). While a cable modem can maintain traffic encryption and authentication keys in volatile memory (RAM), it must store its RSA key pair and X.509 digital certificate in write-once memory.

Although ideally suited for key exchange, public-key algorithms consume more computational resources during encryption and decryption than symmetric algorithms do. It is important to note here that the two-tier key-exchange mechanism employed by BPI+ improves performance considerably by reducing the number of public-key operations performed on cable modems and CMTSs. This is especially crucial for a cable modem, because it prevents the modem from having to perform a private-key decryption each time it exchanges keying material with a CMTS. (Remember from Chapter 2 that private-key RSA operations are approximately 100 times slower than most symmetric operations.)

Table 9-5 details the operation of the Authorization SM. The states of the Authorization SM are listed horizontally along the top edge of the table; events are listed vertically down the left-hand edge of the table. The intersection of a state and event in the table indicates the state that a cable modem transitions to following the event.

TEK State Machine

The TEK SM, which allows a cable modem and CMTS to periodically change encryption and authentication keys, consists of six states and nine events (when including both message transmission and receipt). As with the Authorization SM, each event in a TEK SM triggers a transition to a new state. Following successful authentication and authorization, a cable modem and CMTS maintain

State/Event	Start	Authorize Wait	Authorized	Reauthorize Wait	Authorize Reject Wait	Silent
Provisioned	Authorize Wait					
Authorize Reject		Authorize Reject Wait		Authorize Reject Wait		
Permanent Authorize Reject		Silent		Silent		
Authorize Reply		Authorized		Authorized		
Timeout		Authorize Wait		Reauthorize Wait	Start	
Authorize Grace Timeout			Reauthorize Wait			
Authorize Invalid			Reauthorize Wait	Reauthorize Wait		
Reauthorize			Reauthorize Wait			

Table 9-5 An Authorization State Machine Operational Matrix (CableLabs, 2001a)

two separate TEK SMs for each SAID returned in the *Authorization Reply* by the CMTS. In response to a cable modem's request for keying material, a CMTS creates both TEK SMs and returns them to the cable modem in a *Key Reply* message. The two SMs contain different keying material, and their lifetimes overlap. The keying material contained within the newer of the TEK SMs (TEK_{new}) can be used to encrypt traffic when those keys in the older TEK SM (TEK_{old}) have expired. The overlapping lifetimes help to avoid loss of synchronization following expiration of the keying material in TEK_{old}, since the cable modem and CMTS still share valid encryption keys. During normal operation, a cable modem periodically issues *Key Request* messages to obtain updated keying information from the CMTS. The cable modem issues the request following the expiration of the older of its two TEK SMs, and prior to expiration of the newer SM (see Figure 9-5).

NOTE:
A CMTS uses its TEK_{old} SM to encrypt downstream traffic, and decrypts upstream traffic with either TEK_{old} or TEK_{new}. A cable modem encrypts upstream traffic using its TEK_{new} SM, and decrypts downstream traffic using either TEK_{old} or TEK_{new}.

Table 9-6 demonstrates the operation of a TEK state machine. The states, events, messages, and parameters for a TEK SM are listed in Tables 9-7 through 9-10.

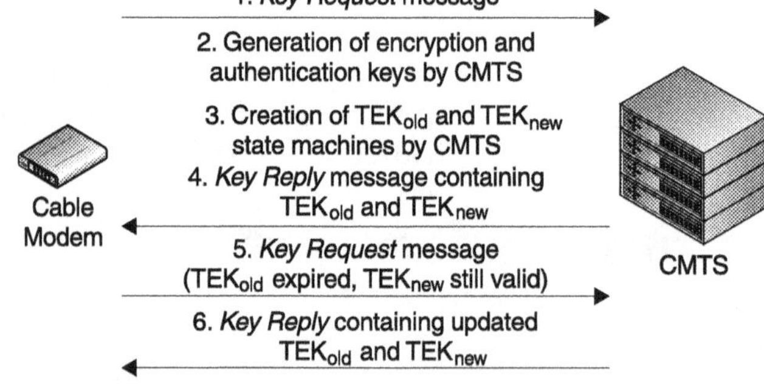
1. Key Request message
2. Generation of encryption and authentication keys by CMTS
3. Creation of TEKold and TEKnew state machines by CMTS
4. Key Reply message containing TEKold and TEKnew
Cable Modem
5. Key Request message (TEKold expired, TEKnew still valid)
CMTS
6. Key Reply containing updated TEKold and TEKnew

Figure 9-5 To initially obtain and update keying material a cable modem and CMTS exchange Key Request and Key Reply messages

State/Event	Start	Operational Wait	Operational Reauthorize Wait	Operational	Rekey Wait	Rekey Reauthorize Wait
Stop		Start	Start	Start	Start	Start
Authorized	Operational Wait					
Authorize Pending		Operational Reauthorize Wait			Rekey Reauthorize Wait	
Authorize Comp.			Operational Wait			Rekey Wait
TEK Invalid				Operational Wait	Operational Wait	Operational Reauthorize Wait
Timeout		Operational Wait			Rekey Wait	
TEK Refresh Timeout				Rekey Wait		
Key Reply		Operational			Operational	
Key Reject		Start			Start	

Table 9-6 A TEK State Machine Operational Matrix (CableLabs, 2001a)

State	Description
Operational	The cable modem has obtained valid keys for SAIDs registered or provisioned by a CMTS.
Operational Reauthorize Wait	The state a TEK SM with invalid keying material occupies while waiting for the Authorization SM to complete a reauthorization.
Operational Wait	The cable modem has sent an initial *Key Request* to the CMTS and is awaiting a reply.
Rekey Authorize Wait	The state a TEK SM with valid keying information occupies while awaiting a response to a request for keying material during a cable modem reauthorization.
Rekey Wait	The cable modem has requested new keying material for its SAIDs following expiration of the TEK Refresh Timeout.
Start	The initial state of the TEK SM.

Table 9-7 States of the BPKM TEK State Machine

Message	Description
Key Reject	Sent from a CMTS to a cable modem. Indicates that the CMTS will not create additional keying material because the SAID is now invalid
Key Reply	Sent from a CMTS to a cable modem in response to a *Key Request*. Includes two sets of keying material for the SAID
Key Request	Sent from a cable modem to a CMTS. A request for keying material for a particular SAID
TEK Invalid	Sent from a CMTS to a cable modem when the CMTS has determined that the cable modem has encrypted upstream traffic using invalid keying material (in other words, the TEK SMs of the cable modem and CMTS are no longer synchronized)

Table 9-8 TEK State Machine Message

NOTE:
All Key Reply *messages are authenticated using a keyed message digest.*

Event	Description
Authorization Complete	Sent from the Authorization SM to a TEK SM in the Operational Reauthorize Wait or Rekey Reauthorize Wait state to clear a previously issued wait state.
Authorization Pending	Generated by the Authorization SM to place a TEK SM in a wait state until successful completion of a reauthorization cycle.
Authorized	Sent by the Authorization SM to a nonactive TEK SM to indicate successful cable modem authorization.
Stop	Issued by the Authorization SM to discontinue use of an active TEK SM and remove all associated SAIDs and keying material from the cable modem's memory.
TEK Invalid	Signals a loss of key synchronization between a cable modem and CMTS.
TEK Refresh Timeout	Indicates that it is time for a TEK SM to refresh its keying material by issuing a *Key Request* message to the CMTS.
Timeout	Marks the expiration of a timer, and usually results in retransmission of a request by a cable modem.

Table 9-9 TEK State Machine Events

Parameter	Description
Operational Wait Timeout	The period that must pass following transmission of a *Key Request* by a cable modem in the Operation Wait state before it can send another *Key Request* message.
Rekey Wait Timeout	The period a cable modem in the Rekey Wait state must wait after sending a *Key Request* before it can issue another *Key Request* message.
TEK Grace Time	The minimum number of seconds prior to expiration of keying material that a cable modem can wait before attempting a rekey.

Table 9-10 TEK State Machine Parameters

BPI+ Key Encryption, Traffic Encryption, and Authentication Algorithms

In BPI+, cryptographic mechanisms prevent eavesdropping, facilitate key exchange and device authentication, and ensure the integrity and authenticity of messages and software upgrade files exchanged between a CMTS and a cable modem. BPI+ supports both 40-bit and 56-bit DES in CBC mode for packet data encryption. Support for 40-bit DES provides backward compatibility with DOCSIS 1.0 BPI. When using 40-bit DES, BPI+ requires communicating devices to mask the 16 leftmost bits of a 56-bit DES key with known values (0's, for instance). CBC mode operation also requires an IV as input when processing the first block of plaintext or ciphertext data. (We discussed the CBC mode of block-cipher operation in detail in Chapter 2.) CMTSs transmit IVs along with keying material in *Key Reply* messages sent to cable modems. During encryption and decryption, if the last block of data contains less than 64 bits of data, a cable modem or CMTS must use *residual termination block processing* as described in the BPI+ specification.

To obtain traffic keying material, a cable modem sends a *Key Request* message to a CMTS. The CMTS responds with a *Key Reply* message containing a 3DES-encrypted TEK, a CBC initialization vector, a key sequence number, and the remaining lifetime for the TEK. To prevent tampering during transit, BPI+ authenticates both messages—as well as *Key Reject* and *TEK Invalid* messages—with a 20-byte HMAC digest, calculated with SHA-1. The one-way, collision-free nature of the SHA-1 digest causes great difficulty for an attacker who wishes to modify the contents of a message without detection. The CMTS derives the authentication key used in HMAC generation from the cable modem's authorization key. Section 7.4 of the BPI+ specification (CableLabs, 2001a) describes the process for deriving the various BPI+ keys from an authorization key. In addition to the HMAC usage outlined in BPI+, the DOCSIS 1.1 Radio

Frequency Interface specification (CableLabs, 2001b) uses this same message authentication key to ensure the integrity of *Dynamic Service Addition Request, Dynamic Service Change Request,* and *Dynamic Service Deletion Request* messages during service provisioning.

As with message authentication keys, the CMTS derives all KEKs and TEKs from the cable modem's authorization key. In a *Key Reply* message, the CMTS secures the TEK by 3DES encrypting it using the currently registered KEK. 3DES operates in *encrypt-decrypt-encrypt (EDE)* mode using two 56-bit DES keys. According to (Stallings, 1999), 3DES EDE operations take this form:

$$Ciphertext = E_{k1}(D_{k2}(E_{k1}(Plaintext)))$$

and

$$Plaintext = D_{k1}(E_{k2}(D_{k1}(Ciphertext)))$$

Here, $k1$ and $k2$ are 56-bit DES keys and E( ) and D( ) are encryption and decryption operations, respectively. Each individual DES encryption and decryption runs in electronic codebook (ECB) mode.

The 3DES KEK must be periodically refreshed to prevent an attacker from obtaining enough ciphertext to mount a cryptanalytic attack. This is especially true when encrypting 40-bit DES keys, as the predictable masking of the original 56-bit DES key improves the likelihood of successful cryptanalysis.

TIP:

In Chapter 2, we learned that block ciphers operating in ECB mode are highly susceptible to cryptanalytic attack since they lack a feedback mechanism (that is, they do not conceal plaintext patterns). This poses a significant threat when encrypting large quantities of data containing patterns (for example, multiple e-mail messages). In the case of BPI+, ECB mode poses far less of a threat because it is only used to encrypt small amounts of random data. In fact, ECB mode is ideally suited for encrypting keying material, such as 56-bit DES keys. Combining ECB mode operation with an integrity mechanism (such as an HMAC) prevents block replay attacks. (For a discussion of ECB mode operation, refer to Chapter 2.)

During authorization, a cable modem and a CMTS negotiate a common set of cryptographic algorithms. The cable modem first transmits a list of supported algorithms to the CMTS via an *Authorization Request* message. The CMTS then chooses a cipher suite that it shares in common with the cable modem, and informs the cable modem of its selection within its *Authorization Reply* message. In support of future security requirements, the creators of BPI+ have anticipated the need for stronger cryptographic algorithms than those currently specified by BPI+ (DES, for instance). Through a process known as *security capabilities selection,* BPI+ supports the negotiation of algorithms other than those included within the specification. Table 9-11 summarizes the cryptographic algorithms employed by BPI+.

NOTE:

Although DOCSIS 1.1 BPI+ includes security capabilities selection, cable modem vendors typically implement traffic-encryption algorithms using hardware. A cable modem shipping solely with DES-encryption hardware will not be able to take advantage of the AES algorithm unless it runs in software. The limited processing power of many embedded devices, including cable modems, makes software-based traffic encryption infeasible. However, this functionality does allow a single software image to operate on multiple cable modems, each implementing a different traffic encryption algorithm in hardware.

Algorithm	Function
DES	Encryption and decryption of traffic exchanged between a CMTS and a cable modem Supports both 40- and 56-bit DES running in CBC mode with 64-bit blocks
3DES	Encryption and decryption of TEKs Operates in EDE mode using two 56-bit DES keys Individual DES encryptions and decryptions run in ECB mode with 64-bit blocks
SHA-1	Creation and verification of HMACs for authenticating *Key Request* and *Key Reply* messages Generation and verification of X.509 certificate signatures (in conjunction with the RSA public-key algorithm)
RSA	Encryption and decryption of keying material (such as authorization keys) exchanged between a CMTS and a cable modem Generation and verification of X.509 certificate signatures (in conjunction with the SHA-1 message digest) Verification of signed software upgrade files

Table 9-11 Cryptographic Algorithms Employed by BPI+

NOTE:
In addition to SHA-1, DOCSIS 1.1 uses MD5 for message integrity. However, the use of MD5 is outside the scope of BPI+.

DOCSIS 1.1 BPI+ X.509 Certificate Usage and PKI Hierarchies

The BPI+ authorization process makes use of digital certificates during cable modem authentication. Each cable modem contains both an RSA key pair and an X.509v3 certificate. A cable modem supplies its certificate to a CMTS within an *Authorization Request* message. In response to the request for authorization, a CMTS authenticates a cable modem based on the information contained within its (valid) manufacturer-installed certificate. Following authentication, the CMTS uses the public key in the certificate to exchange keying material with the cable modem. The cable modem then uses its private key to decrypt the keying material contained within the CMTS's *Authorization Reply message. The use of digital certificates within BPI+ requires a hierarchy for issuing certificates and managing the trust relationships between the issuing CAs.*

NOTE:
This section requires basic knowledge of digital certificates and public-key infrastructure (PKI). For a refresher, refer to Appendix B of this text.

BPI+ Cable Modem Certificate Hierarchy

BPI+ employs a three-level certificate hierarchy as illustrated in Figure 9-6. At the top level of the hierarchy sits the DOCSIS Root Certification Authority. The DOCSIS Root CA generates its own RSA public/private key pair, and issues a single, self-signed certificate binding its identity to its public key. It uses its private key to sign subordinate CA certificates issued to cable modem manufacturers. The manufacturers, in turn, use the private keys corresponding to the public keys in their X.509 certificates to sign end-entity cable modem certificates during manufacturing. Each manufacturer is responsible for either administering its own CA or outsourcing CA administration to a third-party PKI provider. The process and protocols used by a manufacturer to request and generate cable modem certificates during manufacturing is also left up to the manufacturer. RSA Security, Inc., offers a full suite of products and services that allow manufacturers to operate and maintain their own manufacturing CAs and to embed DOCSIS security into their devices. For more information on RSA Security's DOCSIS solution, visit http://www.rsasecurity.com/services/industry-specific/docsis.html and http://www.rsasecurity.com/products/bsafe/broadband.html.

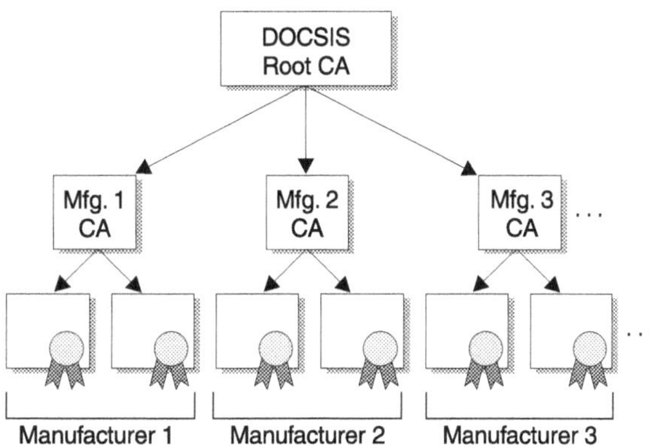

Figure 9-6 DOCSIS 1.1 BPI+ uses a three-level device certificate hierarchy to establish trust relationships between the DOCSIS Root CA, subordinate manufacturer CAs, and cable modems.

NOTE:
In addition to storing its RSA key pair and digital certificate in write-once memory, every cable modem maintains a copy of its manufacturer's CA certificate and the DOCSIS Root CA public key in nonvolatile RAM.

While CableLabs retains control over each of its root CAs (for DOCSIS, PacketCable, and other CableLabs initiatives), it does not actually operate the DOCSIS Root CA. In 2001, CableLabs outsourced this responsibility to Verisign, a third-party PKI service provider. Verisign now hosts the DOCSIS Root CA private key and offers a manufacturer CA certificate-signing service on behalf of CableLabs. To request a CA certificate, a manufacturer generates a PKCS #10–formatted certificate request, and submits the request to Verisign. Once a month, Verisign holds a certificate-signing ceremony where it issues certificates to those manufacturers who have obtained the appropriate approvals from CableLabs. For more information regarding Verisign's signing service, visit http://www.verisign.com/products/cable/signing.html. If you wish to view the DOCSIS Root CA certificate, you may download the certificate from http://www.verisign.com/products/cable/root.html. On a Microsoft Windows–based computer, you can display the certificate by double-clicking the downloaded file (see Figure 9-7). (You may have to change the file extension to .cer before Windows recognizes the file type.)

Manufacturer CAs

To produce cable modems that comply with BPI+, each manufacturer must operate a CA for issuing end-entity device certificates to cable modems. A typical configuration for a manufacturer CA consists of a server running CA software and at least one client for requesting cable modem certificates (see Figure 9-8). The CA may reside on the manufacturing floor or in a centrally managed facility; but regardless of the location, the manufacturer's RSA private key *must* remain secure. The client application is responsible for generating an RSA key

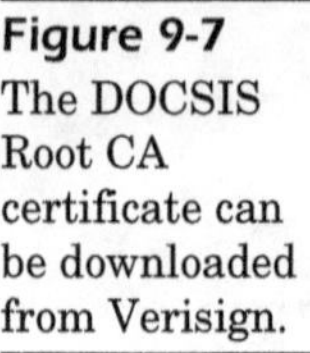

Figure 9-7 The DOCSIS Root CA certificate can be downloaded from Verisign.

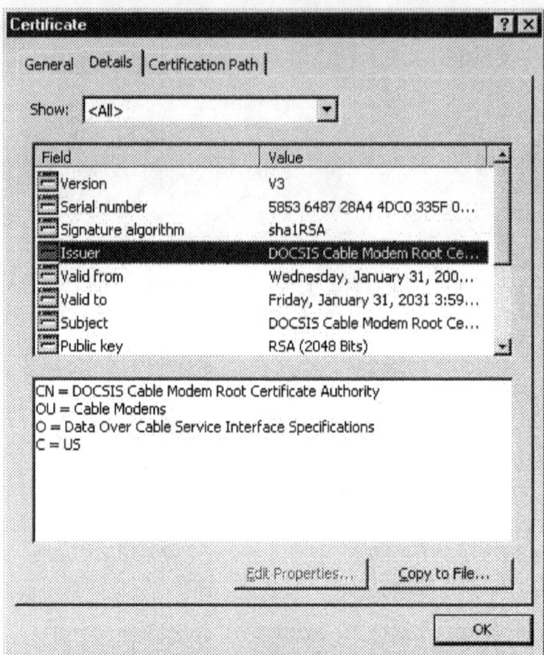

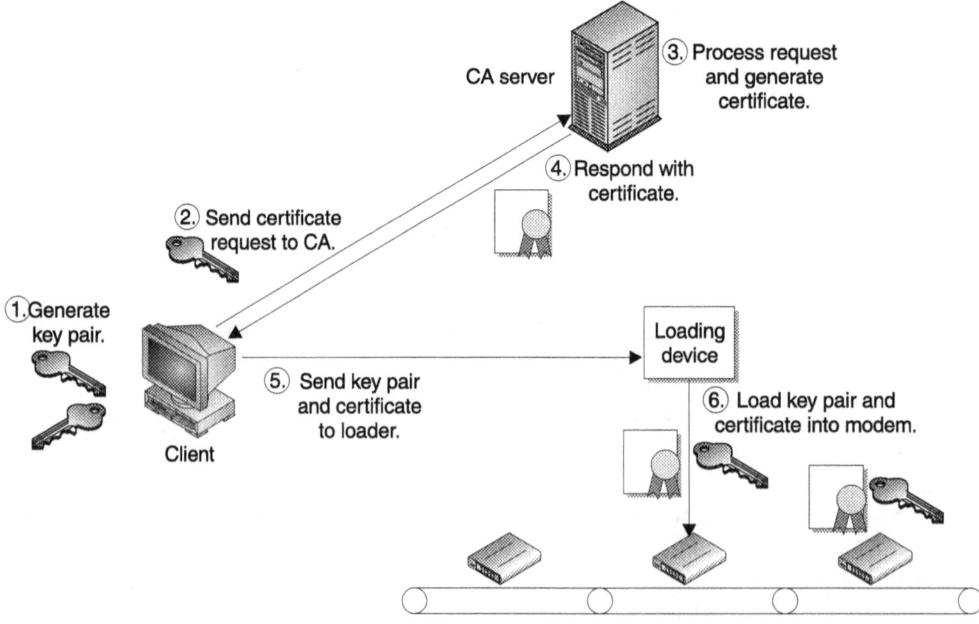

Figure 9-8 A typical manufacturer CA configuration consists of a CA server and at least one certificate request client.

pair for each cable modem and supplying the public key to the CA in a certificate request. Once the CA issues the certificate, the client application parses the response, and passes the certificate and key pair off to a device that loads them into the cable modem's write-once memory.

To improve production times or facilitate production within geographically dispersed manufacturing facilities (see Figure 9-9), a manufacturer may run multiple certificate-request clients. This solution provides scalability beyond the single-client case. When using this configuration, however, steps should be taken to prevent an intruder from intercepting and modifying certificate requests sent from a remote manufacturing facility to the CA. Methods for securing certificate requests might include the use of PKCS #10 or another certificate request protocol that performs integrity checking, or the transmission of certificate requests over a dedicated VPN connection.

TIP:

Since RSA key-pair generation is one of the most computationally expensive operations in the certificate request and generation process, a manufacturer may wish to run multiple clients even within the same manufacturing facility. All clients can request certificates from a single manufacturing CA.

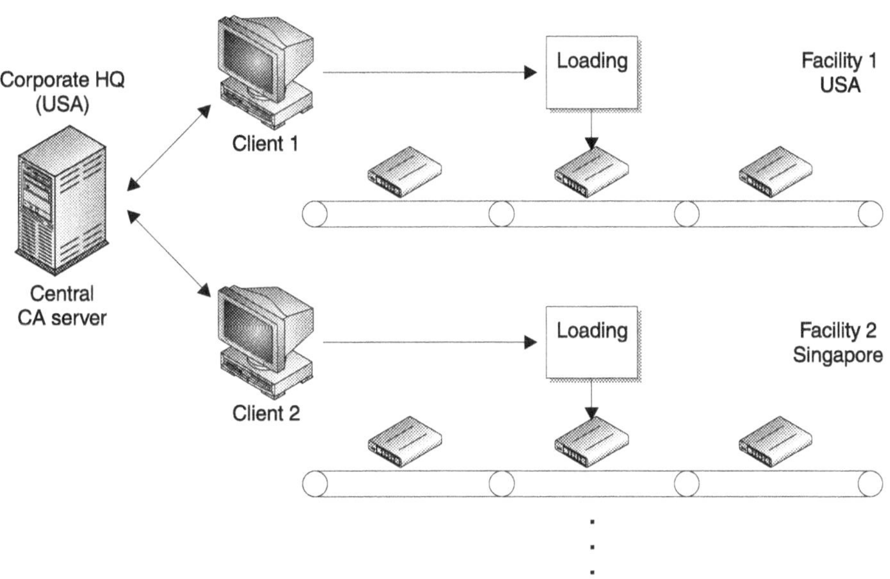

Figure 9-9 A manufacturer may run a single, centrally located CA and a certificate-request client at multiple remote manufacturing sites.

A manufacturer may wish to run multiple CAs, each housed in a different manufacturing facility. In a distributed manufacturing environment, such as that depicted in Figure 9-10, the use of multiple CAs offers higher performance, greater scalability, and more redundancy than a single CA. By keeping certificate requests on the local network, roundtrip request and response times are reduced, resulting in more efficient cable modem production. This approach is also more secure than transmitting requests over a public network such as the Internet. The use of multiple CAs permits the manufacturer to continue producing certificates in the event of a CA failure. If the manufacturer does not wish to request a second manufacturer CA certificate from the DOCSIS Root CA, it may use its existing RSA key pair and CA certificate for more than one manufacturing CA.

CAUTION:
When operating multiple CAs, the manufacturer must take steps to guarantee that the CAs do not issue certificates containing identical serial numbers.

For improved security, some manufacturers may generate CA private keys within *hardware security modules (HSMs)*. These devices are designed to pro-

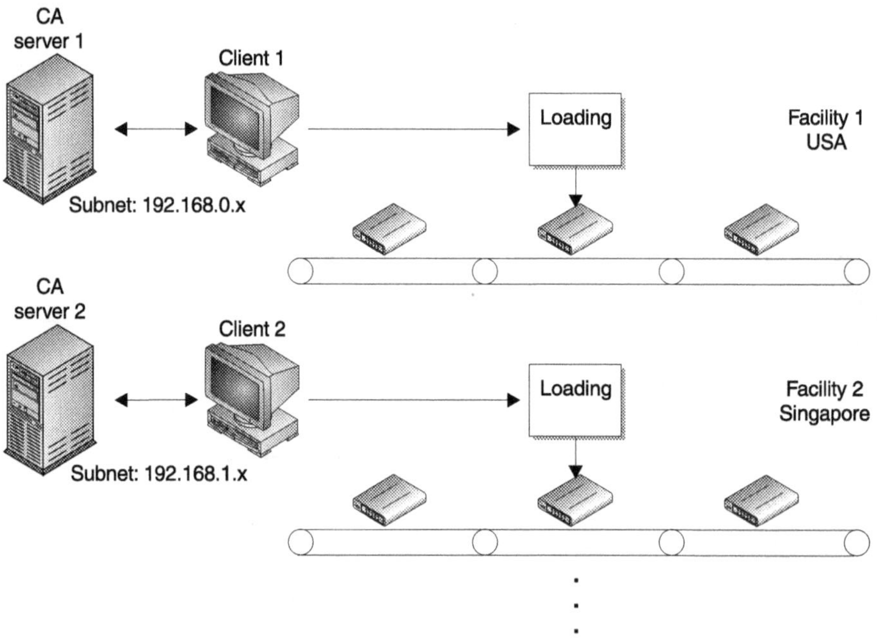

Figure 9-10 In the case of multiple manufacturing facilities, a manufacturer may wish to run more than one CA.

tect keying material by preventing an intruder from physically extracting the keys. In other words, the keys are generated, stored, and destroyed within, and only within, the HSM. Use of an HSM may prevent a manufacturer from duplicating the private key of one manufacturing CA for use by another. To accomplish this task, the manufacturer must generate the original CA private key using an HSM that supports *key backup*. Key backup allows the private key generated by one HSM to be securely copied to another HSM. The Chrysalis Luna CA[3] is an example of an HSM that supports key backup.

In addition to the three-tier device certificate hierarchy, BPI+ defines a second hierarchy for the distribution of code signing certificates. This hierarchy consists of two levels (see Figure 9-11). At the top of the hierarchy lies the DOCSIS Root CA (the same one used for issuing manufacturer CA certificates), and at the bottom rests the manufacturer and MSO code verification certificates. (We'll discuss code verification certificates further in the following section.)

NOTE:
CableLabs has its own CVC for signing manufacturer code images that have passed its DOCSIS 1.1 certification process.

BPI+ Certificate Formats

BPI+ specifies the use of X.509 version 3 certificates for the DOCSIS Root CA, subordinate manufacturer CAs, and cable modems. All certificates must conform to the X.509 standard, as well as the certificate profile specified in RFC 2459 (Housley et al., 1999). BPI+ requires all certificates to be signed using RSA with SHA-1, and restricts the possible values for a number of certificate fields. Appendix B discusses in detail each field of an X.509 certificate. Tables 9-12, 9-13, and 9-14 list the contents of the DOCSIS Root CA, the manufacturer CA, and the cable modem certificates, respectively. (The brackets indicate optional values.)

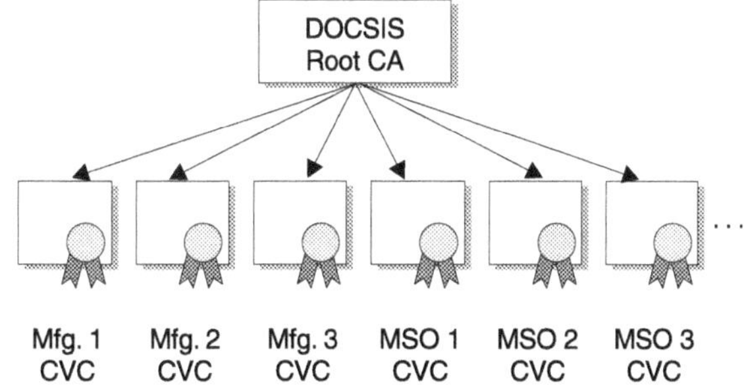

Figure 9-11 BPI+ includes a second certificate hierarchy, stemming from the same DOCSIS Root, for manufacturer and MSO code signing certificates.

Field	Value(s)
Version	3
Serial number	The serial number of the DOCSIS Root CA certificate
Signature algorithm ID	PKCS #1 RSA signature using SHA-1 (ASN.1 OID 1.2.840.113549.1.1.5)
Issuer	A distinguished name of the following format (self-signed): countryName = US organizationName = Data Over Cable Service Interface Specifications organizationalUnitName = Cable Modems commonName = DOCSIS Cable Modem Root Certification Authority
Validity period	Begins on or before the date that the DOCSIS Root CA begins issuing manufacturer CA certificates, and can be periodically reissued.
Subject	A distinguished name of the following format: countryName = US organizationName = Data Over Cable Service Interface Specifications organizationalUnitName = Cable Modems commonName = DOCSIS Cable Modem Root Certification Authority
Subject public-key information	Public key of the DOCSIS Root CA and corresponding algorithm identifier (OID 1.2.840.113549.1.1)
Issuer unique ID	Not present in BPI+ certificates
Subject unique ID	Not present in BPI+ certificates
Extensions	Certificate extensions are optional in the DOCSIS Root CA certificate, and all extensions must conform to RFC 2459. According to the BPI+ specification, the DOCSIS Root CA certificate may contain any noncritical extensions and must contain the following critical extensions: Basic Constraints with the CA parameter set to true and the path length constraint set to one Key Usage with the following bit settings: Certificate signing bit must be turned on. CRL signing bit may be turned on. All other bits should be turned off.

Table 9-12 DOCSIS Root CA Certificate Fields

Field	Value(s)
Version	3
Serial number	A unique serial number issued by the DOCSIS Root CA.
Signature algorithm ID	PKCS #1 RSA signature using SHA-1 (ASN.1 OID 1.2.840.113549.1.1.5)
Issuer	A distinguished name of the following format: countryName = US organizationName = Data Over Cable Service Interface Specifications organizationalUnitName = Cable Modems commonName = DOCSIS Cable Modem Root Certification Authority
Validity period	Begins when the certificate is issued by the DOCSIS Root CA, and may be periodically reissued.
Subject	A distinguished name of the following format: countryName = <country of manufacturer> [stateOrProvinceName = <state/province>] [localityName = <city>] organizationName = <company name> organizationalUnitName = DOCSIS [organizationalUnitName = <manufacturing location>] commonName = <company name> Cable Modem Root Certification Authority
Subject public-key information	Public key of the manufacturing CA and corresponding algorithm identifier (OID 1.2.840.113549.1.1)
Issuer unique ID	Not present in BPI+ certificates
Subject unique ID	Not present in BPI+ certificates
Extensions	Certificate extensions are optional in manufacturer CA certificates. All extensions must conform to RFC 2459. According to the BPI+ specification, manufacturer CA certificates may contain the following extensions: Basic Constraints with the CA parameter set to true and the path length constraint set to zero (may be marked as critical) Key Usage with the following bit settings (may be marked as critical): Certificate signing bit must be turned on. CRL signing bit may be turned on. All other bits should be turned off. Any other noncritical extensions

Table 9-13 Subordinate Manufacturer CA Certificate Fields

Field	Value(s)
Version	3
Serial number	A unique serial number issued by a manufacturing CA; every cable modem certificate issued by the same manufacturer must contain a unique serial number.
Signature algorithm ID	PKCS #1 RSA signature using SHA-1 (ASN.1 OID 1.2.840.113549.1.1.5)
Issuer	A distinguished name of the following format: countryName = <country of manufacturer> [stateOrProvinceName = <state/province>] [localityName = <city>] organizationName = <company name> organizationalUnitName = DOCSIS [organizationalUnitName = <manufacturing location>] commonName = <company name> Cable Modem Root Certification Authority
Validity period	Begins with the manufacturing date of the cable modem and extends beyond the expected operational lifetime of the cable modem (at least 20 years)
Subject	A distinguished name of the following format: countryName = <country of manufacturer> organizationName = <company name> organizationalUnitName = <manufacturing location> commonName = <serial number> commonName = <MAC address>
Subject public-key information	768- or 1024-bit RSA public key of the cable modem and corresponding algorithm identifier (OID 1.2.840.113549.1.1)
Issuer unique ID	Not present in BPI+ certificates
Subject unique ID	Not present in BPI+ certificates
Extensions	Certificate extensions are optional in cable modem certificates, and all extensions must conform to RFC 2459. Cable modem certificates may contain *only* noncritical extensions. According to the BPI+ specification, when using the Key Usage extension, the key agreement and key encipherment bits must be turned on, the certificate and CRL signing bits must be turned off, and all other bits should be turned off.

Table 9-14 Cable Modem End-Entity Certificate Fields

An *object identifier (OID)* is a series of integer values representing an ASN.1 (Abstract Syntax Notation One) object. The IETF uses a decimal notation to separate the components of an OID. The portions are arranged hierarchically from left to right. An organization with the OID 1.2.840 is allowed to issue OIDs of the format 1.2.840.*x*, where *x* is another integer value. Using the more descriptive ITU-T notation, the RSA signature OID of 1.2.840.113549.1.1.5 corresponds to {iso(1) member-body(2) us(840) rsadsi(113549) pkcs(1) pkcs-1(1) 5}. The organization named in the OID—RSA Data Security, Inc. (rsadsi)—defined its own OIDs for the PKCS series of standards. In this case, the OID refers to PKCS #1. (See Kaliski, 1993 for a description of ASN.1.)

NOTE:
Cable modem certificates cannot be reissued, therefore they must contain a validity period extending the anticipated life of the cable modem.

In support of secure software upgrades (discussed in the section entitled "Signed Software Upgrade Verification"), BPI+ includes an additional certificate format for *code verification certificates (CVCs)*. The DOCSIS Root CA issues these certificates to cable modem manufacturers and MSOs for use in signing software upgrades. Table 9-15 describes the format of a BPI+ CVC.

Certificate Validation on the CMTS

Before a CMTS accepts a cable modem's certificate as proof of identity, it performs a number of checks to determine the validity of the certificate. Certificate validation involves examining certain fields and extensions within the certificate, verifying the certificate's signature, and chaining the certificate up to a trusted certification authority (the DOCSIS Root CA). Cable modem certificate validation succeeds when the following criteria are met:

- The certificate chains up to the DOCSIS Root CA or is marked as trusted within the CMTS.
- The CMTS can verify the certificate signature.
- The current date and time are within the validity period specified by the certificate.
- The certificate is not contained within a hot list.
- The subject identifier within the certificate matches the MAC address of the cable modem.
- The key usage extension within the certificate has the appropriate bits enabled for key agreement and key encipherment, and has the bits for certificate and CRL signing disabled.

Field	Value(s)
Version	3
Serial number	A unique serial number issued by the DOCSIS Root CA.
Signature algorithm ID	PKCS #1 RSA signature using SHA-1 (ASN.1 OID 1.2.840.113549.1.1.5)
Issuer	A distinguished name of the following format: countryName = US organizationName = Data Over Cable Service Interface Specifications organizationalUnitName = Cable Modems commonName = DOCSIS Cable Modem Root Certificate Authority
Validity period	Begins when issued by the DOCSIS Root CA, and may be periodically reissued
Subject	A distinguished name of the following format: countryName = <country of subject> organizationName = <subject code-signing agent> organizationalUnitName = DOCSIS commonName = Code Verification Certificate
Subject public-key information	1024-, 1536-, or 2048-bit RSA public key of the cable modem manufacturer or MSO and corresponding algorithm identifier (OID 1.2.840.113549.1.1)
Issuer unique ID	Not present in BPI+ CVCs
Subject unique ID	Not present in BPI+ CVCs
Extensions	CVCs must contain only the Extended Key Usage extension, which must be marked as critical. The extension should indicate that the CVC is for code signing. Additional extensions or an Extended Key Usage extension marked as noncritical or missing results in certificate validation failure.

Table 9-15 Manufacturer and MSO Code-Verification Certificate (CVC) Fields

A CMTS can cache certificate validation results by calculating a thumbprint for each validated cable modem and manufacturer CA certificate. A *thumbprint* is simply the result of passing a certificate through a message digest. Instead of validating a certificate each time a CMTS receives an initial *Authorization Request* message, the CMTS may refer to its list of calculated thumbprints to determine whether the certificate and its corresponding certificate chain have been previously validated. When compared to the time spent validating even a short certificate chain, the calculation and comparison of a certificate thumbprint offers an efficient alternative.

Certificate Revocation and Hot Lists

A manufacturer may generate and distribute *certificate revocation lists (CRLs)* to MSOs identifying revoked manufacturer CA and cable modem certificates. However, BPI+ does not define the CRL format or the process used by the manufacturer and MSO to exchange revocation information. Although a manufacturer and MSO may exchange CRLs, BPI+ includes no built-in support for revocation lists. Instead, BPI+ requires that all CMTSs support the use of *hot lists*. Hot lists are maintained by the MSOs, and contain revoked, or otherwise untrusted, manufacturer CA and cable modem certificates. Before a CMTS authorizes a cable modem, it must verify that the modem is not included in the most recent hot list. The protocols and procedures associated with management and installation of hot lists are beyond the scope of BPI+.

TFTP Configuration Files

During the provisioning process, every DOCSIS cable modem downloads a configuration file from a TFTP (Trivial File Transfer Protocol) server maintained by the MSO who owns the cable network. The configuration file contains numerous parameters, many of which relate directly to BPI+ initialization and operation. Table 9-16 contains a list of the BPI+ parameters that can appear in a configuration file.

Each configuration file contains a CMTS *MIC (message integrity check)* field for ensuring the integrity of the file. The CMTS MIC calculation involves the generation of a keyed message digest over the contents of the configuration file using HMAC-MD5. The CMTS and provisioning server (the server responsible for generating the configuration file) exchange a shared secret *authentication string*, which the CMTS uses as the key in its HMAC-MD5 calculation. When a cable modem receives a configuration file from a CMTS, it updates its configura-

Parameter	Range	Default Value
Authorization Lifetime	1–6,048,000 seconds	7 days (604,800 seconds)
TEK Lifetime	1–604,800 seconds	12 hours (43,200 seconds)
Authorize Wait Timeout	1–30 seconds	10 seconds
Reauthorize Wait Timeout	1–30 seconds	10 seconds
Authorization Grace Time	1–6,047,999 seconds	10 minutes (600 seconds)
Operational Wait Timeout	1–10 seconds	10 seconds
Rekey Wait Timeout	1–10 seconds	10 seconds
TEK Grace Time	1–302,399 seconds	1 hour (3,600 seconds)
Authorize Reject Wait Timeout	1–600 seconds	60 seconds
SA Map Wait Timeout	1–10 seconds	1 second
SA Map Max Retries	0–10 attempts	4

Table 9-16 BPI+ TFTP Configuration File Parameters

tion settings. It then extracts the MIC from the configuration file, and forwards it back to the CMTS in a *Registration Request (REG-REQ)* message along with a list of its current configuration settings. Upon receipt of the REG-REQ, the CMTS computes an HMAC-MD5 digest over the shared authentication string and the appropriate configuration setting fields in the request message and compares the result with the CMTS MIC provided by the cable modem. If the values match, then the REG-REQ is accepted; otherwise, the CTMS returns a registration rejection notice to the cable modem (CableLabs, 2001b).

Signed Software Upgrade Verification

To facilitate firmware upgrades, BPI+ supports remote downloading of code images by cable modems deployed in subscribers' homes and offices. Automated remote software upgrades eliminate the need for technicians or subscribers to manually install code updates, and lowers maintenance costs by reducing supports calls and on-site visits associated with the upgrade process. Additionally, an MSO can upgrade a network in a fraction of the time required to manually perform the upgrade and ensure that the upgrade process is completed successfully on each modem.

Although remote software upgrades offer many benefits, they provide a potential target for attackers. If there are no appropriate countermeasures in place, a malicious user might force a cable modem to download and install a rogue code update. Pushing malicious code to a single cable modem may only disrupt service for a single user. However, a large-scale attack involving many cable modems could potentially affect an entire network. Take, for instance, the case where a rogue update causes a cable modem to generate spurious network traffic. While a single malfunctioning modem might not noticeably affect network performance, imagine the effects of dozens or even hundreds of modems working together in a DoS attack. An attack of this magnitude could slow a best-effort IP-based network to a crawl. To prevent such attacks, BPI+ requires that all software upgrade files be signed by the issuing manufacturer, and optionally signed by the MSO.

A BPI+ signed software upgrade, as shown here, is a PKCS #7–compliant data structure consisting of at least one PKCS #7 detached signature, a list of TLV encoded options, and a binary code image.

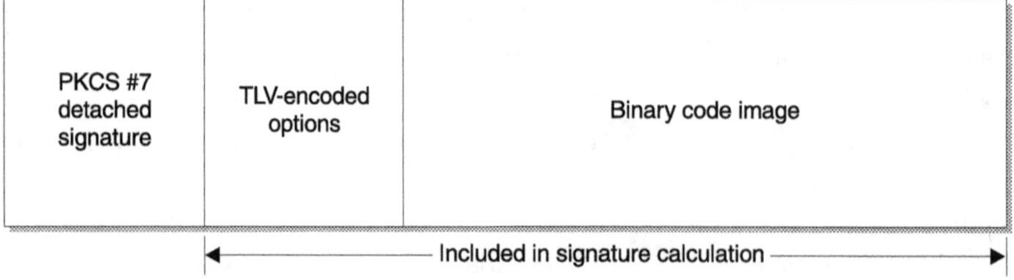

TLV encoding provides a method for describing the format of data as a collection of fields with defined types, lengths, and values. Section 4.2.2 of

(CableLabs, 2001a) defines the TLV encoding scheme employed by BPI+. All software upgrade files must follow the *Distinguished Encoding Rules (DERs)* for ASN.1 encoding. In addition to a firmware code image, a signed software upgrade might also contain an updated DOCSIS Root CA public key and updated manufacturer certificates to replace those stored in the cable modem. Table 9-17 represents the structure of a BPI+ signed-software upgrade. All software upgrade files contain the following:

- The cable modem manufacturer's *Code Verification Signature (CVS)*
- The manufacturer's *Code Verification Certificate (CVC)*—the name in the certificate must match the manufacturer name stored in the cable modem
- The CVS and CVC of an MSO (optional)
- A mandatory list of download parameters
- The DOCSIS Root CA public key (optional)
- Manufacturer certificate(s) (optional)
- A code image

Code File	Description
PKCS #7 Digital Signature{	
ContentInfo	
contentType	SignedData
SignedData()	EXPLICIT signed-data content value; includes CVS and X.509 CVC
}	
SignedContent{	
DownloadParameters {	Mandatory TLV format (Type 28) defined in section 4.2.2.28 (CableLabs, 2001a). (Length is zero if there are no sub-TLVs.)
RootCAPublicKey()	Optional TLV for the Root CA Public Key for CVC verification, formatted according to the RSA-Public-Key TLV format (Type 4) defined in section 4.2.2.4 (CableLabs, 2001a).
MfgCerts()	Optional TLV for one or more DER-encoded Manufacturer Certificate(s) each formatted according to the CA-Certificate TLV format (Type 17) defined in section 4.2.2.17 (CableLabs, 2001a).
}	
CodeImage()	Upgrade code image
}	

Table 9-17 The Contents of a PKCS #7 Signed Software Upgrade (CableLabs, 2001a)

NOTE:
The fields containing the DOCSIS Root CA public key and the manufacturer CA certificates are separate from the code image in the PKCS #7 structure, which makes it possible for an MSO to change these values without modifying the code image.

Table 9-18 lists the DER-encoded contents of the PKCS #7 digital signature portion of a BPI+ signed software upgrade.

PKCS #7 Field	Description
Signed Data {	
version	Version = 1
digestAlgorithmIdentifier	SHA-1
contentInfo	
contentType	Data (SignedContent is concatenated at the end of the PKCS#7 structure)
certificates {	DOCSIS Code Verification Certificate (CVC)
mfgCVC	Required for all code files
msoCVC	OPTIONAL; required for MSO co-signatures
} *end certificates*	
SignerInfo {	
MfgSignerInfo {	REQUIRED for all code files
version	Version = 1
issuerAndSerialNumber	From the singer's certificate
issuerName	Distinguished name of the certificate issuer
CountryName	US
organizationName	Data Over Cable Service Interface Specifications
organizational UnitName	Cable Modems
commonName	DOCSIS Cable Modem Root Certificate Authority
certificateSerialNumber	From CVC; Integer, 8 octets
digestAlgorithm	SHA-1
authenticatedAttributes	
contentType	Data; contentType of signedContent
signingTime	UTCTime (GMT), YYMMDDhhmmssZ
messageDigest	Digest of the content as defined in PKCS #7 (RSA Laboratories, 1993)
digestEncryptionAlgorithm	rsaEncryption

Table 9-18 PKCS #7 Detached Signature Format Employed by BPI+ (CableLabs, 2001a)

PKCS #7 Field	Description
encryptedDigest	
} *end mfg signer info*	
MsoSignerInfo {	OPTIONAL; required for MSO co-signatures
version	Version = 1
issuerAndSerialNumber	From the signer's certificate
issuerName	Distinguished name of the certificate issuer
CountryName	US
organizationName	Data Over Cable Service Interface Specifications
organizationUnitName	Cable Modems
commonName	DOCSIS Cable Modem Root Certificate Authority
certificateSerialNumber	From CVC; Integer, 8 octets
digestAlgorithm	SHA-1
authenticatedAttributes	
contentType	Data; contentType of signedContent
signingTime	UTCTime (GMT), YYMMDDhhmmssZ
messageDigest	Digest of the content as defined in PKCS #7 (RSA Laboratories, 1993)
digestEncryptionAlgorithm	rsaEncryption
encryptedDigest	
} *end mso signer info*	
} *end signer info*	
} *end signed data*	

Table 9-18 PKCS #7 Detached Signature Format Employed by BPI+ (CableLabs, 2001a) *(continued)*

Generation and Verification of Signed Software Upgrade Files

The process for generating signed software upgrade files is relatively simple. First, the manufacturer generates and tests the code file. The manufacturer then bundles the code file with the appropriate options fields and certificates, and signs the content using the private key associated with its CVC. Following signing, the manufacturer sends the code update to an MSO who validates the image by verifying the signature. If the signature verifies correctly, the MSO has the option of co-signing the image before distributing it to the cable modems on its network.

Prior to installing a software upgrade, a cable modem must verify the manufacturer's signature and, if present, the optional signature of the MSO to ensure that the code file originated from a trusted source and that it has not been altered in transit. A cable modem must follow these steps when verifying a signed software upgrade:

1. Ensure that the signing time contained within the manufacturer's PKCS #7 signature is greater than or equal to the **cvcAccessStart** time stored in the cable modem, and within the validity start and end time provided in the manufacturer's CVC.
2. Ensure that the manufacturer's CVC is valid. (In other words, it must contain a manufacturer name matching that stored in the cable modem, have a validity start time greater than or equal to the **cvcAccessStart** time stored on the cable modem, and have the appropriate extended key usage values.)
3. Ensure that the CVC signature verifies correctly, using the DOCSIS Root CA public key.
4. Ensure that the manufacturer's signature (see Table 9-18) generated over the SignedContent portion of the software upgrade verifies correctly. The cable modem must be capable of verifying RSA signatures with modulus sizes of 1024, 1526, or 2048 bits.
5. In the presence of an MSO signature, repeat steps 1–4 above.
6. If signature verification is successful, install the code image and update the values for the manufacturer and optional MSO time-varying controls (**codeAccessStart** and **cvcAccessStart**) stored in the modem's nonvolatile memory.

These steps do not have to be completed in the order shown here, but the cable modem must complete successfully all steps before installing the code image.

NOTE:

*The manufacturer and MSO use the time-varying controls—**codeAccessStart** and **cvcAccessStart**—to prevent a cable modem from installing an outdated code update or accepting an old CVC. The code image must contain a signature with a signing time greater than or equal to the **codeAccessStart** time, and the manufacturer's CVC (and optionally the MSO's CVC) must have a validity start time greater than or equal to the **cvcAccessStart** time. The values for these fields must not decrease over time.*

The code update process for BPI+ does not support CRLs. Instead, when a cable modem receives an updated CVC within a configuration file or via SNMP, it must update its cvcAccessStart *variable to indicate the new validity period for the updated CVC. This variable prevents the cable modem from accepting an outdated CVC while verifying a signed software upgrade.*

References

CableLabs. 29 August, 2001. *DOCSIS Baseline Privacy Plus Interface Specification (SP-BPI+-I07-010829)*. Cable Television Laboratories.

CableLabs. 29 August, 2001. *DOCSIS Radio Frequency Interface Specification (SP-RFIv1.1-I07-010829)*.

National Institute of Standards and Technology. April, 1982. *Federal Information Processing Standards Publication 140-1, Security Requirements for Cryptographic Modules*.

Housley, R., W. Ford, W. Polk, and D. Solo. January, 1999. *RFC 2459: Internet X.509 Public Key Infrastructure Certificate and CRL Profile*. Internet Engineering Task Force.

Kaliski, Burton S., Jr. November, 1993. *A Layman's Guide to a Subset of ASN.1, BER, and DER*. RSA Laboratories.

RSA Laboratories. 1 November, 1993. *PKCS #7 v1.5: Cryptographic Message Syntax Standard*.

Stallings, William. 1999. *Cryptography and Network Security: Principles and Practice*. Upper Saddle River, New Jersey: Prentice-Hall.

References

Chapter 10

Securing Real-Time Multimedia: PacketCable Security

As a natural extension to DOCSIS 1.1 cable modem networks, CableLabs and its members realized the need for a scalable architecture supporting the delivery of real-time multimedia applications over two-way cable systems. In 1997, CableLabs began the PacketCable initiative to develop a set of interface specifications for the delivery of real-time multimedia traffic with varying levels of QoS over packet-based networks. While IP telephony will likely serve as the first practical application for PacketCable networks, potential PacketCable services include voice and video conferencing, interactive gaming, and other real-time streaming multimedia applications.

As mentioned briefly in Chapter 5, PacketCable operates at or above the Network layer and includes the necessary QoS functionality to support constant bit rate data services. Although PacketCable includes QoS functionality of its own, it must also interact with DOCSIS 1.1 for end-to-end QoS. (Remember from Chapter 4 that end-to-end QoS requires all network layers to support the necessary QoS mechanisms.)

NOTE:
The use of IP-based packet-switching networks for the delivery of QoS-enabled real-time services provides two primary advantages over cell relay (ATM): cost and ubiquity. IP-based networks are less expensive and far more common than ATM-based networks, and they are more practical for delivering services to home and small- to medium-sized business subscribers. See Chapter 4 for a list of the advantages and disadvantages for packet-switching and cell-relay networks.

PacketCable combines many new and existing standards and technologies into a complete solution providing end-to-end support for signaling, media transport at various levels of QoS, security, provisioning, billing, and network management functions (CableLabs, 2002). PacketCable includes functional specifications for the following technologies:

- Audio/video codecs and media stream mapping
- Audio server protocol
- Call Management Server (CMS) signaling
- Dynamic QoS
- Electronic surveillance
- Event messaging
- Interdomain QoS
- Internet Signaling Transport Protocol (ISTP)
- Management event mechanisms
- Management Information Base (MIB) specifications
- Media Terminal Adapter (MTA) device provisioning
- Network-based call signaling
- Public Switched Telephone Network (PSTN) gateway call signaling
- Security

NOTE:
Each PacketCable specification is denoted by a version number based on the functionality it includes. PacketCable 1.0 covers the basic intra-domain architecture of a PacketCable network, which includes security. (The term "intra-domain" describes all network components within a single service provider's network.) PacketCable 1.1 introduces electronic surveillance and primary line-capable services. ("Primary line" refers to a communications service providing virtual constant availability as a replacement for a telephone line.) PacketCable 1.2 includes inter-domain QoS ("inter-domain" means from one provider's PacketCable network to another) and call management server (CMS) signaling.

In addition to these specifications, PacketCable includes a number of technical reports addressing the PacketCable architectural framework, network call signaling, operations support systems, call flow architectures, and primary line service capabilities. For more information regarding these specifications, visit http://www.packetcable.com/specifications.html.

The PacketCable architecture, depicted in Figure 10-1, encompasses far more network elements and protocols than DOCSIS 1.1. In addition to the elements

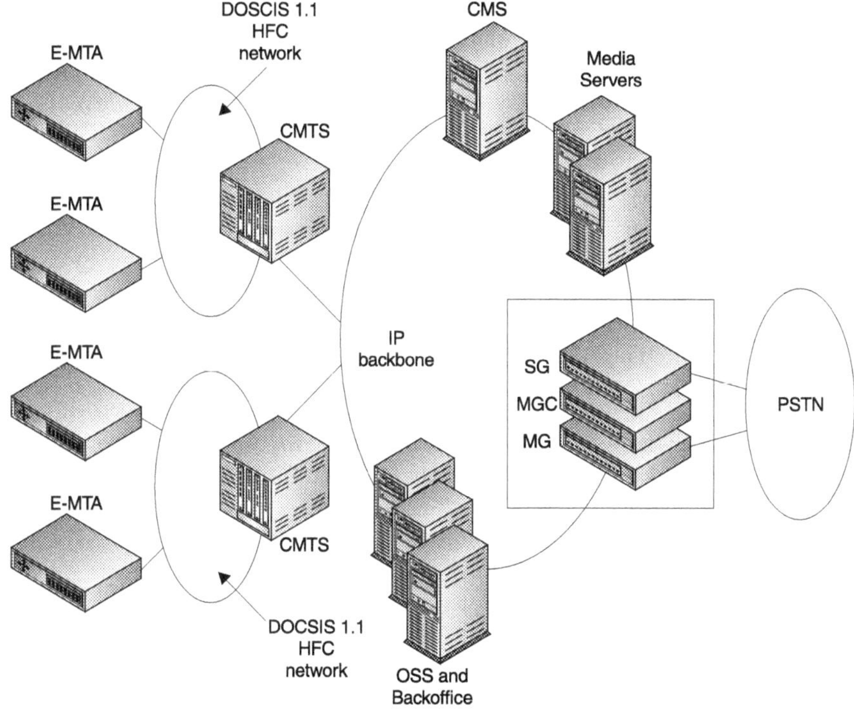

Figure 10-1 At a high level, the PacketCable network architecture consists of over half a dozen functional elements.

comprising a DOCSIS hybrid fiber-coax (HFC) network, the PacketCable environment consist of MTAs, CMSs, various media and signaling gateways for interfacing to PSTN networks, operation support systems (for billing, provisioning, and configuration of subscriber devices), electronic surveillance systems, and security servers. These components communicate using a wide variety of transport, signaling, and security protocols. Table 10-1 offers a brief definition of each component within a PacketCable network, Table 10-2 lists the subcomponents, and Table 10-3 summarizes the protocols used on the various network interfaces.

Component	Description
Border router	A router that sits on the network boundary between two service providers. Border routers are responsible for aggregating network traffic sent between the two networks and for establishing Resource Reservation Protocol (RSVP) resource reservations for inter-domain QoS.
CM	*Cable modem* A component of the underlying DOCSIS 1.1 HFC access network (see Chapter 9). Cable modems and MTAs coexist within the same device.
CMS	*Call management server* Controls audio call setup, termination, and other call signaling operations. CMSs are also involved in dynamic QoS establishment and consist of two primary components: *call agents (CAs)* and *gate controllers (GCs)*. A CMS may also contain an optional *media player controller (MPC)* and a *media gateway controller (MGC)* (see Table 10-2).
CMTS	*Cable modem termination system* A component of the underlying DOCSIS 1.1 HFC access network that communicates directly with cable modems (see Chapter 9).
DF	*Delivery function* A server used in electronic surveillance (in other words, wiretapping). A DF server collects *call data (CD)*, which are signaling messages that contain the cipher suite and keying material used to secure a communication session, and *call content (CC)*, which is the actual media stream. It then forwards the CD and CC to a *collection function* server, at which point the call may be reconstructed from the CD and CC by an authorized call interception agency.
EBP	*Exterior border proxy* A Session Initiation Protocol (SIP)[1] proxy that routes traffic between service provider domains.

Table 10-1 Elements Within a PacketCable Network

Component	Description
Edge router	A router that sits between an access network and a service provider's internal network. A CMTS routes traffic destined for its provider's internal network through an upstream edge router (which may be a component of the CMTS or a separate device within the network). Edge routers also perform admission control (see "IP Integrated Services Architecture" in Chapter 4), per-flow RSVP resource reservation, and aggregation of RSVP flows entering the provider's network.
KDC	*Key distribution center* A trusted third-party Kerberos server that issues authentication tickets to various PacketCable network components. In the PacketCable 1.0 architecture, MTAs are the only devices that obtain tickets from the KDC. In the PacketCable 1.2 architecture, the CMS and EBP obtain tickets for inter-domain call signaling.
Media servers	Network elements that process media streams in support of voice communication service options.
MG	*Media gateway* Provides a media interface to the PSTN and converts a PacketCable media stream into a format that is appropriate for a PSTN network.
MGC	See Table 10-2.
MP	*Media player* A media resource server that interprets commands from an MPC. An MP delivers voice announcements (for misdialed numbers, numbers no longer in service, and so on) within media streams to MTAs and accepts/reports user inputs.
MPC	See Table 10-2.
MTA	*Media terminal adapter* A subscriber device that contains the necessary media stream signaling/transport protocols and audio CODECs required for voice-over-IP (VoIP) communication. MTAs include both network and voice device (i.e., telephone) interfaces.
OSS	*Operation support systems* Network servers required for configuration, provisioning, and management of subscriber devices and other network components (Dynamic Host Configuration Protocol or DHCP, Trivial File Transfer Protocol or TFTP, billing systems, and provisioning servers).
RKS	*Record keeping server* Collects and correlates event messages related to customer billing.

Table 10-1 Elements Within a PacketCable Network *(continued)*

Component	Description
SG	*Signaling gateway* Provides a signaling interface to the PSTN. An SG sits on the border between a PacketCable network and a Signaling System 7 (SS7) network used by a PSTN.

[1] The Session Initiation Protocol (SIP) is a control protocol for establishing, modifying, and terminating sessions between one or more communicating protocols.

Table 10-1 Elements Within a PacketCable Network *(continued)*

NOTE:
PacketCable Security currently addresses only embedded MTAs, or E-MTAs (those MTAs contained within a DOCSIS 1.1 cable modem). A future specification will define the operation of standalone MTAs (S-MTAs).

Subcomponent	Description	Required?
CA	*Call agent* Provides network intelligence, manages call states, and controls media gateways	Yes
GC	*Gate controller* Coordinates all QoS authorization and control functions for packet-based voice communication	Yes
MGC	*Media gateway controller* Maintains the call state information specific to media gateways	No
MPC	*Media player controller* Initiates and manages media player announcement services	No

Table 10-2 Required and Optional Subcomponents of a Call Management Server

Protocol	Description
CMSS	*CMT-to-CMS Signaling* Used to exchange signaling information between call management servers.
COPS	*Common Open Policy Service Protocol* Provides policy control for QoS signaling and management of provisioned network resources.
DiffServ + Aggregated RSVP	Provides differentiated QoS services for packet-based networks. See Chapter 4.
DOCSIS 1.1	The Link layer (or Data Link layer when using the OSI Reference Model) protocol used by DOCSIS HFC networks. DOCSIS 1.1 is discussed in Chapter 9.
Gate coordination messages	Provide acknowledgment and synchronization information for dynamic QoS between a CMS and CMTS.
IntServ + RSVP	Provides integrated QoS services for packet-based networks. See Chapter 4.
ISTP	*Internet Signaling Transport Protocol* Provides a signaling interconnection between an MGC and an SG on a PacketCable network.
Kerberos	A network authentication protocol used to exchange keying information and obtain access to network servers. Kerberos is discussed in Chapter 6.
NCS	*Network-based Call Signaling* Used to communicate call signaling information between MTAs and CMSs.
PKCROSS	*Public-Key Cryptography for Cross-Realm Authentication* Allows KDCs to authenticate each other during the establishment of inter-realm keys. These keys are later used to negotiate cross-realm tickets.
PKINIT	*Public-Key Cryptography for Initial Authentication* Specifies the use of public-key cryptography and digital certificates during initial Kerberos authentication.

Table 10-3 Protocols Employed by PacketCable Network Interfaces

Protocol	Description
RADIUS	*Remote Authentication Dial-In User Service* A network authentication system originally designed for dial-in environments. RADIUS is now used by many systems for remote authentication. (Note: PacketCable doesn't actually use RADIUS but employs RADIUS-formatted messages.)
RTCP	*Real-Time Control Protocol* The control protocol used in conjunction with RTP (see Chapter 4).
RTP	*Real-time Transport Protocol* A transport protocol for providing QoS to real-time media streams over packet-based networks (see Chapter 4).
SNMP	*Simple Network Management Protocol* A network control and monitoring protocol for communicating status information between an SNMP agent and a centralized management console. Network devices (agents) may be configured remotely via commands sent from an SNMP manager to an agent. PacketCable only supports the use of SNMPv3.
TCAP/IP	*Transaction Capabilities Application Protocol* A protocol within the SS7 stack used to query a service control point for routing numbers associated with dialed phone numbers and to validate calling card transactions. A discussion of SS7 and other PSTN technologies is beyond the scope of this book.
TFTP	*Trivial File Transfer Protocol* A simplified version of the File Transfer Protocol (FTP) used to copy files quickly from one host to another (uses host-based authentication).
TGCP	*Trunking Gateway Control Protocol* A control protocol used in centralized call control architectures to interface between a VoIP network and a PSTN.
UDP	*User Datagram Protocol* A connectionless and unreliable Transport layer protocol that runs on top of the Internet Protocol.

Table 10-3 Protocols Employed by PacketCable Network Interfaces *(continued)*

Overview of PacketCable Security

The security concerns addressed by DOCSIS 1.1 BPI+ apply equally to PacketCable networks. Service providers must incorporate security mechanisms into their PacketCable networks to prevent eavesdropping, unauthorized access to services, and Denial-of-Service (DoS) attacks. In this section, we introduce the PacketCable Security specification (CableLabs, 2002). This specification addresses the threats present in PacketCable networks and outlines the security and key management protocols used on each PacketCable interface. Before we begin our discussion of the various security and key management protocols, we introduce the primary threats that exist on PacketCable networks:

- Theft of service or unauthorized levels of QoS via MTA cloning, network server impersonation, subscription fraud, nonpayment for services rendered, falsification of billing records, or direct attacks against a transport or signaling protocol
- Disclosure of sensitive information (a voice conversation, for example) to an eavesdropper or cloned/impersonated network element, and offline cryptanalysis of data captured by an attacker using a network monitoring utility
- Loss of privacy due to the capture of keying material on a compromised network server
- Disclosure of sensitive information in signaling messages that may be used for marketing purposes or to identify and locate a caller
- Disruption of network service through direct manipulation of transport and signaling protocols and MTA configuration files, device impersonation, and unauthorized access to network elements

The interfaces within a PacketCable network communicate using different transport and signaling protocols. Most network components use a combination of security and key management protocols to secure communication over the various interfaces of the network. While PacketCable does employ DOCSIS 1.1 BPI+ as its Link layer protocol over HFC networks, PacketCable has a number of security requirements that cannot be met by BPI+ (end-to-end security, for instance). All security protocols used by PacketCable operate at the Network layer or Application layer, as illustrated in Figure 10-2. (Remember that the Application layer of the TCP/IP Reference Model corresponds to layers 5–7 of the OSI Reference Model.)

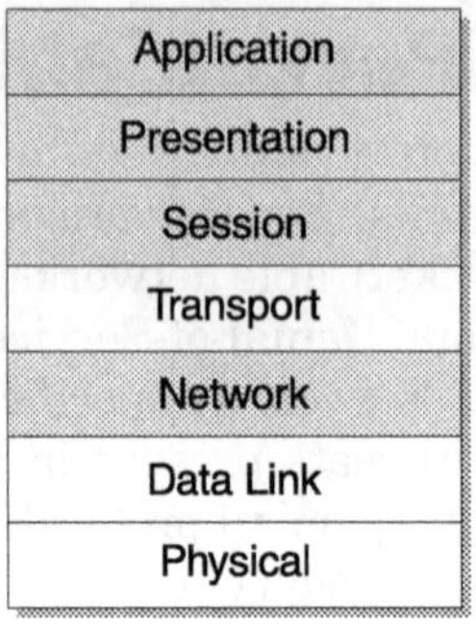

Figure 10-2 Security protocols within PacketCable operate at the Network layer or within the confines of an application (the Application layer).

Table 10-4 summarizes the threats associated with each PacketCable interface as presented in (CableLabs, 2002). The security services required by each interface to counter these threats are listed in Table 10-5, along with the security and key management protocols employed by each interface. In Table 10-5, we use the convention [Security Service?, Required] to denote the security services required on each interface. The convention uses the following symbols:

- A = authentication
- C = confidentiality
- I = integrity
- AC = access control
- r = required
- n = not required or optional

For example, the value [A, r] indicates that authentication is required, and the value [C, n] signifies that confidentiality is optional.

Interface	Protocol	Threats
Device and Service Provisioning		
MTA–DHCP	DHCP	DoS attacks resulting in improper MTA configuration
MTA–SNMP Manager	SNMPv3	DoS attacks resulting in improper MTA configuration or unavailability of provisioning equipment
MTA–Provisioning Server	TFTP	DoS attacks resulting in improper MTA configuration or configuration using parameters from an old TFTP configuration file
Quality-of-Service Signaling		
CM–CMTS	DOCSIS 1.1	See Chapter 9
CMTS–CMS	Gate Coordination Messages	Service theft and DoS

Table 10-4 PacketCable Interfaces and Associated Threats

Interface	Protocol	Threats
CMS Gate Controller–CMTS	COPS	Theft or modification of QoS, DoS, and leakage of customer information
CMS–RSVP router, RSVP router–RSVP router	IntServ + RSVP	Theft or modification of QoS and DoS
DiffServ router–DiffServ router	DiffServ + Aggregated RSVP	Theft or modification of QoS
Billing System Interfaces		
CMS–RKS	Radius	Service theft, DoS, falsification of billing records, and leakage of customer information and provider communication patterns
CMTS–RKS	Radius	Service theft; DoS; falsification of billing records; and leakage of customer information, provider communication patterns, QoS data, and network performance figures
MGC–RKS	Radius	Service theft, DoS, falsification of billing records, and leakage of customer information and provider communication patterns
Call Signaling		
MTA–CMS	NCS	Impersonation of an MTA or CMS, modification to signaling messages (may result in improperly dialed phone numbers), and leakage of customer information and dialed phone numbers
CMS–CMS, CMS–EBP, EBP–EBP	CMSS	Impersonation of a CMS or SIP Proxy, modification to signaling messages, and leakage of customer information and dialed phone numbers
PSTN Gateway Interface		
MGC–MG	TGCP	Impersonation attacks against an MG or MGC, modification of TGCP signaling messages, and leakage of customer information (such as a dialed phone number)
MGC–SG	TCAP, ISTP	Impersonation of MGC or SG, tampering of ISTP call signaling messages, leakage of customer information (such as dialed phone numbers)
CMS–SG	TCAP	DoS attacks caused by tampering of TCAP messages and leakage of customer information (such as dialed numbers) in TCAP messages
Media Stream		
MTA–MTA, MTA–MG	RTP	Eavesdropping of RTP message content, DoS, replay attacks, and leakage of identity information pertaining to one or more communication endpoints
MTA–MTA, MTA–MG	RTCP	DoS attacks against RTCP to disrupt media stream transmission, leakage of identity information pertaining to one or more communication endpoints

Table 10-4 PacketCable Interfaces and Associated Threats *(continued)*

Interface	Protocol	Threats
Audio Server Services		
MPC–MP	TGCP	Impersonation of an MPC or MP to play obscene or inappropriate messages, tampering of signaling messages to allow obscene or inappropriate messages, and leakage of service-related information or the identity of communication endpoints from TGCP signaling messages
MTA–MP	RTP/RTCP	Eavesdropping of message content, DoS, replay attacks, tampering of call content messages to allow obscene or inappropriate messages, and leakage of identity information pertaining to one or more communication endpoints
Electronic Surveillance		
CMS–DF, CMTS–DF, MGC-DF, DF–DF (event interfaces)	UDP	Service theft, DoS, and leakage of customer information and communication patterns
CMTS–DF, MG-DF, DF–DF (call content interfaces)	UDP	Modification of media stream content and unauthorized eavesdropping by non-government entities

Table 10-4 PacketCable Interfaces and Associated Threats *(continued)*

NOTE:
Event interfaces *are used for call signaling, whereas* call content interfaces *are used for media stream transport.*

NOTE:
It is difficult to make changes to voice signals without detection, so in most cases, integrity is optional for media stream transport.

For a more detailed handling of the security mechanisms used on the various PacketCable interfaces, refer to Section 7 of CableLabs, 2002.

IPSec

Many of the interfaces listed in Table 10-5 use IPSec for encryption, integrity, and message authentication. IPSec operates at the Network layer and includes two protocols, Authentication Header (AH) and Encapsulating Security Payload (ESP), that offer varying levels of protection for IP packets. The AH protocol can authenticate the source and ensure the integrity of an entire IP packet, including the IP header. However, it offers no encryption mechanisms. The ESP protocol, on the other hand, can be used to encrypt the upper-layer data encapsulated within an IP packet. It can also authenticate the source and ensure the integrity of the data, but it cannot authenticate the IP header. Both IPSec protocols operate in two modes: transport and tunnel. PacketCable employs only the ESP protocol in transport mode for the following reasons:

- Authentication of IP headers does not significantly improve security within PacketCable networks (CableLabs, 2002).
- Communication between components in PacketCable networks requires end-to-end encryption and authentication that spans the entire path between the source and destination. Only transport mode offers end-to-end security.

Interface	[Required Security Services]: Security Protocol/Key Management Protocol
Device and Service Provisioning	
MTA–DHCP	[A, n], [C, n], [I, n], [AC, n]: Securing DHCP is the responsibility of the service provider, and is not specified within PacketCable security.
MTA–SNMP Manager	[A, r], [C, n], [I, r], [AC, r]: SNMPv3 security/Kerberized Key Management (initial authentication via Kerberos + PKINIT).
MTA–Provisioning Server (TFTP Server)	[A, r], [C, n], [I, r], [AC, r]: Hash and optional encryption/keying material sent over SNMPv3 (with SNMPv3 security enabled).[2]
Quality-of-Service Signaling	
CM–CMTS	[A, r], [C, r], [I, r], [AC, r]: BPI+/BPKM.
CMTS–CMS	[A, r], [C, r], [I, r], [AC, r]: IPSec/IKE (with pre-shared keys).
CMS Gate Controller–CMTS	[A, r], [C, n], [I, r], [AC, n]: Radius authentication using an MD5-based MAC/CMS key distribution using COPS messages.
CMS–RSVP router, RSVP router–RSVP router	[A, r], [C, n], [I, r], [AC, r]: RSVP integrity object, which uses HMAC-MD5 for authentication and integrity. Access control provided via pre-shared keys.
DiffServ router–DiffServ router	[A, n], [C, n], [I, n], [AC, r]: DiffServ Code Point (DSCP) authorization.[3]
Billing System Interfaces	
CMS–RKS	[A, r], [C, r], [I, r], [AC, r]: IPSec/IKE (with pre-shared keys).
CMTS–RKS	[A, r], [C, r], [I, r], [AC, r]: IPSec/IKE (with pre-shared keys).
MGC–RKS	[A, r], [C, r], [I, r], [AC, r]: IPSec/IKE (with pre-shared keys).
Call Signaling	
MTA–CMS	[A, r], [C, r], [I, r], [AC, r]: IPSec/ Kerberized Key Management (initial authentication via Kerberos + PKINIT).
CMS–CMS, CMS–EBP, EBP–EBP	[A, r], [C, r], [I, r], [AC, r]: IPSec/ Kerberized Key Management (initial authentication via Kerberos).

Table 10-5 Security Services and Protocols Employed by PacketCable Interfaces

Interface	[Required Security Services]: Security Protocol/Key Management Protocol
PSTN Gateway Interface	
MGC–MG	[A, r], [C, r], [I, r], [AC, r]: IPSec/IKE (with pre-shared keys).
MGC–SG	[A, r], [C, r], [I, r], [AC, r]: IPSec/IKE (with pre-shared keys).
CMS–SG	[A, r], [C, r], [I, r], [AC, r]: IPSec/IKE (with pre-shared keys).
Media Stream	
MTA–MTA, MTA–MG (via RTP)	[A, r], [C, r], [I, o], [AC, o]: Application layer security (cipher + HMAC)/CMS-based key management.[4]
MTA–MTA, MTA–MG (via RTCP)	[A, r], [C, r], [I, r], [AC, o]: Application layer security (cipher + HMAC)/CMS-based key management.
Audio Server Services	
MPC–MP	[A, r], [C, r], [I, r], [AC, r]: IPSec/IKE (with pre-shared keys).
MTA–MP	[A, r], [C, r], [I, r], [AC, o]: Application layer security (cipher + HMAC) /CMS-based key management (for RTP and RTCP).
Electronic Surveillance	
CMS–DF, CMTS–DF, MGC-DF (event interfaces)	[A, r], [C, r], [I, r], [AC, r]: IPSec/IKE (with pre-shared keys).
CMTS–DF, MG-DF (call content interfaces)	[A, r], [C, r], [I, o], [AC, o]: Application layer security (cipher + HMAC)/CMS-based key management.[5]
DF-DF (event interfaces)	[A, r], [C, r], [I, r], [AC, r]: IPSec/IKE (with digital signatures and X.509 certificates).
DF–DF (call content interfaces)	[A, r], [C, r], [I, o], [AC, o]: Application layer security (cipher + HMAC)/CMS-based key management.[6]

[2] The authentication and optional encryption services for the MTA–Provisioning Server interface are applied directly to configuration files and not to the network interface itself. The provisioning server generates a hash over each configuration file by passing the file through a message digest. It then transmits the hash to an MTA within an *SNMP Set* command during MTA provisioning. Once the MTA downloads the configuration file from the TFTP server, it uses the hash to authenticate the file. The provisioning server is responsible for building the configuration file and deciding whether to encrypt it.

[3] DSCP authorization is used to control access to QoS resources on DiffServ routers.

[4] CMS-based key management exchanges keying information within call signaling messages.

[5] In this case, the CMTS and DF server do not directly apply the security mechanisms. The content exchanged between the CMTS and the DF server has already been secured using RTP/RTCP Application layer security over an MTA–MTA or MTA–MG interface.

[6] Similar to the CMTS–DF call content interface, the DF–DF call content interface applies no security mechanisms of its own. It relies on the traffic being previously secured using RTP/RTCP Application layer security.

Table 10-5 Security Services and Protocols Employed by PacketCable Interfaces *(continued)*

Table 10-6 provides a list of the IPSec encryption transforms and key sizes supported by PacketCable. The supported IPSec authentication algorithms are listed in Table 10-7. Refer to Chapter 6 for a complete handling of IPSec and the ESP protocol.

In addition to encryption, integrity, and authentication, IPSec includes protection against replay attacks by monitoring the sequence numbers of all incoming IP packets. IPSec assumes that any packet containing a sequence number that falls outside of a window of valid sequence numbers is part of a replay attack, and it counters by dropping the packet. PacketCable requires IPSec anti-replay protection to be enabled at all times.

Internet Key Exchange

One of the IPSec key management protocols supported by PacketCable is Internet Key Exchange (IKE). PacketCable uses IKE on interfaces that do not include a large number of connections, and in situations when the communication endpoints are known ahead of time (CableLabs, 2002). As discussed in Chapter 6, IKE key management consists of two distinct phases. For Phase 1 IKE negotiations, PacketCable supports authentication utilizing pre-shared keys and digital signatures. When using digital signatures, both parties must exchange X.509 certificates to obtain the public keys required for signature verification.

Encryption Transform	Key Size	Required?	Description
ESP_3DES	192 bits	Yes	3DES in CBC mode
ESP_RC5	128 bits	No	RC5 in CBC mode
ESP_IDEA	128 bits	No	IDEA in CBC mode
ESP_CAST	128 bits	No	CAST in CBC mode
ESP_BLOWFISH	128 bits	No	Blowfish in CBC mode
ESP_AES	128 bits	No	AES in CBC mode using a 128-bit block size
ESP_NULL	0 bit	Yes, but only on interfaces that do not require confidentiality	No encryption

Table 10-6 IPSec Encryption Transforms Supported by PacketCable

Authentication Algorithm	Key Size	Required?	Description
HMAC-MD5	128 bits	Yes	MD5 HMAC
HMAC-SHA	160 bits	Yes	SHA-1 HMAC

Table 10-7 IPSec Authentication Algorithms Supported by PacketCable

NOTE:
PacketCable does not define a method for generating and exchanging pre-shared secrets. As a result, communicating parties should ensure that they use appropriate entropy (that is, sources of randomness) when generating shared secrets and that they exchange the shared secrets (or the passwords used to generate the shared secrets) securely.

During Phase 2 IKE negotiations, communicating parties use information obtained during a Phase 1 exchange to establish ESP security associations (SAs) securely. These SAs describe the security services—the IPSec protocol, encryption and authentication transforms, keying material, and key lifetimes—for the connection. All IPSec keying material can be derived from a shared Phase 2 secret using the one-way function described in RFC 2409 (Harkins and Carrel, 1998) or using a Diffie-Hellman exchange. The PacketCable specification specifically prohibits use of Diffie-Hellman to generate keying material due to its associated performance penalties.

NOTE:
IKE key management operations are performed asynchronous to any call-signaling messages. As a result, they do not incur additional delays during communication setup (CableLabs, 2002).

SNMPv3 Security

PacketCable employs SNMPv3 Security to protect network management messages exchanged between an SNMP manager and SNMP agents. Currently, the only SNMPv3 interface specified by PacketCable exists between an MTA and a provisioning server. Table 10-8 and Table 10-9 list the SNMPv3 encryption and authentication algorithms supported by PacketCable.

Encryption Transform	Key Size	Required?	Description
SNMPv3_DES	64 bits	Yes	DES in CBC mode
SNMPv3_NULL	0 bit	Yes	No encryption

Table 10-8 SNMPv3 Encryption Transforms Supported by PacketCable

ASNMPv3 Encryption Transforms Supported by PacketCableuthentication Algorithm	Key Size	Required?	Description
SNMPv3_HMAC-MD5	128 bits	Yes	MD5 HMAC
SNMPv3_HMAC-SHA-1	160 bits	No, but should be supported	SHA-1 HMAC

Table 10-9 SNMPv3 Authentication Algorithms Supported by PacketCable

PacketCable's Use of Kerberos

This section provides a high-level overview of the Kerberos functionality employed by PacketCable and discusses how PacketCable uses Kerberos for key management within the IPSec and SNMPv3 protocols. Much of the terminology used in this section was introduced in the discussion of Kerberos in Chapter 6. A typical Kerberos authentication session can be divided into three phases (see Figure 10-3):

- Phase 1 (this phase is optional in PacketCable):

1. A Kerberos client sends an *AS Request* message to the Authentication Server (AS) component of the KDC to obtain a ticket granting ticket (TGT).
2. The AS responds with an *AS Reply* message containing the TGT.

- Phase 2:

3. The client uses the TGT to obtain an application server ticket by sending a *TGS Request* message to the Ticket Granting Server (TGS) component of the KDC.
4. The TGS responds with a *TGS Reply* message containing the application server ticket for the particular server the client wishes to access.

- Phase 3:

5. The client uses the server ticket received in step 4 to authenticate to the application server. It does so by sending an *AP Request* message to the server.
6. The application server responds with an *AP Reply*. The client and server may then generate security parameters from information contained within the *AP Request* and *AP Reply* messages.

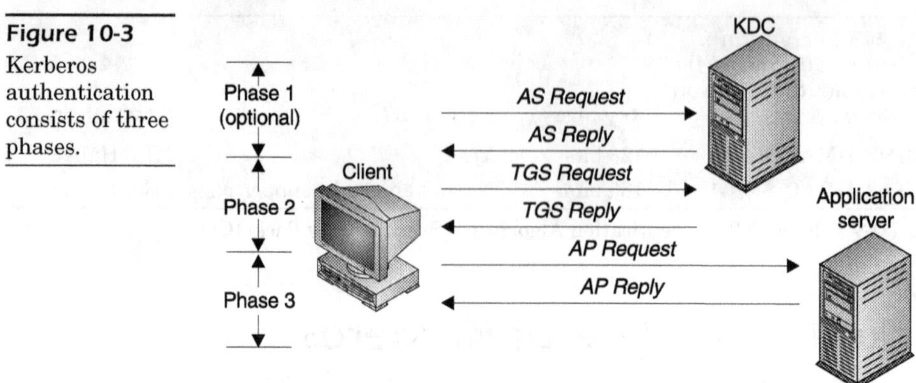

Figure 10-3 Kerberos authentication consists of three phases.

NOTE:

As we will discuss later in "Kerberized Key Management for IPSec and SNMPv3," messages exchanged within the third phase of a Kerberos authentication session may facilitate key management within other security protocols.

The Kerberos V5 specification (Neuman et al., 2000) defines all three authentication phases, along with the *AS Request, AS Reply, TGS Request, TGS Reply, AP Request,* and *AP Reply* messages. CableLabs, 2002 contains the *profiles* (necessary fields and values) for each of the six message types as required by PacketCable. Unless PKINIT is used, client authentication in Phase 1 makes use of a secret value (a symmetric client key) shared between the client and the KDC and follows the steps listed in the section entitled "Kerberos Authentication" in Chapter 6.

Instead of authenticating via a shared secret, an MTA *must* use PKINIT to obtain a TGT during a Phase 1 exchange or an application server ticket during a Phase 2 exchange. When using PKINIT, the MTA authenticates itself to the KDC using public-key cryptography and a digital signature. Whether or not an MTA obtains a TGT prior to requesting an application server ticket is specific to the MTA implementation. (Within PacketCable, an MTA can directly request an application ticket without first obtaining a TGT by skipping Phase 1.) However, the use of TGTs offers an important advantage over directly requesting application server tickets. By first requesting a TGT, the MTA is not required to authenticate using a PKINIT exchange each time the MTA wants to obtain a Phase 2 application server ticket. This avoids repeated use of computationally expensive public-key operations. An MTA can use a single TGT to request multiple application server tickets. To obtain a Kerberos ticket, the MTA sends a *PKINIT Request* message to the KDC containing the following elements:

- The Kerberos principal name of the client.
- The principal name of the KDC or the application server that the client wishes to contact (an application server principal name is used only when an MTA decides to skip the Phase 1 exchange).
- The current time (used to prevent replay attacks).
- A random nonce generated by the client.
- The client's Diffie-Hellman public value (used for key agreement).
- An RSA digital signature (using SHA-1) generated by the MTA.
- The MTA's X.509v3 digital certificate, which contains the public key needed to verify the MTA's signature. The request must also contain the certificate of the MTA manufacturer.
- An MD5 checksum generated over a portion of the request.

NOTE:
Only the MTA-to-CMS interface currently utilizes Kerberos with PKINIT.

Upon receiving the request, the KDC must validate the certificate chain up to and including the MTA Root CA certificate, verify the signature contained within the request, and check the validity of the MD5 checksum. If all these checks pass, the KDC responds with a *PKINIT Reply* message that includes the following:

- The TGT or application server ticket
- The X.509v3 certificate of the KDC along with the certificates of the local system operator CA (if one exists) and the service provider CA
- The KDC's Diffie-Hellman public value
- The nonce supplied by the client
- An RSA digital signature (using SHA-1) generated by the KDC
- A session key and key validity period (encrypted using the Diffie-Hellman algorithm)

Upon receiving the reply, the MTA must check the nonce value, validate the certificate chain between the KDC and the CableLabs Service Provider Root CA, and verify the KDC's signature. For a more detailed description of the PKINIT Request and PKINIT Reply messages, see Neuman et al., 2001. Table 10-10 lists the Kerberos principal identifier formats as required by PacketCable.

Table 10-10 Kerberos Principal Identifier Formats Used by PacketCable

Component	Kerberos Principal Identifier Format
KDC	krbtgt/<Kerberos realm>@<Kerberos realm>
CMS	cms/<FQDN of CMS>@<Kerberos realm>
Provisioning Server	mtaprovsrvr/<FQDN of server>@<Kerberos realm>
MTA	mta/<FQDN of MTA>@<Kerberos realm>
Services	<service name>/<FQDN of host>

NOTE:
In a PacketCable Kerberos principal identifier, the Kerberos realm name must appear in uppercase characters; all other components of a Kerberos principal identifier, including fully qualified domain names (FQDNs), must be in all lowercase characters. The Kerberos realm for each component is usually the domain name of the service provider (that is, MSO123.COM) and corresponds to the administrative domain of a single KDC.

The *PKINIT Request* and *PKINIT Reply* messages are carried within *pre-authenticator fields* inside *AS Request* and *AS Reply* messages, respectively. A KDC uses the pre-authenticator fields to actively authenticate the MTA prior to issuing a Kerberos ticket. Alternatively, an MTA could authenticate itself following the issuance of a Kerberos ticket by proving that it could decrypt the contents of the ticket. However, this mode is not supported by PacketCable on MTA interfaces because it may aid in a cryptanalytic attack in which an attacker requests tickets using other users' identities to try to obtain those users' client keys.

NOTE:
An MTA should save all Kerberos tickets within nonvolatile memory so that they can be reused even after the MTA has been powered down. Kerberos tickets must not be valid for more than seven days following the data of issuance of the ticket.

Kerberized Key Management for IPSec and SNMPv3

The third phase of a Kerberos authentication session may be used to provide key management for security protocols, such as IPSec and SNMPv3. To facilitate the negotiation of security parameters between a Kerberos client and an application server, PacketCable defines a number of new message types—*Wake Up, Security Parameter Recovered,* and *Rekey*—in addition to the *AP Request* and *AP Reply* messages already introduced. The contents of all five message types are listed in Table 10-11. CableLabs, 2002 describes two Phase 3 key management exchanges that employ a combination of these messages to establish security parameters for IPSec and SNMPv3: the *AP Request*/*AP Reply* and *Rekey* exchanges. A client and an application server can use either exchange to refresh or renew security parameters periodically. *AP Request*/*AP Reply* exchanges follow these steps, as shown in Figure 10-4:

1. The application server sends a *Wake Up* message to the client to initiate a key management exchange. This step is optional and requires that the application server (a CMS) and the Kerberos client (an MTA) have already established IPSec security associations (SAs) for the connection. A CMS may use a *Wake Up* message to negotiate new SAs after previous SAs have expired.
2. The client replies by sending an *AP Request* message to the application server (when a Kerberos client receives a *Wake Up* message, it must transmit an *AP Request*). Even if the client does not receive a *Wake Up* message from the application server, it will still send an *AP Request* message on its own to establish new SAs prior to the expiration of the existing ones. The *AP Request* message *may* contain a randomly generated subkey for use in deriving security parameters for the connection.
3. The application server verifies the HMAC contained within the client's *AP Reply*. If the HMAC verifies correctly, the application server responds with an *AP Reply* message containing a random subkey for use in deriving security parameters. Upon receiving the *AP Reply,* the client must verify the included HMAC. If either HMAC verification fails, the corresponding message is discarded.
4. If the Boolean ACK-required (acknowledgment-required) flag within the *AP Reply* is set to true, the client must respond with a *Security Parameter Recovered* message. This indicates to the application server that the client received the *AP Reply* message and that it successfully established a set of security parameters.

Figure 10-4 Kerberos clients and application servers within PacketCable may use AP Request/AP Reply exchanges to negotiate security parameters for IPSec and SNMPv3 sessions.

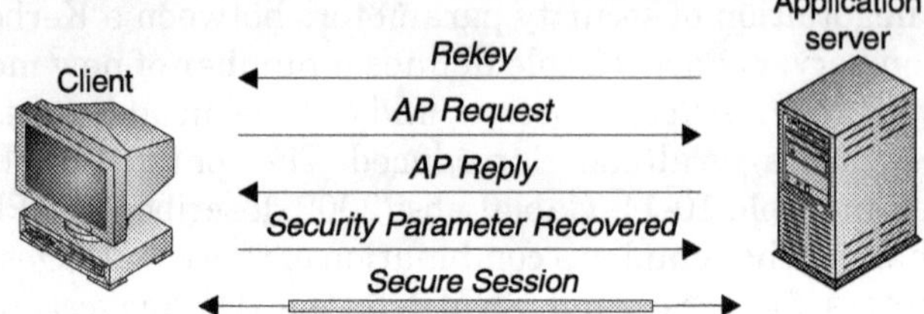

Message Type	Contents
AP Reply	• *Key management message ID* A 1-byte field containing the value 0x03 • *DOI (Domain of Interpretation)* a 1-byte field indicating the protocol for which the security parameters are being negotiated (IPSec = 1, SNMPv3 = 2) • *Protocol version* A 1-byte field identifying the parent protocol (the value is 0x10 for PacketCable) • *KRB_AP_REP* A DER encoded Kerberos message as defined within Neuman et al., 2000 • *Application-specific data* Additional information that the server must communicate to the client to describe the protocol for which security parameters are being negotiated (IPSec or SNMPv3) • *The selected cipher suites to be used within the target protocol (IPSec or SNMPv3)* Includes authentication and encryption algorithms • *Security parameter lifetime* A 4-byte field indicating when the security parameters generated during this key management session will expire in seconds from the current time (used only in conjunction with IPSec—the security parameters for SNMP do not expire) • *Grace period* A 4-byte field specifying the number of seconds prior to the expiration of the security parameters that the client can begin requesting new parameters (used only with IPSec) • *Re-establish flag* A 1-byte field containing a Boolean value that, when set to "TRUE" (1), indicates that a new set of security parameters must be negotiated prior to the expiration of the old parameters; when set to "FALSE" (0), the old parameters must be allowed to expire before requesting new ones • *ACK-required flag* A 1-byte field containing a Boolean value that, when set to "TRUE" (1), indicates that the *AP Reply* requires an acknowledgment (a *Security Parameter Recovered* message) • *SHA-1 HMAC* A 20-byte HMAC (using a SHA-1 hash of the session key) calculated over the contents of the message excluding the HMAC field itself

Table 10-11 Phase 3 Kerberized Key Management Message Formats

Message Type	Contents
AP Request	• *Key management message ID* A 1-byte field containing the value 0x02 • *DOI* Same as in *AP Reply* message • *Protocol version* Same as in *AP Reply* message • *KRB_AP_REQ* A DER-encoded Kerberos message as defined within Neuman et al., 2000; when replying to a *Rekey* message, this data structure contains the subkey • *Server-nonce* A 4-byte field containing the random binary string supplied by the application server in a *Wake Up* or *Rekey* message • *Application-specific data* Same as in *AP Reply* message • *A list of cipher suites supported by the client* Includes authentication and encryption algorithms • *Re-establish flag* A 1-byte field containing a Boolean value that, when set to "TRUE" (1), indicates that the client is attempting to renegotiate a new set of security parameters before the old ones expire • *SHA-1 HMAC* Same as in *AP Reply* message
Rekey (only used in conjunction with IPSec)	• *Key management message ID* A 1-byte field containing the value 0x05 • See the description for the *AP Reply* message • *Protocol version* Same as in *AP Reply* message • *Server-nonce* A 4-byte field containing a random binary string • The Kerberos principal name of the application server (see Table 10-10) • *Timestamp* A string indicating the current UTC time in the format *YYMMDDhhmmssZ* • *Application-specific data* Same as in *AP Reply* message • A list of cipher suites supported by the server • *Security parameter lifetime* Same as in *AP Reply* message • *Grace period* Same as in *AP Reply* message • *Re-establish flag* Same as in *AP Reply* message • *SHA-1 HMAC* Same as in *AP Reply* message, except the HMAC key is derived from the last *AP Reply* message sent by the server
Security Parameter Recovered	• *Key management message ID* A 1-byte field containing the value 0x04 • *DOI* Same as in *AP Reply* message • *Protocol version* Same as in *AP Reply* message • *SHA-1 HMAC* A 20-byte HMAC (using the subkey in the *AP Reply*) calculated over the contents of the previous *AP Reply* message
Wake Up	• *Key management message ID* A 1-byte field containing the value 0x01 • *DOI* Same as in *AP Reply* message • *Protocol version* Same as in *AP Reply* message • *Server-nonce* A 4-byte field containing a random binary string • The Kerberos principal name of the application server (see Table 10-10)

Table 10-11 Phase 3 Kerberized Key Management Message Formats *(continued)*

NOTE:
Prior to issuing an AP Request *message, the Kerberos client must acquire a server ticket from the KDC, which it uses to authenticate to the application server.*

The *Rekey* exchange offers better performance than the *AP Request / AP Reply* exchange, and the former may be used when the client and application server already share a server authentication key (a server authentication key is a 20-byte SHA-1 hash of the Kerberos session key shared between the client and server). *Rekey* exchanges (see Figure 10-5) may be used only in the establishment of security parameters for IPSec, and such exchanges must follow these steps:

1. The application server transmits a *Rekey* message to the Kerberos client. This message replaces the *Wake Up* message within the AP *Request / AP* Reply exchange.
2. The client verifies the HMAC contained within the *Rekey* message; if the verification fails, the message is discarded. The client then generates a random subkey, which it uses to establish security parameters for the connection. It sends the subkey to the application server in an *AP Request* message.
3. The application server establishes a corresponding set of security parameters once it acquires the subkey contained within the client's *AP Request*.

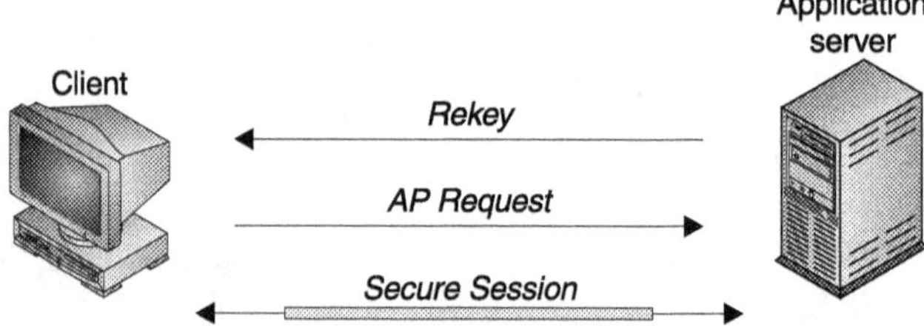

Figure 10-5 The Rekey exchange provides similar functionality to an AP Request/AP Reply exchange but offers better performance when the client and application server already share a server authentication key.

NOTE:
All key management messages are exchanged over UDP using port 1293 on both the client and the application servers.

Following a Phase 3 Kerberos exchange, a client and server use the application-specific data and subkey(s) obtained in the Phase 3 exchange to negotiate either IPSec security associations (SAs) or encryption and authentication keys for SNMPv3. When Kerberized key management is used in conjunction with IPSec, the client creates a single IPSec subkey by XORing the optional subkey contained within its *AP Request* message with the one included in the *AP Reply* message it receives from the application server. The server performs the same XORing operation to obtain an identical IPSec-specific subkey. The client and server then derive the encryption and authentication keys for inbound and outbound SAs from the IPSec-specific subkey. If the client's *AP Request* message did not include a subkey, the client and server simply use the subkey contained within the *AP Reply* for SA negotiation.

Table 10-12 lists the application-specific data and subkey requirements for IPSec and SNMPv3 security parameter derivation. CableLabs, 2002, further describes the process for deriving IPSec and SNMPv3 encryption and authentication keys.

	Domain of Interpretation	Application-Specific Data	Protocol-Specific Subkey
IPSec	Set to 1	• *AP Request* message Security parameter index (SPI) for the client's inbound security association (SA) • *AP Reply* and *Rekey* messages SPI for the server's inbound SA	Equal to the XOR of the 46-byte subkey within the *AP Request* and the 46-byte subkey within the *AP Reply*

Table 10-12 Application-Specific Data and Subkey Requirements for AP Request, AP Reply, and Rekey Messages Used in Conjunction with Kerberized IPSec and SNMPv3

	Domain of Interpretation	Application-Specific Data	Protocol-Specific Subkey
SNMPv3	Set to 2	• *AP Request* message A concatenation of SNMP agent-specific attributes (see CableLabs, 2002, for the detailed list of attributes) • *AP Reply* message A concatenation of SNMP manager-specific attributes (see CableLabs, 2002, for the detailed list of attributes)	Must be a 46-byte value

Table 10-12 Application-Specific Data and Subkey Requirements for AP Request, AP Reply, and Rekey Messages Used in Conjunction with Kerberized IPSec and SNMPv3 *(continued)*

Cross-Realm Operation

There may be times when a client within one Kerberos realm must exchange security parameters with an application server in another realm. Prior to negotiating security parameters, the client must authenticate to the server using a process known as *cross-realm authentication,* which was introduced in Chapter 6. For cross-realm authentication, PacketCable supports the use of PKCROSS (Hur et al., 2001). Instead of using a pre-shared key to authenticate a KDC to another in a different realm, PKCROSS specifies the use of public-key encryption. For a client in Realm 1 and an application server in another realm, Realm 2, cross-realm authentication using PKCROSS follows these steps:

1. The client in Realm 1 requests a TGT by sending an *AS Request* message to its local KDC.
2. The KDC responds with an *AS Reply* message containing the TGT.
3. The client uses the TGT to obtain a cross-realm TGT for the KDC in Realm 2. It does so by sending a *TGS Request* message to the local KDC.
4. The KDC in Realm 1 authenticates itself to the KDC in Realm 2 using PKCROSS and requests a cross-realm TGT in lieu of the client using a *PKCROSS Request* message.
5. The KDC in Realm 2 responds with a *PKCROSS Reply* containing the cross-realm TGT.

6. The KDC in Realm 1 transmits the cross-realm TGT to the client within a *TGS Reply* message.
7. The client uses the cross-realm TGT obtained in step 6 to request an application server ticket from the remote KDC in Realm 2. It does so by sending a *TGS Request* message to the remote KDC.
8. The remote KDC responds with a *TGS Reply* message that includes the server ticket.
9. The client authenticates itself to the application server in Realm 2 by sending an *AP Request* message directly to the server.
10. The application server responds with an *AP Reply*.

Once the Kerberos client receives an application server ticket from the remote KDC in Phase 2, it may contact the remote application server directly for all Phase 3 security parameter negotiations. PacketCable uses Kerberos + PKCROSS for cross-realm authentication in CMS-to-CMS communication sessions that employs CMSS.

Securing RTP and RTCP

Chapter 4 discussed the QoS requirements for various multimedia applications. Currently, the only application defined for PacketCable networks is voice telephony. While it doesn't require large amounts of bandwidth, telephony is a rigid, real-time application with low tolerance to signal delay and variation (in fact, rigid, real-time applications are the least tolerant of all multimedia applications). Voice transmission typically requires 8–64 Kbps, depending on the compression and sampling techniques used. End-to-end latency greater than 300 ms becomes intolerable for interactive voice-only communication, and jitter of greater than 20 ms results in sufficient signal degradation to hinder an intelligible conversation. To support the delivery of rigid, real-time applications over packet-based networks, PacketCable defines the use of the Real-time Transport Protocol (RTP) and the Real-Time Control Protocol (RTCP). Both protocols were introduced in Chapter 4.

PacketCable uses RTP for transporting MTA-to-MTA, MTA-to-MP, and MTA-to-PSTN traffic. (MTA-to-PSTN exchanges occur between an MTA and a media gateway.) For end-to-end privacy, integrity, and authentication of data transmitted over RTP, PacketCable supports the encryption and authentication algorithms listed in Tables 10-13 and 10-14. Figure 10-6 illustrates the format of an encrypted and authenticated RTP message. PacketCable encrypts only the payload of the message but authenticates both the header and payload using an optional MAC. The variable-length payload of the RTP message contains compressed audio (a *voice packet*) or signaling information related to establishing a telephone call (an *event packet*).

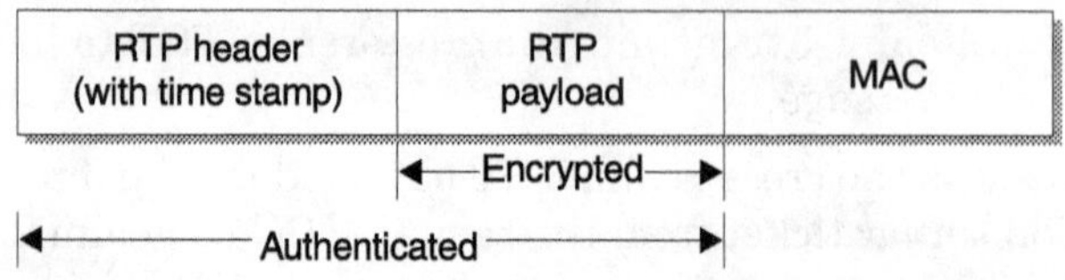

Figure 10-6 During RTP message encoding, PacketCable encrypts the variable-length header and payload of an RTP message and appends an optional MAC for integrity and origin authentication.

CAUTION:

The real-time nature of RTP traffic requires both communication endpoints to maintain strict timing requirements during signal encoding, decoding, and playback. These requirements may be difficult to meet when using a stream cipher, such as RC4. Although stream ciphers are fast, they are often difficult to synchronize in constant-bit-rate applications. Stream ciphers are better suited to asynchronous applications that send and receive data as quickly as possible without regard to timing. The preferred algorithm for encrypting RTP media streams is AES.

Encryption Transform	Key Size	Required?	Description
RTP_AES	128 bits	Yes	AES in CBC mode with 128-bit block size
RTP_XDESX_CBC	192 bits	No	DESX-XEX[7] in CBC mode
RTP_DES_CBC_PAD	128 bits	No	DES in CBC mode with a random pad
RTP_3DES_CBC	128 bits (effectively 112 bits)	No	3DES in EDE-CBC mode
RTP_RC4	128 bits	No	RC4 stream cipher

[7] XDESX (also known as DESX and DESX-XEX) is a slightly modified version of DES that eliminates some of the weaknesses present in the DES algorithm. A discussion of these weaknesses is beyond the scope of this text.

Table 10-13 Supported RTP Encryption Transforms

Authentication Algorithm	Key Size	Required?	Description
AUTH_NULL	0 bit	Yes	No authentication
RTP_MMH_2	Variable	Yes	2-byte MMH HMAC
RTP_MMH_4	Variable	Yes	4-byte MMH HMAC

Table 10-14 Supported RTP Authentication Algorithms

NOTE:
The Multilinear Modular Hash (MMH) *HMAC used to calculate the message authentication codes appended to RTP messages is a high-performance MACing algorithm that can be implemented efficiently within digital signal processors. A detailed discussion of MMH is outside the scope of this text. (For more information on MMH, see Section 9.8 of CableLabs, 2002).*

Before an MTA can encode or decode RTP messages, it must derive encryption and authentication keys from an end-to-end secret it shares with the receiving MTA, MG, or MP. MTAs exchange end-to-end secrets with MGs, MPs, and other MTAs via trusted third-party call-management servers. The keys are transported within network call-signaling messages. We describe this process further later in the chapter in "Key Management for RTP and RTCP." The sender and recipient must maintain two sets of encryption and authentication keys—one for inbound traffic, and the other for outbound traffic. CableLabs, 2002, discusses the process for generating the end-to-end shared secret and using the secret to derive encryption and authentication keys.

When decoding an RTP message, the recipient should first check a timestamp contained within the header of the message. The timestamp indicates the position of the RTP payload in the original media stream. If the timestamp falls outside a window of acceptable values, the message is discarded. (Remember that in the case of rigid, real-time applications, it is better to discard a message that arrives too late than to halt or delay playback while waiting for the message.) Following the timestamp check, the recipient verifies the MAC computed over the message and decrypts the payload using its inbound authentication and encryption keys, respectively. If the MAC verification fails, the recipient discards the RTP message.

Using RTCP, communicating parties exchange information about the current delivery conditions for a connection. They use this information to adjust playback settings dynamically to meet predefined QoS requirements. Encrypted

RTCP messages take the form shown in Figure 10-7 and consist of the following fields:

- **Sequence number** A 4-byte field starting with zero for the first transmitted RTCP message and incremented by one for each subsequent message.
- **IV** The initialization vector required by the CBC mode block cipher used to encrypt the RTCP message. The length of the IV is equal to the block size of the cipher.
- **Variable** The length of the encrypted RTCP message.
- **MAC** The message authentication code computed over the sequence number, IV, and the encrypted RTCP message. The length of the MAC depends on the algorithm chosen.

Tables 10-15 and 10-16 list the supported encryption and authentication algorithms for use in securing RTCP messages.

Encryption Transform	Key Size	Required?	Description
AES-CBC	128 bits	Yes	AES in CBC mode with 128-bit block size
XDESX-CBC	192 bits	No	DESX-XEX in CBC mode
DES-CBC-PAD	128 bits	No	DES in CBC mode with a random pad
3DES-CBC	128 bits (effectively 112 bits)	No	3DES in EDE-CBC mode

Table 10-15 Supported RTCP Encryption Transforms

Authentication Algorithm	Key Size	Required?	Description
HMAC-SHA1-96	168 bits	Yes	SHA-1 HMAC
HMAC-MD5-96	128 bits	No	MD5 HMAC

Table 10-16 Supported RTCP Authentication Algorithms

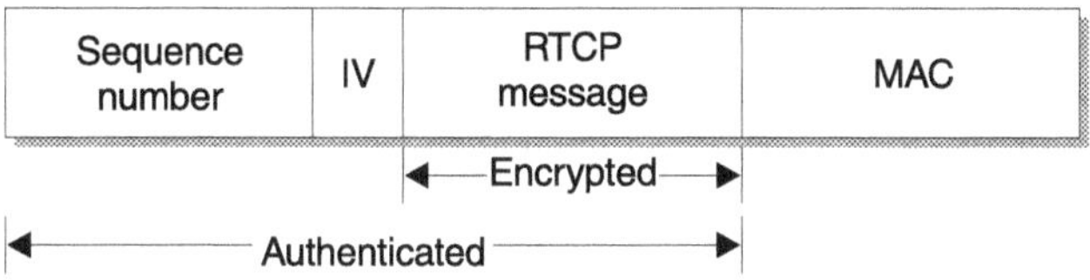

Figure 10-7 An encrypted RTCP message consists of a sequence number, initialization vector, encrypted RTCP message, and a message authentication code.

The process for encoding encrypted RTCP messages follows these steps:

1. The sender generates a random IV.
2. The sender encrypts the RTCP message using the selected block cipher and the IV generated in Step 1.
3. The sender concatenates the current sequence number, IV, and encrypted message.
4. The sender calculates the MAC over the combined sequence number, IV, and encrypted message, and appends the MAC to the end of the message.

RTCP decoding follows these steps:

1. The recipient generates a MAC from the sequence number, IV, and encrypted RTCP message, and then compares the result to the MAC contained within the message. The message is dropped if the MACs do not match.
2. The recipient verifies the sequence number contained within the message by making sure that it falls within a window of acceptable values. The message is dropped if the sequence number falls outside the current sliding window.
3. The recipient decrypts the message using the included IV.

Key Management for RTP and RTCP

MTAs and MPs establish end-to-end shared secrets for MTA-to-MTA and MTA-to-MP communication using network-based call signaling (NCS) messages; MTAs and MGs negotiate shared secrets using the Trunking Gateway Control Protocol (TGCP). These devices do not exchange keying material directly; instead, they use third-party call management servers as proxies. PacketCable protects all MTA-to-CMS, CMS-to-CMS, and MG-to-MGC NCS messages using IPSec, which provides encryption, integrity, authentication, and anti-replay protection.

Once an IPSec session has been established, an MTA and a CMS, or an MG and an MGC, can securely exchange keying material via NCS. The PacketCable specification describes the key management procedures for MTA-to-MTA communication using RTP/RTCP, summarized here:

1. The calling MTA (MTA_0) sends a *NTFY* message to its local CMS (CMS_0). Note that we use the term "local" to infer that the MTA and CMS lie within the same service provider domain and the term "remote" to indicate that they belong to different domains.
2. CMS_0 responds with a *CRCX* message containing a list of available RTP and RTCP cipher suites for the local connection.
3. MTA_0 generates a 46-byte end-to-end shared secret value, $Secret_0$, and a 46-byte pad value, Pad_1, and establishes inbound keys for RTP/RTCP based on its preferred cipher suites and $Secret_0$. It then sends an *ACK* message back to CMS_0 that includes the following:
   - MTA_0's choice of RTP and RTCP cipher suites (listed in order of decreasing preference), based on those provided by CMS_0 for the local connection.
   - The end-to-end values $Secret_0$ and Pad_1. MTA_1 later used $Secret_0$ and Pad_1 to derive outbound and inbound RTP/RTCP keys, respectively.
   - A connection ID and endpoint for MTA_0. The combination (Connection ID, Endpoint) uniquely identifies the connection.

NOTE:
Before sending the acknowledgment message to CMS_0, MTA_0 must enable inbound and outbound RTP and RTCP security in preparation to begin receiving encrypted traffic from MTA_1.

---

4. CMS_0 sends an *INVITE* message to the remote CMS (CMS_1) specifying the cipher suites and end-to-end secret/pad value chosen by MTA_0.
5. CMS_1 sends a *CRCX* message to the destination MTA (MTA_1) containing a list of available cipher suites for the local connection and the cipher suites and end-to-end secret/pad value chosen by MTA_0.
6. MTA_1 generates a 46-byte end-to-end shared secret value, $Secret_1$, and may generate a 46-byte pad value, Pad_0. It establishes inbound keys for RTP/RTCP based on its preferred cipher suites, $Secret_1$, and Pad_1, and derives outbound RTP/RTCP keys based on its preferred cipher suites,

$Secret_0$, and Pad_0 (if MTA1 generates Pad_0). MTA_1 then sends an *ACK* message back to CMS_1 that includes the following:

- MTA_1's choice of RTP and RTCP cipher suites (listed in order of preference) based on the intersection of the cipher suites required by CMS_1 for the local connection and those requested by MTA_0 for the remote connection. MTA_1 may select cipher suites other than those preferred by MTA_0, but must not try to send data to MTA_0 until MTA_0 has been informed of the selection.
- A 46-byte end-to-end secret and a 46-byte pad value generated by MTA_1 that it will use to derive the encryption and authentication keys for its inbound traffic.
- A connection ID and endpoint for MTA_1.

NOTE:
Before sending the acknowledgment message to CMS_1, MTA_1 must enable inbound and outbound RTP and RTCP security in preparation to begin receiving encrypted traffic from MTA_0.

---

7. CMS_1 forwards the acknowledgement to CMS_0 within a *200 OK* message. The acknowledgment contains the cipher suites and end-to-end secret/pad value selected by MTA_1.
8. CMS_0 sends a final *MDCX* message to MTA_0 providing the following:
   - The connection ID and endpoint for MTA_1
   - The end-to-end secret and pad value chosen by MTA_1
   - A new list of cipher suites if CMS_0 wishes to modify the list of cipher suites previously provided to MTA_0
9. Upon receiving the *MDCX* message, MTA_0 must generate new inbound RTP/RTCP keys if the message contains the pad value, Pad_0, or if the message indicates an alternate selection of cipher suites by MTA_1. If Pad_0 is present, MTA_0 derives the keys from $Secret_0$, Pad_0, and the cipher suites chosen by MTA_1. If the message does not contain Pad_0, but indicates an alternate cipher suite selection, MTA_0 generates new inbound RTP/RTCP keys using only the new cipher suites and $Secret_0$. Finally, MTA_0 establishes outbound RTP/RTCP keys based on the agreed upon cipher suites, $Secret_1$, and Pad_1.

NOTE:
Upon receiving the final message from CMS_0, MTA_0 must determine the appropriate encryption and authentication keys for processing incoming RTP and RTCP messages based on the end-to-end secret/pad value supplied by MTA_1. MTA_0 must also be prepared to begin exchanging encrypted RTP and RTCP messages with MTA_1.

Figure 10-8 illustrates the process described in the preceding example. The NCS messages used in the exchange are defined in CableLabs, 2001a. MTA-to-MG communication uses an identical process. However, the MG contacts an MGC instead of a CMS.

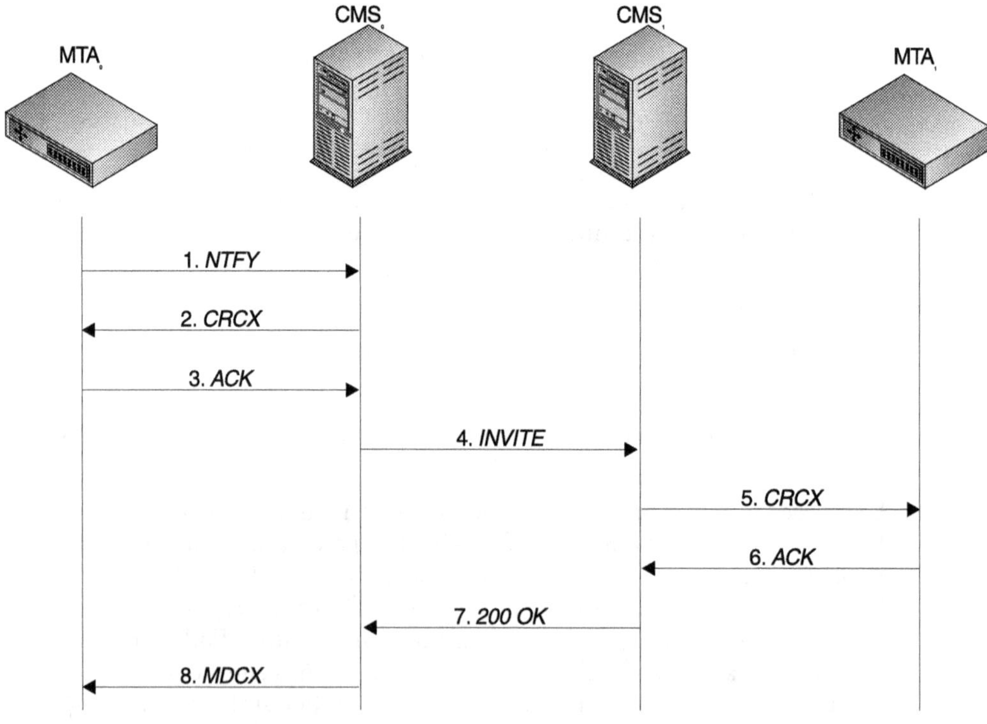

Figure 10-8 RTP and RTCP key management communicates end-to-end shared secrets via NCS messages.

PacketCable Security Certificate Usage and PKI Hierarchies

Like DOCSIS 1.1 BPI+, PacketCable employs X.509v3 digital certificates. The various components of a PacketCable network use these certificates for authentication, key exchange, and verification of software upgrade files. Every PacketCable MTA must contain an RSA key pair along with a certificate binding its MAC address to its public key, both of which are installed during manufacturing. To manage the issuance of device certificates, PacketCable also includes an MTA device certificate hierarchy that closely resembles that used by DOCSIS 1.1 BPI+ (see Figure 10-9). At the top of the hierarchy sits the MTA Root CA. CableLabs issues and maintains the self-signed MTA Root CA certificate and uses the corresponding private key to sign MTA manufacturer CA certificates. The manufacturer CA, in turn, issues device certificates to MTAs during manufacturing. We discussed this process in detail in "Manufacturer CAs" in Chapter 9. Tables 10-17 through 10-19 list the fields within the certificates of the MTA device certificate hierarchy. (Fields within brackets indicate optional values.)

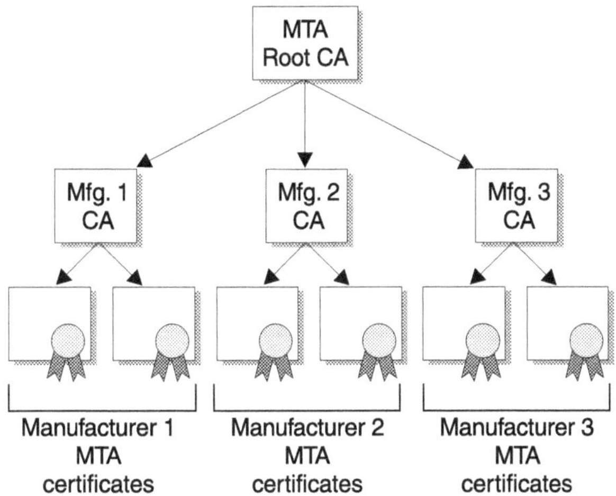

Figure 10-9 The PacketCable MTA device certificate hierarchy closely resembles that used by DOCSIS 1.1 BPI+ for cable modem authentication.

NOTE:
Within an MTA, the MTA Root CA and manufacturer CA certificates are stored within nonvolatile (but not permanent write-once) memory. To update these certificates, a manufacturer and service provider must include new certificates within secure software upgrade files. (See "Secure Software Upgrades," later in this chapter, for more information.)

In addition to the version, serial number, signature algorithm ID, issuer name, validity period, subject name, and subject public-key information fields, PacketCable certificates may contain certificate extensions. The following extensions must be used as indicated by CableLabs, 2002:

- **Subject key identifier** This extension allows multiple public keys that belong to the same subject to be distinguished during the certification process. The subject key identifier extension must be included within *all* PacketCable CA certificates and is marked as noncritical.
- **Authority key identifier** This extension allows multiple public keys that belong to a single CA to be differentiated when verifying a certificate signature. The authority key identifier extension must be included within *all* PacketCable certificates, except for root CA certificates. The value for the authority key identifier of a certificate must match the subject key identifier within the certificate of the issuing CA. This extension is marked as noncritical.
- **Key usage** Restricts the usage of the public key contained within the certificate. *All* PacketCable CA certificates must contain the key usage extension with the certificate signing and CRL signing bits enabled. End-entity certificates may contain this extension. This extension must be marked as critical in CA certificates and should be marked as critical when included within device certificates and other end-entity certificates.
- **Basic constraints** Indicates whether the subject identified within the certificate is a CA, and if so, lists the maximum depth for a certification path passing through that CA. *All* PacketCable CA certificates must include the basic constraints extension, and it must be marked as critical.

PacketCable certificates may include any other X.509v3-compliant extension, but they must be marked as noncritical. For a more detailed discussion of certificate extensions, see Appendix B.

Field	Value(s)
Version	3
Serial number	The serial number of the MTA Root CA certificate
Signature algorithm ID	PKCS #1 RSA signature using SHA-1 (ASN.1 OID 1.2.840.113549.1.1.5)
Issuer	A distinguished name of the following format (self-signed): countryName = US organizationName = CableLabs organizationalUnitName = PacketCable commonName = PacketCable Root Device Certificate Authority
Validity period	Begins on or before the date that the MTA Root CA begins issuing manufacturer CA certificates and should never be reissued. The lifetime of the MTA Root CA certificate is 20+ years.
Subject	A distinguished name of the following format: countryName = US organizationName = CableLabs organizationalUnitName = PacketCable commonName = PacketCable Root Device Certificate Authority
Subject public key information	RSA public key of the MTA Root CA with a modulus length of 2048 bits and corresponding algorithm identifier (OID 1.2.840.113549.1.1.1).
Issuer unique ID	Not present in PacketCable certificates.
Subject unique ID	Not present in PacketCable certificates.
Extensions	Must contain the following extensions: **Key usage** Marked as critical with certificate and CRL signing bits enabled **Subject key identifier** Marked as noncritical **Basic constraints** Marked as critical with CA bit set to true and path length constraint of one May contain any other extensions as long as they are marked as noncritical.

Table 10-17 MTA Root CA Certificate Format

Field	Value(s)
Version	3
Serial number	A unique serial number issued by the MTA Root CA.
Signature algorithm ID	PKCS #1 RSA signature using SHA-1 (ASN.1 OID 1.2.840.113549.1.1.5)
Issuer	A distinguished name of the following format: countryName = US organizationName = CableLabs organizationalUnitName = PacketCable commonName = PacketCable Root Device Certificate Authority
Validity period	Begins when the certificate is issued by the MTA Root CA and may be periodically reissued. The lifetime of a manufacturer CA certificate is 20 years.
Subject	A distinguished name of the following format: countryName = <country of manufacturer> organizationName = <company name> [stateOrProvinceName = <state/province>] [localityName = <city>] organizationalUnitName = PacketCable [organizationalUnitName = <manufacturer's facility>] commonName = <company name> PacketCable CA
Subject public-key information	RSA public key of the manufacturing CA with a modulus length of 2048 bits and corresponding algorithm identifier (OID 1.2.840.113549.1.1.1).
Issuer unique ID	Not present in PacketCable certificates.
Subject unique ID	Not present in PacketCable certificates.
Extensions	Must contain the following extensions: **Key usage** Marked as critical with certificate and CRL signing bits enabled **Subject key identifier** Marked as noncritical **Authority key identifier** Marked as noncritical **Basic constraints** Marked as critical with CA bit set to true and path length constraint of zero May contain any other extensions as long as they are marked as noncritical.

Table 10-18 MTA Manufacturer CA Certificate Format

Field	Value(s)
Version	3
Serial number	A unique serial number issued by a manufacturing CA; every MTA certificate issued by the same manufacturer must contain a unique serial number.
Signature algorithm ID	PKCS #1 RSA signature using SHA-1 (ASN.1 OID 1.2.840.113549.1.1.5).
Issuer	A distinguished name of the following format: countryName = <country of manufacturer> organizationName = <company name> [stateOrProvinceName = <state/province>] [localityName = <city>] organizationalUnitName = PacketCable [organizationalUnitName = <manufacturer's facility>] commonName = <company name> PacketCable CA
Validity period	Begins with the manufacturing date of the MTA and extends beyond the expected operational lifetime of the device (at least 20 years).
Subject	A distinguished name of the following format: countryName = <country of manufacturer> organizationName = <company name> [stateOrProvinceName = <state/province>] [localityName = <city>] organizationalUnitName = PacketCable [organizationalUnitName = <product name>] [organizationalUnitName = <manufacturer's facility>] commonName = <MAC address>
Subject public-key information	RSA public key of the MTA with a modulus length of 1024, 1536, or 2048 bits and corresponding algorithm identifier (OID 1.2.840.113549.1.1.1).
Issuer unique ID	Not present in PacketCable certificates.
Subject unique ID	Not present in PacketCable certificates.
Extensions	Must contain the following extension: **Authority key identifier** Marked as noncritical May contain the key usage extension marked as critical with the digital signature and key encipherment bits enabled. May contain any other extensions as long as they are marked as noncritical.

Table 10-19 MTA Device Certificate Format

In addition to the MTA device certificate hierarchy, PacketCable includes an additional PKI hierarchy for service provider end-entity certificates, as illustrated in Figure 10-10. These certificates are issued to network servers and other equipment within a service provider's network. The top level of the hierarchy includes the CableLabs Service Provider Root CA. The Service Provider Root CA issues certificates to service provider CAs, which then issue certificates to local system operator CAs. A service provider CA corresponds to the top of the PKI hierarchy within a single provider's domain. (A domain consists of all entities under the administrative control of a single provider.) The local system operator CAs are responsible for issuing certificates to network servers, such as KDCs, and other network equipment. In this hierarchy, the local system operator CAs are optional; a service provider may use its service provider CA to directly issue end-entity certificates. Tables 10-20 through 10-24 list the fields within the certificates of the CableLabs Service Provider certificate hierarchy. (Fields within brackets indicate optional values.)

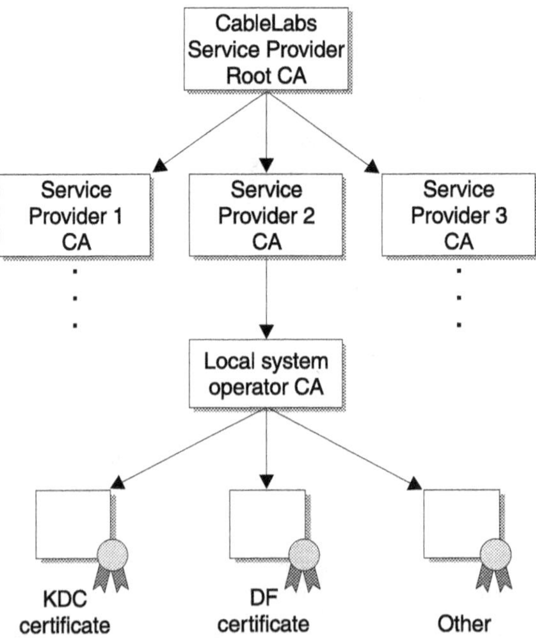

Figure 10-10 PacketCable defines a service provider certificate hierarchy for issuing end-entity certificates to network servers and other equipment within service provider networks.

NOTE:
The only end-entity certificates currently defined by the PacketCable specification are key distribution center and delivery function certificates.

Field	Value(s)
Version	3
Serial number	The serial number of the Service Provider Root CA certificate.
Signature algorithm ID	PKCS #1 RSA signature using SHA-1 (ASN.1 OID 1.2.840.113549.1.1.5).
Issuer	A distinguished name of the following format (self-signed): countryName = US organizationName = CableLabs commonName = CableLabs Service Provider Root CA
Validity period	Begins on or before the date that the Service Provider Root CA begins issuing service provider CA certificates and should never be reissued. The lifetime of the Service Provider Root CA is 20+ years.
Subject	A distinguished name of the following format: countryName = US organizationName = CableLabs commonName = CableLabs Service Provider Root CA
Subject public-key information	RSA public key of the Service Provider Root CA with a modulus length of 2048 bits and corresponding algorithm identifier (OID 1.2.840.113549.1.1.1).
Issuer unique ID	Not present in PacketCable certificates.
Subject unique ID	Not present in PacketCable certificates.
Extensions	Must contain the following extensions: **Key usage** Marked as critical with the certificate and CRL signing bits enabled **Subject key identifier** Marked as critical **Basic constraints** Marked as critical with the CA bit set to true May contain any other extensions as long as they are marked as noncritical.

Table 10-20 CableLabs Service Provider Root CA Certificate Format

Field	Value(s)
Version	3
Serial number	A unique serial number issued by the Service Provider Root CA.
Signature algorithm ID	PKCS #1 RSA signature using SHA-1 (ASN.1 OID 1.2.840.113549.1.1.5).
Issuer	A distinguished name of the following format: countryName = US organizationName = CableLabs commonName = CableLabs Service Provider Root CA
Validity period	Begins when the certificate is issued by the Service Provider Root CA and may be periodically reissued. The lifetime of a service provider CA certificate is 20 years.
Subject	A distinguished name of the following format: countryName = <country of service provider> organizationName = <company name> commonName = <company name> CableLabs Service Provider, CA
Subject public-key information	RSA public key of the service provider CA with a modulus length of 2048 bits and corresponding algorithm identifier (OID 1.2.840.113549.1.1.1).
Issuer unique ID	Not present in PacketCable certificates.
Subject unique ID	Not present in PacketCable certificates.
Extensions	Must contain the following extensions: **Key usage** Marked as critical with certificate and CRL signing bits enabled **Subject key identifier** Marked as noncritical **Authority key identifier** Marked as noncritical **Basic constraints** marked as critical with CA bit set to true and path length constraint of one May contain any other extensions as long as they are marked as noncritical.

Table 10-21 Service Provider CA Certificate Format

Field	Value(s)
Version	3
Serial number	A unique serial number issued a service provider CA.
Signature algorithm ID	PKCS #1 RSA signature using SHA-1 (ASN.1 OID 1.2.840.113549.1.1.5).
Issuer	A distinguished name of the following format: countryName = <country of service provider> organizationName = <company name> commonName = <company name> CableLabs Service Provider, CA
Validity period	Begins when the certificate is issued by a service provider CA and may be periodically reissued. The lifetime of a local system operator CA certificate is 20 years.
Subject	A distinguished name of the following format: countryName = <country of local system> organizationName = <company name> organizationalUnitName = <local system name> commonName = <company name> CableLabs Local System CA
Subject public-key information	RSA public key of the local system CA with a modulus length of 1024, 1536, or 2048 bits and corresponding algorithm identifier (OID 1.2.840.113549.1.1.1).
Issuer unique ID	Not present in PacketCable certificates.
Subject unique ID	Not present in PacketCable certificates.
Extensions	Must contain the following extensions: **Key usage** Marked as critical with the certificate and CRL signing bits enabled **Subject key identifier** Marked as noncritical **Authority key identifier** Marked as noncritical **Basic constraints** Marked as critical with the CA bit set to true and a path length constraint of zero May contain any other extensions as long as they are marked as noncritical.

Table 10-22 Local System CA Certificate Format

Field	Value(s)
Version	3
Serial number	A unique serial number issued by a service provider CA or a local system operator CA.
Signature algorithm ID	PKCS #1 RSA signature using SHA-1 (ASN.1 OID 1.2.840.113549.1.1.5).
Issuer	A distinguished name of the following format: countryName = <country of service provider> organizationName = <company name> commonName = <company name> CableLabs Service Provider CA or countryName = <country of local system> organizationName = <company name> organizationalUnitName = <local system name> commonName = <company name> CableLabs Local System CA
Validity period	Begins when the certificate is issued by a service provider or local system operator CA and may be periodically reissued. The lifetime of a KDC certificate is 20 years.
Subject	A distinguished name of the following format: countryName = <country of provider> organizationName = <company name> [organizationalUnitName = <local system name>] organizationalUnitName = CableLabs Key Distribution Center commonName = <DNS name>
Subject public-key information	RSA public key of the KDC with a modulus length of 1024, 1536, or 2048 bits and corresponding algorithm identifier (OID 1.2.840.113549.1.1.1).
Issuer unique ID	Not present in PacketCable certificates.
Subject unique ID	Not present in PacketCable certificates.
Extensions	Must contain the following extensions: **Authority key identifier** Marked as noncritical **Subject alternate name** Marked as noncritical and set to the Kerberos principal name of the KDC. (See Chapter 6 for a discussion of Kerberos and principal names.) May contain the key usage extension marked as critical with the digital signature bit enabled. May contain any other extensions as long as they are marked as noncritical.

Table 10-23 Key Distribution Center Certificate Format

Field	Value(s)
Version	3
Serial number	A unique serial number issued by a service provider CA or a local system operator CA.
Signature algorithm ID	PKCS #1 RSA signature using SHA-1 (ASN.1 OID 1.2.840.113549.1.1.5).
Issuer	A distinguished name of the following format: countryName = <country of service provider> organizationName = <company name> commonName = <company name> CableLabs Service Provider CA or countryName = <country of local system> organizationName = <company name> organizationalUnitName = <local system name> commonName = <company name> CableLabs Local System CA
Validity period	Begins when the certificate is issued by a service provider or local system operator CA and may be periodically reissued. The lifetime of a DF certificate is 20 years.
Subject	A distinguished name of the following format: countryName = <country of provider> organizationName = <company name> [organizationalUnitName = <local system name>] organizationalUnitName = PacketCable Electronic Surveillance commonName = <IP address>
Subject public-key information	RSA public key of the DF server with a modulus length of 2048 bits and corresponding algorithm identifier (OID 1.2.840.113549.1.1.1).
Issuer unique ID	Not present in PacketCable certificates.
Subject unique ID	Not present in PacketCable certificates.
Extensions	Must contain the following extensions: **Authority key identifier** Marked as noncritical **Subject alternate name** Marked as noncritical and set to the DNS name of the DF server May contain the key usage extension marked as critical with the digital signature bit enabled. May contain any other extensions as long as they are marked as noncritical.

Table 10-24 DF Certificate Format

NOTE:
DF certificates are used for authentication during Phase 1 IKE negotiations between DF servers within a single service provider's domain.

The final PKI hierarchy used by PacketCable is that for code verification certificates (CVCs). PacketCable uses these certificates for the verification of signed software upgrade files. The CVC PKI hierarchy and certificate structure employed by embedded PacketCable MTAs is identical to that of DOCSIS 1.1 BPI+ (CableLabs, 2001b).

NOTE:
Because the current version of the PacketCable Security specification addresses only MTAs embedded within DOCSIS 1.1 cable modems, PacketCable relies completely on the secure software upgrade functionality provided by DOCSIS 1.1 BPI+.

PacketCable Certificate Validation

Certificate validation within PacketCable entails the validation of entire certificate chains. The certificate chains include all the certificates—up to and including a root CA certificate—required to build trust in an MTA device certificate or a service provider equipment certificate. For instance, the certificate chain for a provider's key distribution center may take the form *[CableLabs Service Provider Root CA certificate + Service Provider CA certificate + Local System Operators certificate + KDC certificate]*. For the last certificate in the chain to pass validation, all certificates preceding it in the certificate chain must verify correctly (they must contain valid signatures and validity periods, for example). We discuss certificate validation in greater detail in Appendix B.

NOTE:
A common step during certificate validation is to check whether the certificate has been revoked prior to its scheduled expiration date. Revocation occurs for a number of reasons, including key or certificate compromise, certificate reissuance, change of affiliation or employment status, or another unexpected occurrence. This prevents an attacker—and even the legitimate owner of the certificate—from using the certificate to gain unauthorized access to information or resources. PacketCable currently does not support certificate revocation.

Those devices that use certificates for authentication exchange their certificate chains during an authentication session. In most cases, the root certificate of the hierarchy does not appear in the certificate chains exchanged over the wire.

Instead, each device maintains its own copy of the MTA Root CA certificate and/or the CableLabs Service Provider Root CA certificate. This implies that all devices must know the identity of the root CAs ahead of time. If a root CA certificate does appear in the chain, it must not differ from the copy maintained by the device except for possibly the serial number, validity period, and the signature value. Certificates containing modifications in any other fields must fail validation.

NOTE:
In PacketCable, the validity period of a certificate need not fall within the validity period of the certificate of the entity that issued it. This allows the certificates of the MTA Root CA and subordinate manufacturer CAs to be reissued without invalidating the certificates of MTAs contained within permanent write-once memory. According to CableLabs, 2002, MTA certificates cannot be replaced or renewed.

Physical Protection of Keying Material

To prevent cloning, PacketCable requires that MTA vendors deter access to all keying material within an MTA. Each MTA must maintain an RSA key pair within permanent write-once memory, and store IPSec encryption and authentication keys and Kerberos session keys within persistent memory. None of the keying material should be extractable via any of the MTA interfaces (most important, the interface to the cable network), and all MTAs should meet FIPS 140-1 Level 1 security requirements for embedded devices.

NOTE:
FIPS 140-1 Level 1 security requirements address the protection of keying material in multiple-chip standalone modules. This includes the use of production-grade chips, metal or hard plastic enclosures, and a sealant for covering exposed circuitry (FIPS140-1, 1982).

The security of keying material stored within an MTA is contingent upon the owner preventing unauthorized physical access to the device. Although an attacker cannot obtain keys by querying the network interfaces of an MTA, he or she may be able to extract keying material by opening the device and tapping into its circuitry. This keying material can then be used to create a cloned device. Section 10.1 of CableLabs, 2002, describes active and passive attacks involving cloned modems. In summary, an *active clone* participates in network exchanges, while a *passive clone* monitors the traffic of its parent device (the parent device is the legitimate MTA belonging to a paying subscriber).

Secure Software Upgrades

PacketCable mandates the use of signed code files for upgrading MTAs deployed in the field. The current version of the PacketCable specification addresses only MTAs embedded within DOCSIS 1.1. BPI+-compliant cable modems. As such, PacketCable relies on the code upgrade verification provided by DOCSIS 1.1 BPI+ (CableLabs, 2001). A future PacketCable specification is likely to address the case of standalone MTAs. These devices must implement software upgrade verification on their own using the appropriate manufacturer and service provider code verification certificates (CVCs). See Chapter 9 for more information on secure software upgrades.

References

CableLabs. December 21, 2001. *PacketCable Network-Based Call Signaling Protocol Specification (PKT-SP-EC-MGCP-I04-011221)*. Cable Television Laboratories.

CableLabs. August 29, 2001. *DOCSIS Baseline Privacy Plus Interface Specification (SP-BPI+-I07-010829)*. Cable Television Laboratories.

CableLabs. January 16, 2002. *PacketCable Security Specification (PKT-SP-SEC-I05-020116)*. Cable Television Laboratories.

FIPS140-1. April, 1982. *Federal Information Processing Standards Publication 140-1, Security Requirements for Cryptographic Modules*. National Institute of Standards and Technology.

Harkins, D., and D. Carrel. November 1998. *RFC 2409: The Internet Key Exchange (IKE)*. The Internet Engineering Task Force.

Hur, Matthew, Brian Tung, Tatyana Ryutov, Clifford Neuman, Gene Tsudik, Ari Medvinsky, and Bill Sommerfeld. November 8, 2001. *Public Key Cryptography for Cross-Realm Authentication in Kerberos*. The Internet Engineering Task Force.

Neuman, Clifford, John Kohl, and Theodore Ts'o. November 24, 2000. *The Kerberos Network Authentication Service (V5)*. The Internet Engineering Task Force.

Neuman, Clifford, Brian Tung, Mathew Hur, Ari Medvinsky, Sasha Medvinsky, John Wray, and Jonathan Trostle. November 28, 2001. *Public Key Cryptography for Initial Authentication in Kerberos*. The Internet Engineering Task Force.

Chapter 11

Securing Interactive Television: DVB MHP Security

For decades, the one-way nature of broadcast television services restricted viewer interaction to flipping dials and clicking buttons on remote controls. Consequently, viewers could do little to tailor their viewing experience. The broadcasting stations and television service providers decided what to play and when to play it. If viewers didn't like what was playing, their only options were to change the channel or find something else to do. Early efforts at enhancing broadcast television resulted in modest improvements over existing services. Since the mid-1970s, providers in many European countries have offered *teletext* services for distributing textual information—such as program listings, news, and weather—to television sets or set-top boxes with teletext decoders. Teletext systems transmit data within the vertical blanking interval of a television signal (the nonviewable portion of the signal that the TV uses to reposition its electron beam between successive frames). Closed-captioning also uses this technique to transmit text captions for the hearing impaired.

With the advent of *pay-per-view (PPV)* television, cable and satellite television providers offered subscribers the ability to select and pay for movies, sporting events, concerts, and other shows on a per-view basis. Subscribers chose PPV events from a list of currently playing titles, and telephoned their requests into their provider. The provider then charged each subscriber's account, and gave him or her access to the next showing of the event. Each PPV title played in a continuous loop on a particular channel, so a viewer chose a time that best fit his or her schedule, and waited for the next show. Current PPV systems enable users to request a PPV title directly from their set-top box; the set-top box contacts the provider via a bi-directional cable or satellite system, or via a dial-up connection, to transmit billing related information. Both teletext and PPV extend the functionality of broadcast television, but are far from interactive.

True *interactive television*, or *iTV,* combines the content of broadcast television with the interactivity of a personal computer. Unlike unidirectional broadcast television, iTV is capable of accepting user input and communicating it back to the provider in real-time. This enables a service provider to address the personal interests and viewing habits of each of its subscribers by tailoring the delivery of content on a subscriber-by-subscriber basis. iTV applications include interactive program guides and television shows, information services, impulse pay-per-view (that is, not having to wait for the next showing), video-on-demand, e-commerce, online gaming, Internet access, and much more.

The Multimedia Home Platform

In Chapter 5, we introduced the *Multimedia Home Platform (MHP)* from the Digital Video Broadcast group. MHP is intended to extend the interactive functionality of DVB to all transmission networks including cable, wireless, and satellite. The MHP specification (Digital Video Broadcasting Project, 2001) defines an *application programming interface (API)* between iTV applications and their execution environment (that is, a set-top box). The MHP 1.1 specification may be downloaded from http://www.mhp.org/technical_essen/pdf_and_other_files/Ts102812.V1.1.1.zip.

As depicted in Figure 11-1, the basic MHP environment consists of an MHP terminal, user input/output (I/O) devices, and one or more MHP resources. The terminal outputs audio and video streams to devices, such as televisions and stereo receivers, and accepts input from remote control units, keyboards, mice, and other input devices. An MHP terminal also includes interfaces for connecting to a broadcast network and a *return channel*. The return channel allows the

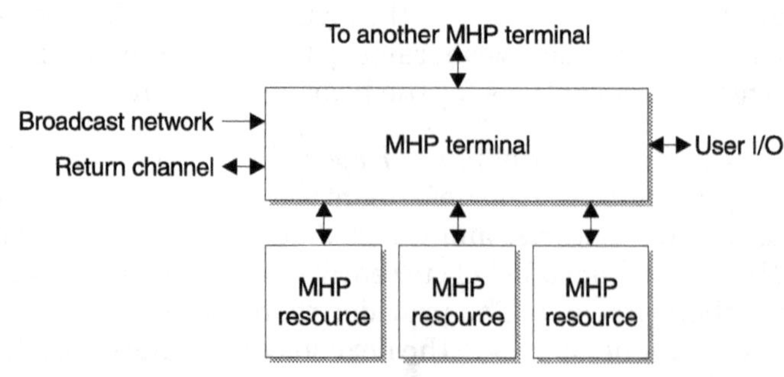

Figure 11-1 The MHP environment consists of one or more MHP terminals, a collection of MHP resources, network access channels, and user input/output devices.

terminal to communicate information back to the provider and to contact other remote entities. Applications running within an MHP terminal have access to the MHP resources, and may direct data and traffic flows to storage repositories.

MHP can be described using a three-layer model consisting of resources, system software, and applications (see Figure 11-2). The resource layer includes media controllers and decoders, tuners, CPU, memory, storage devices, audio and video subsystems, network interfaces, and associated data stores and media streams. Applications cannot directly access these resources. Instead, the applications make calls to the *system software* layer, which acts as a proxy to the resource layer. The system software also includes an *application manager* component for supervising application loading, pausing, stopping, resuming, unloading, and other operations relating to the life cycle of an application. The bulk of the MHP specification addresses the API sitting between the applications and the system software layer, and leaves the remaining details of the two lower layers to the terminal manufacturer. The API provides a transparent interface between MHP applications and the vendor-specific system software, and decouples the applications from the details of the underlying terminal hardware and software. According to (Digital Video Broadcasting Project, 2001), the API provides an abstract view of)

- Media streams and I/O routines
- Commands and events
- Data records and files
- Hardware resources

MHP applications run within a platform known as *DVB-J*. The DVB-J runtime environment includes a Java *virtual machine (VM)* that implements many Java classes. A vendor may include support for both proprietary and MHP interoperable applications within its terminal. Interoperable MHP applications interface to the system software layer via the MHP API. Vendor-specific code, on the other hand, may directly access the system software via proprietary system calls. Whereas MHP interoperable applications function on any MHP terminal, proprietary applications may only operate on a particular vendor's terminal.

Figure 11-2
The MHP hardware/ software model can be described by using three main layers and an intermediate API layer.

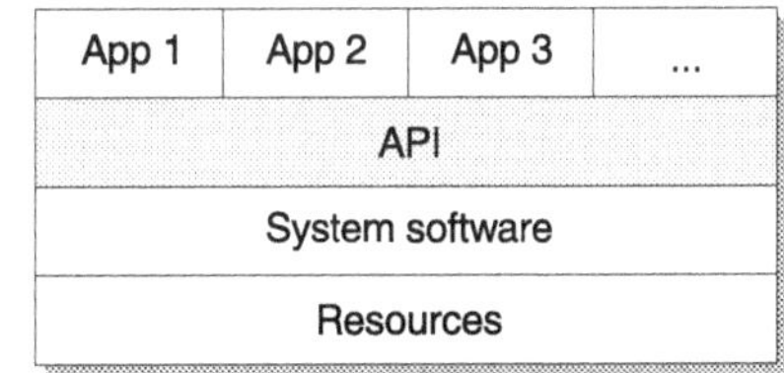

Table 11-1 MHP **application_id** Value Ranges

ID Value	Use
0x0000...0x3fff	Application IDs for unsigned applications
0x4000...0x7fff	Application IDs for signed applications
0x8000...0xfffd	Reserved for future use by DVB
0xfffe	Special wildcard value for signed applications of an organization
0xffff	Special wildcard value for all applications of an organization

All MHP applications are identified by using a 6-byte application identifier consisting of an organization and application ID. The organization ID (**organization_id**) is a unique 32-bit identifier assigned to the organization responsible for the application. The application ID (**application_id**) is a unique 16-bit value indicating the function of the application. Table 11-1 lists the ID ranges for signed and unsigned applications as presented in (Digital Video Broadcasting Project, 2001).

MHP Security Overview

During normal operation, DVB-J applications and plug-ins typically require access to one or more protected MHP resources. Not all applications require access to the same resources, and the level of access granted by the terminal varies from one application to the next. For instance, one application may request access to a media decoder to process an incoming video stream, while another asks to communicate with a remote host via the return channel. An application may even request to launch a separate application, or access data that it does not own. The level of access afforded each application depends on the terminal's security policy and whether or not the source of the application can be trusted. To protect terminal resources from misuse by rogue applications, MHP provides a number of security services including

- Source authentication for MHP applications
- Configurable security policy
- Certificate management
- Return channel privacy and integrity

To prevent the downloading of malicious code, MHP includes security components for authenticating the source of application code and other files downloaded to set-top boxes, televisions with integrated DVB receivers, and multimedia PCs. An MHP terminal must authenticate the source of code files before storing or executing an application that requires access to sensitive MHP

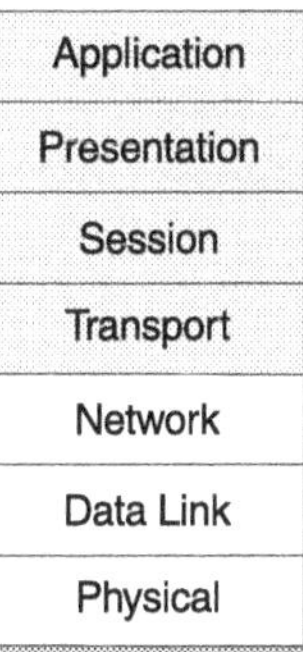

Figure 11-3
MHP security operates at or above the Transport layer.

resources. MHP defines three message types for use in authenticating applications—hash codes, signatures, and certificates—which we discuss in the next section. Following successful authentication, MHP determines the access rights a terminal should grant to the application based on the rights requested by the service provider and those granted by the owner of the MHP terminal. To prevent eavesdropping and data modification on the return channel, MHP employs the industry-standard *Transport Layer Security (TLS)* protocol. MHP security operates at or above the Transport layer of the OSI Network Reference model (see Figure 11-3).

Authentication Messages

MHP uses three message types to communicate authentication information pertaining to an application:

- Hash codes
- Signatures
- Certificates

By using a combination of these messages, MHP individually authenticates each component of an application and efficiently binds all authenticated components into a single trusted application. Hash codes employ message digest operations to create a condensed representation of a subset (for example, files and directories) of an application. The message digest helps ensure the integrity of each component by making changes easy to detect. A single application may employ multiple hash codes to authenticate the contents of its various subdirectories. A single digital signature within the application's root directory verifies the integrity of all hash codes for that application. MHP uses certificate messages to communicate identifying information about the source of the application, and to chain the source to a trusted third party. The terminal uses this information when verifying digital signatures. All authentication messages are encapsulated within files stored in the terminal's local file system. Hash codes, signatures, and certificates are stored within hash files, signature files, and certificate files, respectively. We discuss each of these files types in the following sections. The authentication scheme employed by MHP applies to any hierarchical file system (see Figure 11-4).

CAUTION:

In Chapter 2, we explained that message digests alone do not ensure the integrity of data because a message digest provides no means of cryptographically binding a digest to the original message. An attacker can modify the data, create a new digest, and swap it out with the original digest without detection. To bind the digest to the data, a message digest operation must be combined with a shared secret (as is the case with an HMAC) or some form of encryption. MHP generates a digital signature over all of the hash codes of an application. The private-key encryption required to generate the signature prevents an intruder from modifying the application and swapping the hash codes.

Authenticating applications in this manner offers a key advantage in terms of performance. Whenever a terminal loads a signed application or plug-in from the broadcast network or an external interface, it must authenticate the application code. Not only do MHP terminals have limited processing resources to devote to authentication, but computationally expensive cryptographic operations can greatly increase the load time of an application. The combination of message digests and digital signatures employed by MHP results in efficient authentication. Remember that generating a message digest is far faster than generating a

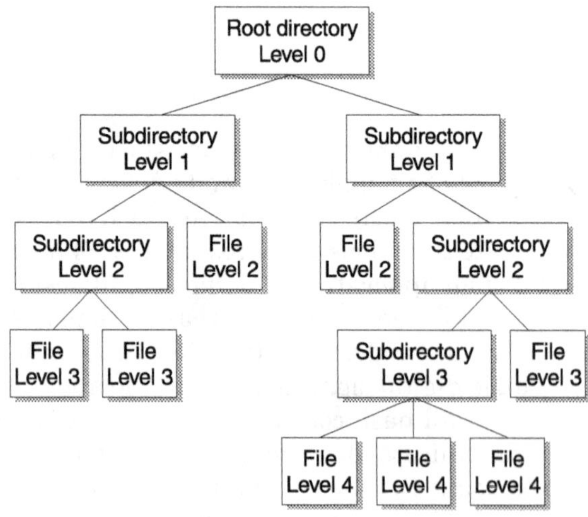

Figure 11-4 A sample hierarchical file system consisting of a root directory, subdirectories, and files.

digital signature. Instead of signing every file and directory within an application, MHP uses fast digest operations to check the integrity of each individual component, and uses a single digital signature to ensure the integrity of all the hash files. The signature also authenticates the source of the application, something that message digests alone cannot do.

Hash Files

Each directory containing objects to be authenticated must include a hash file with the name **dvb.hashfile**. The hash file lists the contents of the directory as a collection of file and directory objects, and includes all objects within the directory except for any MHP signature files and the hash file itself. Also contained within the hash file of an authenticated application is a list of MD5 or SHA-1 message digests calculated over the objects in the order they appear in the hash file. For a file object, the digest calculation includes the entire contents of the file. For a directory object, the contents of the directory's hash file are included in the calculation. Because the digest calculation includes the hash files of all subdirectories, the hash file authenticates all the objects below it in the directory tree. Figure 11-5 demonstrates the placement of hash files within a hierarchical files system.

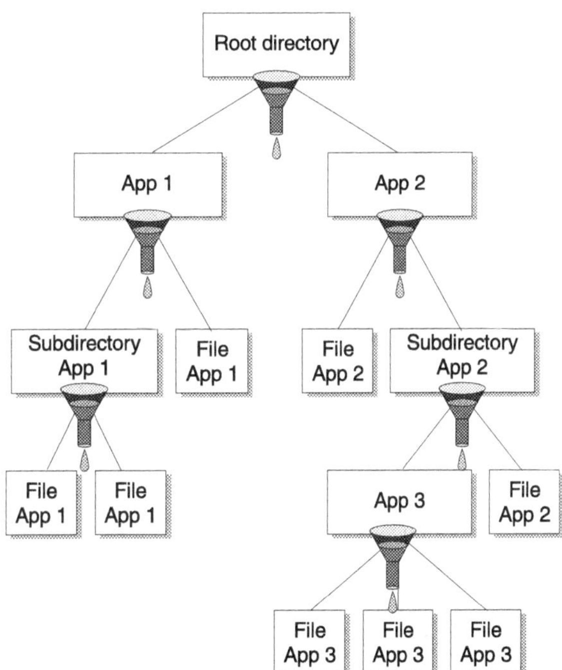

Figure 11-5 The location of hash files within a hierarchical file system

A hash file takes the following form, as specified in (Digital Video Broadcasting Project, 2001):

```
Hashfile {
 specify digest_count
 for each digest {
  specify digest_type
  specify name_count
  for each name {
   specify name_length
   for each byte in the object name {
    list name_byte
                   }
        }
  for each byte in the digest of length digest_length {
   list digest_byte
             }
}
}
```

The components of the hash file have the following meaning:

- **digest_count** A 16-bit value indicating the number of digest values contained within the hash file.
- **digest_type** An 8-bit value specifying the message digest algorithm used to generate the digest for an object (Table 11-2 lists the available algorithms).
- **name_count** A 16-bit value signifying the number of objects represented by a particular digest value. An object can be a file or directory.
- **name_length** An 8-bit value indicating the length of an object name in bytes.
- **name_byte** An 8-bit value containing a single byte of the object name.
- **digtest_length** An integer equal to the number of bytes in the digest; the number of bytes depends on the algorithm chosen.
- **digest_byte** An 8-bit value containing one byte of the MD5 or SHA-1 digest.

Signature Files

The root of a terminal's file system and the root directory of each authenticated application contain one or more signature files. Each signature file includes a sin-

Table 11-2 Supported Message Digest Algorithms for Use in Hash Code Computations

Value	Algorithm
0	Nonauthenticated (0 bytes)
1	A 16-byte MD5 digest
2	A 20-byte SHA-1 digest

gle digital signature generated over the contents of the directory tree up to and including the directory containing the signature file (that is, the root directory and all its subdirectories of an application). The number of signature files depends on the number of entities who have signed the application. The naming convention used by MHP to denote a signature file is **dvb.signaturefile.<x>**, where <x> is a decimal integer value. The possible values for <x> begin with the integer 1 and increment by one for each signature file present in the directory. For example, an application with a root directory containing two signature files—dvb.signaturefile.1 and dvb.signaturefile.2—indicates that the application has been signed by two (and only two) separate entities. Figure 11-6 illustrates the location of signature files within a hierarchical file system.

The signature contained within a signature file is calculated over the hash file located in the same directory. The signature generation process involves creating an MD5 or SHA-1 digest and encrypting the digest using the public-key encryption algorithm specified in the signer's certificate (see Chapter 2 for a discussion of digital signature generation and verification). Each signature file contains a reference to the certificate containing the public key needed to verify the signature, the message digest algorithm used to create the signature, and the signature itself. The ASN.1[1]

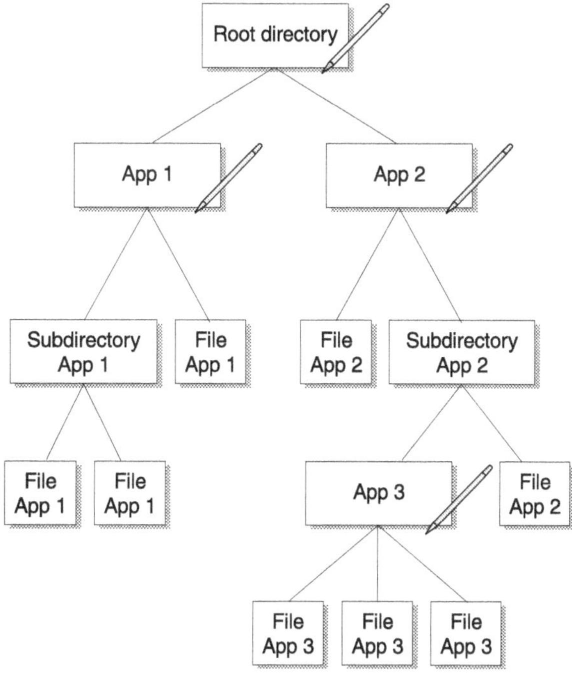

Figure 11-6 The location of signature files within a hierarchical file system

[1] Refer to Kaliski (1993) for more information on ASN.1 DER encoding.

DER-encoded signature structure employed by MHP uses the following syntax as specified in (Digital Video Broadcasting Project, 2001):

```
Signature ::= SEQUENCE {
   certificateIdentifier      AuthorityKeyIdentifier,
   hashSignatureAlgorithm     HashAlgorithmIdentifier,
   signatureValue             BIT STRING  }
```

The components of the signature have the following meanings:

- **certificateIdentifier** Signifies the certificate containing the public key required to verify the signature. See (ITU-T, 1997) for a description of the **AuthorityKeyIdentifier** certificate extension.
- **hashSignatureAlgorithm** Indicates the message digest used to generate the signature. This value identifies either the ASN.1 OID for MD5 (1.2.840.113549.2.5) or SHA-1 (1.3.14.3.2.26).
- **signatureValue** A bit string representing the signature.

Certificate Files

Each directory containing one or more signature files must also include a certificate file for each signature file (see Figure 11-7). The certificate file contains all the information required to build trust in the source of the application. It includes a certificate chain consisting of all certificates in the certification path between the signer of the application and the trusted root certification authority (see the section entitled "MHP X.509 Certificate Usage and PKI Hierarchy" for description of how MHP uses digital certificates). MHP names the certificates files using

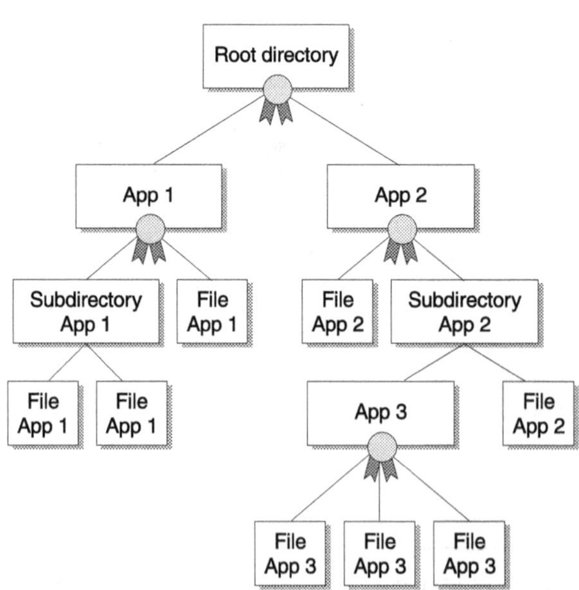

Figure 11-7 The location of certificate files within a hierarchical file system.

the convention **dvb.certificates.<x>**, where the integer value corresponds to the integer value contained within the name of the corresponding signature file.

MHP certificate files take the following format:

```
Certificatefile {
    specify certificate_count
    for each certificate {
        specify certificate_length
        list certificate_contents
    }
}
```

The fields within the certificate file have the following significance:

- **certificate_count** A 16-bit integer value specifying the number of certificates contained within the certificate file.
- **certificate_length** A 24-bit integer value indicating the length of the certificate in bytes.
- **certificate_contents** A variable-length data structure containing a single certificate.

NOTE:
The certificates are listed in the order they appear in the certificate chain between the signer and root CA; the signer's certificate appears first, and the root CA certificate last.

MHP certificates contain public keys for use with the RSA algorithm (OID 1.2.840.113549.1.1). All certificate signatures are generated using RSA with MD5 (OID 1.2.840.113549.1.1.4) or RSA with SHA-1 (OID 1.2.840.113549.1.1.5).

NOTE:
According to Digital Video Broadcasting Project, 2001, the distinguished names within all certificates must be encoded as UTF8 strings. The use of UTF8 encoding facilitates name matching by allowing fast binary comparisons.

The Object Authentication Process

As previously mentioned, the process for authenticating an application uses a combination of hash code, signature, and certificate messages to authenticate the various components of an application. In this section, we demonstrate the authentication process for a single file or directory object. The integrity of an entire file system may be ensured by applying this method to all elements within the file system. For a file, the authentication process follows these steps:

1. The terminal first checks that the hash file located in the same directory as the file to be authenticated contains a reference to the file.

2. The terminal then performs a digest operation by supplying the file as input to the appropriate message digest algorithm as specified within the hash file. The value obtained from the digest operation must match the value stored within the hash file.
3. The terminal ascends the directory tree checking to see that each directory is properly authenticated by its parent directory (that is, the directory above it in the hierarchy). This process involves generating a digest from the hash file of each directory, and verifying that it matches the digest value for the corresponding directory object in the hash file of its parent directory.
4. The terminal stops when it reaches a directory containing one or more signature files. This directory is termed the *root directory* of the authenticated application.
5. To complete the authentication, the terminal must verify the signature contained within the signature file, which requires the terminal to first locate the corresponding certificate file.
6. The terminal validates each certificate in the certificate chain—from the signer to the root CA—contained within the certificate file, and verifies that none of the certificates have been revoked (see the section entitled "Certificate Revocation," later in this chapter, for a discussion of MHP certificate revocation).
7. If all certificates validate successfully and the terminal trusts the root CA, the terminal extracts the public key from the signer's certificate. It uses this key to verify the signature value in the signature file.
8. In the presence of multiple signature files, the terminal must repeat steps 5 to 7 for each signature file located in the application's root directory.

The process for authenticating a directory is almost identical to that for authenticating a file:

1. The terminal first checks that the hash file located in the parent directory contains a reference to the hash file within the directory it is authenticating.
2. The terminal then performs a digest operation by supplying the authenticated directory's hash file as input to the appropriate message digest algorithm as specified within the hash file of the parent directory. The value obtained from the digest operation must match the value stored within the parent directory's hash file.
3. The terminal ascends the directory tree checking to see that each directory is properly authenticated by its parent directory. This process involves generating a digest from the hash file of each directory, and

verifying that it matches the digest value for the corresponding directory object in the hash file of its parent directory.

4. The terminal stops when it reaches a directory containing one or more signature files. This directory is termed the *root directory* of the authenticated application.
5. To complete the authentication, the terminal must verify the signature contained within the signature file, which requires the terminal to first locate the corresponding certificate file.
6. The terminal validates each certificate in the certificate chain—from the signer to the root CA—contained within the certificate file, and verifies that none of the certificates have been revoked.
7. If all certificates validate successfully and the terminal trusts the root CA, the terminal extracts the public key from the signer's certificate. It uses this key to verify the signature value in the signature file.
8. In the presence of multiple signature files, the terminal must repeat steps 5 to 7 for each signature file located in the application's root directory.

The steps do not need to be completed in the order shown, but failure to successfully complete any of the steps listed results in an immediate authentication failure. Applications that fail authentication are treated as unsigned applications.

NOTE:
Section 12.4.4 of the current draft of the MHP v1.1.1 specification (Digital Video Broadcasting Project, 2001) eliminates the signature verification mentioned in step 5 above. The author believes that this is an unintentional error in the specification, and has included the step here for completeness.

The aforementioned process enables an MHP terminal to authenticate quickly a single file or a small collection of objects (for example, a single application) within a hierarchical directory structure. This process is *not* as efficient for verifying the integrity of a complete file system, because it can potentially involve a very large number of digest calculations. A more efficient approach for checking the integrity of a file system might be to create a single compressed image of the file system, pass the image through a message digest, and encrypt the digest output using a private key. The result would be a single signature generated over the entire file system.

MHP X.509 Certificate Usage and PKI Hierarchy

The authentication process outlined by the MHP specification requires the use of X.509v3 digital certificates. Certificates provide a mechanism for service providers and MHP terminals to communicate the identity information necessary to verify and authenticate signed applications. However, they require a PKI for managing certificate issuance and revocation, and for establishing a certification path between end-entity certificates and a trusted root CA. At the time of this writing, the MHP specification lacks a defined PKI hierarchy, such as the three-tiered model employed by DOCSIS 1.1 BPI+. Nevertheless, the presence of certificate chains within certificate files implies support for an arbitrary PKI hierarchy chaining up to one or more root CAs. Table 11-3 outlines the format of an MHP digital certificate, and Table 11-4 lists the X.509v3 extensions that may appear in each certificate. Appendix B provides a quick review of digital certificates and public-key infrastructure concepts.

Field	Value(s)
Version	3
Serial number	A unique serial number generated by the issuing CA
Signature algorithm ID	MHP supports two signature algorithms: RSA signature using SHA-1 (ASN.1 OID 1.2.840.113549.1.1.5) RSA signature using MD5 (ASN.1 OID 1.2.840.113549.1.1.4)
Issuer	At least the commonName of the issuer
Validity period	Begins with the date of issue and extends for a period determined by the issuing CA
Subject	Minimum fields for Root CA and subordinate CA certificates: commonName countryName Minimum fields for end-entity (or *leaf*) certificates: commonName countryName organizationName, which takes the form <text_name>.organization_id
Subject public-key information	RSA public key of the subject (ASN.1 OID 1.2.840.113549.1.1)
Issuer unique ID	Not present in MHP certificates
Subject unique ID	Not present in MHP certificates
Extensions	See Table 11-4

Table 11-3 The Format of an MHP Digital Certificate

Extension	Broadcast	Implementation	Criticality
Authority key identifier	Optional	Optional	Noncritical
Subject key identifier	Optional	Optional	Noncritical
Key usage	Mandatory	Mandatory	If marked critical: the digital signature bit shall be set within end-entity certificates; the certificate signing bit shall be set within CA certificates.
Private key usage period	Optional	Optional	Noncritical
Certificate policies	Optional	Optional	Noncritical
Policy mappings	Optional	Optional	Noncritical
Subject alternative name	Mandatory	Optional	Noncritical
Issuer alternative name	Mandatory	Optional	Noncritical
Subject directory attributes	Optional	Optional	Noncritical
Basic constraints	Optional	Mandatory	May be marked critical
Name constraints	Optional	Mandatory	May be marked critical
Policy constraints	Optional	Optional	Noncritical
Extended key usage	Optional	Optional	Noncritical
CRL distribution points	Optional	Optional	Noncritical

Table 11-4 X.509v3 Extension Usage Within the MHP Environment

MHP digital certificates may contain a number of X.509v3 extensions. MHP makes a distinction between certificates distributed with an application over a broadcast channel and those stored along with an application in an MHP terminal. For purposes of clarity, we term these certificates broadcast and implementation certificates, respectively. Both certificate types have slightly different requirements for extension usage, as presented in Table 11-4 (Digital Video Broadcasting Project, 2001). Mandatory fields are fields that must be included within broadcast and implementation certificates. An MHP terminal must be capable of processing all mandatory extensions. Broadcast certificates may include any of the optional fields, which a terminal may ignore. A terminal will treat a certificate as invalid if it contains unrecognized extensions marked as critical.

Storage and Management of Root Certificates

All MHP terminals maintain within persistent storage a list of valid root CA certificates, which the terminal vendor installs during manufacturing. In the event that a root certificate is retired or revoked, a service provider must be capable of updating the certificate lists contained within all terminals attached to its network. Fortunately, MHP incorporate such a mechanism for disseminating updated certificate information. By using a *Root Certificate Management Message (RCMM),* a service provider can transmit a list of certificates within an MPEG transport stream sent over a broadcast channel. Each RCMM contains multiple signatures generated using the private keys corresponding to the certificates being replaced; the number of signatures is equal to the quantity of root certificates appearing in the RCMM. Before updating the root certificate list, a terminal must verify all signatures attached to the RCMM. The use of multiple signatures, each generated using a different private key, prevents a terminal from accepting an RCMM signed by an attacker who has managed to compromise only a single root CA private key or certificate. The file naming convention for an RCMM is **dvb.rcmm**.

Certificate Revocation

There are a handful of situations, such as key compromise and change of operational status or group affiliation, in which a certification authority may revoke an MHP certificate prior to its scheduled expiration. Revoked certificates may not be used by a terminal to authenticate an application or to establish a TLS session with a remote host. To prevent an MHP terminal from trusting a revoked certificate, a service provider may distribute *certificate revocation lists (CRLs).* A terminal uses a CRL to check the revocation status of all certificates in the chain contained within a certificate file prior to executing an application or trusting a TLS server.

The provider may distribute CRLs by using one of two methods. The first method involves transmitting the CRL over the return channel. The MHP specification suggests this as a viable approach for disseminating CRLs for TLS certificates, but not for broadcast application certificates. For those terminals lacking a return channel interface, the provider must distribute CRLs via an MPEG transport stream sent over a broadcast channel. This is the preferred method for updating revocation information pertaining to broadcast applications, as it scales to a much larger number of terminals than the return channel method. The latter approach also allows distribution of CRLs even when the terminal is not connected via its return channel.

CRLs issued by subordinate CAs are contained within files that follow the naming convention **dvb.crl.x**, where the value x corresponds to the integer value contained within the name of the certificate file used to authenticate the CRL. MHP includes a second naming convention for files containing CRLs issued by root CAs: **dvb.crl.root.x**. Because all terminals maintain a list of trusted root certificates, the value x is only used to distinguish between multiple CRL files within the same directory. A terminal must cache CRLs locally within persistent storage. According to (Digital Video Broadcasting Project, 2001), local caching improves both the efficiency and security of status checking. CRL files stored within a terminal's file system must be located in a subdirectory by the name of **dvb.crl** off the root directory of the file system. MHP uses the X.509 CRL format defined by (ITU-T 1997).

Application Security Policy

Access rights vary from application to application depending on the rights requested by the service provider and those allowed by the user of an MHP terminal. Signed and unsigned applications also have different access to resources. In general, a terminal provides an unsigned application with very limited access, because the terminal cannot authenticate the source or verify the integrity of the application. For signed applications, a service provider can request additional access to MHP resources by including a signed *permission request file* with the application. Within a permission request file, a provider can request to access files within the local file system, control the life cycle of an application, modify user settings and preferences, access the return channel, communicate with remote entities, and more.

NOTE:
DVB-J applications request access to resources via Java method calls. Any attempt to access a resource results in one of the security exceptions listed in Section 11.10 of (Digital Video Broadcasting Project, 2001). The terminal grants access to the requested resource by responding to the security exception.

Permission Request File

Permission request files are *XML* (*Extensible Markup Language*) documents whose *document type definition (DTD)* is defined within the current MHP specification. (Note: A DTD describes the syntax of an XML document.) Permission request files sit in the same directory as the application for which the provider is requesting permissions, and are named using the convention **dvb.<application name>.perm**. The value <application name> indicates the name of the first file in the application. Permission files include an application identifier containing the ID of the organization who issued the application, as well as an application ID within one of the ranges specified in Table 11-1. A terminal uses the application identifier to associate an application with the data it can access and the other applications it can execute on the terminal. A permission request file may also contain a list of credentials granting an application access to resources belonging to another organization. Table 11-5 lists the permissions that a provider can request using a permission request file. Table 11-6 indicates the default permissions for signed and unsigned applications.

NOTE:
No additional permissions are granted if the XML file cannot be parsed.

Permission	Description
File Access	Specifies a level of access to persistent storage
CA API	Controls access to Java-based conditional access methods
Application Lifecycle Control Policy	Limits the ability of an application to launch and control other applications
Return Channel Access Policy	Determines whether an application can access the return channel
Tuning Access Policy	Specifies whether an application can modify tuner settings using the Tuning API
Service Selection Policy[2]	Controls an application's ability to select a new service
Media API Access Policy	Specifies access to a Java-based media player via the Java Media Framework API
Inter-application Communication Policy	Determines whether an application can communicate with another application via the inter-application communication API
User Setting and Preferences Access Policy	Indicates the user settings and preferences, such as language, parental rating, fonts sizes, and the like, that an application can read and write
Network Permissions	Specifies access to remote network hosts along with the actions that may be performed over the return channel
Dripfeed Permissions	Determines whether an application can access the drip feed mode of an MPEG-2 video decoder; drip feed mode allows an application to progressively feed an MPEG-2 video decoder with small chunks of a video data stream (as opposed to the entire stream all at once)
Privileged Runtime Code Extension Permission	Limits an application's ability to perform runtime code extension (see (Digital Video Broadcasting Project, 2001) for details)
Application Storage	Controls an application's ability to store an application to an MHP terminal
Non-CA Smart Card Access	Indicates whether the application can access nonconditional access smart cards; smart cards generally store credentials (for example, keys and certificates) or contain microprocessors for performing cryptographic operations

[2] Service selection may require access to the return channel. If access has not been granted to the application to access the return channel, the service selection operation will fail.

Table 11-5 Permissions That May Appear in Permission Request Files.

Permission	Default Value
File Access	Unsigned application—no access Signed application—no access unless dictated by a permission request file Signed applications granted access by a permissions request file adhere to the following policy: An application owns everything within its root directory and subdirectories (that is, the files it creates) An application may grant read-only, write-only, and read-write access to other applications for the files it owns An application can access a file if it has been granted the appropriate application, organization, or world access permissions
Conditional Access API	Unsigned application—cannot access the CAModule.buyEntitlements, CAModule.openMessageSession, CAModuleManager.addMMIListener, CAModule.queryEntitlements, and CAModule.listEntitlements methods Signed application—same as an unsigned application, unless a provider overrides the default permission using a permission request file
Application Lifecycle Control Policy	Unsigned application—can launch and control the life cycle of any application operating within the same service; cannot control the life cycle of applications it has not launched Signed application—same as an unsigned application, unless a provider overrides the default permission using a permission request file
Return Channel Access Policy	Unsigned application—no access to the return channel Signed application—same as an unsigned application, unless a provider overrides the default permission using a permission request file
Tuning Access Policy	Unsigned application—no access to the Tuner API Signed application—same as an unsigned application, unless a provider overrides the default permission using a permission request file
Service Selection Policy	Unsigned application—may *not* select new services Signed application—may select *any* new service
Media API Access Policy	Access rights are determined by the DVB-J environment; all applications operate within a "sandbox" that controls access to processing and memory resources

Table 11-6 Default MHP Permissions

Permission	Default Value
Inter-application Communication Policy	Unsigned application—may communicate with other unsigned applications, but may *not* communicate with a signed application via the inter-application communication API Signed application—may communicate with *any* other signed application operating within the same service
User Setting and Preferences Access Policy	Unsigned application—can read user language, parental rating, default font size, and country code; cannot read or write to any other settings or preferences Signed application—same as an unsigned application, unless a provider overrides the default permission using a permission request file Signed applications granted access by a permissions request file may read and/or write to all user settings and preferences
Network Permissions	Unsigned application—cannot access remote network hosts Signed application—the permission request file contains a list of remote hosts the application can contact and the actions the application can perform
Dripfeed Permissions	Unsigned application—no access to drip feed mode Signed application—same as an unsigned application, unless a provider overrides the default permission using a permission request file
Privileged Runtime Code Extension Permission	Unsigned application—cannot perform runtime code extension Signed application—signed applications with no permissions request file can perform runtime code extension using string data from any source; those with a permission request file must adhere to the internal and external string sources described in the file
Application Storage	Unsigned application—not granted permission to store applications Signed application—same as an unsigned application, unless a provider overrides the default permission using a permission request file
Non-CA Smart Card Access	Unsigned application—no access to smart cards Signed application—same as an unsigned application, unless a provider overrides the default permission using a permission request file

Table 11-6 Default MHP Permissions *(continued)*

Return Channel Security

Return channel security within the MHP environment is provided by the TLS protocol (Dierks and Allen, 1999). MHP uses a subset of the total functionality offered by TLS, and only requires a terminal to implement the client-side component of the protocol. It is not necessary for a terminal to support any version of Secure Sockets Layer (SSL) although a vendor may wish to include this support to communicate with web servers running SSLv2 and SSLv3. Table 11-7 lists the TLS cipher suites supported by MHP for use in securing the return channel.

NOTE:
3DES EDE112 mode is identical to the two-key triple DES operation described in Chapter 9.

Before establishing a TLS session, a terminal must authenticate the remote server to which it is connecting. This requires the terminal to validate each certificate within the certificate chain provided for the server. The chain must also terminate with the certificate of a trusted root CA. A service provider may distribute one or more TLS root certificates along with an application. By combining an application with a series of trusted TLS root certificates, a provider can restrict the list of servers to which a terminal may connect using that application. The provider transmits each TLS certificate within a file by the name of **dvb.tls.organization_id.application_id.x**, where x is used to distinguish between multiple TLS root certificate files within the same directory. The organization and application IDs identify the source and function of the application. In the absence of a list of TLS root certificates, an application can call Sun's *Java*

Cipher Suite	Key Exchange	Cipher	Message Digest
TLS_NULL_WITH_NULL_NULL	NULL	NULL	NULL
TLS_RSA_WITH_NULL_MD5	RSA	NULL	MD5
TLS_RSA_WITH_NULL_SHA	RSA	NULL	SHA-1
TLS_RSA_EXPORT_WITH_DES40_CBC_SHA	RSA EXPORT	DES40_CBC	SHA-1
TLS_RSA_WITH_DES_CBC_SHA	RSA	DES_CBC	SHA-1
TLS_RSA_WITH_3DES_EDE_CBC_SHA	RSA	3DES_EDE_CBC	SHA-1
DVB RSA WITH 3DES EDE112 SHA	RSA	3DES EDE112 CBC	SHA-1

Table 11-7 Tls Cipher Suites Supported By Mhp

Secure Sockets Extension (JSSE) API to retrieve a certificate chain for the server. All MHP-compliant DVB-J implementations must support the JSSE 1.0.2 API for TLS on the return channel. For more information on Sun's JSSE API, visit http://java.sun.com/products/jsse/.

Supported Java Security Classes

MHP-compliant DVB-J VMs must support numerous security-related Java classes from the java.awt, java.io, java.lang, java.net, java.security, java.util, javax.net, and javax.security packages from Sun Microsystems. These classes pertain mostly to security policy configuration, privilege management, certificate processing, and security on the return channel. A detailed discussion of these classes is beyond the scope of this text; however, we list the supported classes in Table 11-8. For more information on these classes, refer to Section 11.8 of (Digital Video Broadcasting Project, 2001).

NOTE:
MHP-compliant DVB-J implementations must follow the JSSE 1.0.2 API from Sun Microsystems for TLS on the return channel.

Package	Classes
java.security	AccessControlContext, AccessControlException, AccessController, AllPermission, BasicPermission, CodeSource, GeneralSecurityException, Guard, InvalidKeyException, Key, KeyException, KeyFactory, NoSuchAlgorithmException, NoSuchProviderException, Permission, PermissionCollection, Permissions, Policy, Principal, PrivilegedAction, PrivilegedActionException, PrivilegedExceptionAction, ProtectionDomain, PublicKey, SignatureException, SecurityPermission, SecureClassLoader, and UnresolvedPermission
java.security.spec	EncodedKeySpec, KeySpec, X509EncodedKeySpec
java.security.cert	Certificate, CertificateEncodingException, CertificateException, X509Certificate, CertificateExpiredException, and CertificateNotYetValidException
JSSE classes	javax.net, javax.net.ssl, and javax.security.cert
Other	java.io.FilePermission, java.io.SerializablePermission, java.lang.RuntimePermission, java.util.PropertyPermission, java.net.SocketPermission, and java.awt.AWTPermission

Table 11-8 Java Security Classes Supported by an Mhp-Compliant Dvb-J Implementation

References

Dierks, T., and C. Allen. 1999. *RFC 2246: The TLS Protocol Version 1.0*. The Internet Engineering Task Force.

Digital Video Broadcasting Project. November, 2001. *Digital Video Broadcasting (DVB) Multimedia Home Platform (MHP) Specification 1.1 (ETSI TS 102 812 V1.1.1)*. European Telecommunications Standards Institute.

ITU-T. June, 1997. *ITU-T Recommendation X.509: Information Technology—Open Systems Interconnect—The Directory: Authentication Framework*. International Telecommunication Union.

Kaliski, Burton S., Jr. November, 1993. *A Layman's Guide to a Subset of ASN.1, BER, and DER*. RSA Laboratories.

Chapter 12

Design Scenarios

Now that you have the knowledge and tools to design a successful broadband security infrastructure, it's time to apply what you have learned. In this chapter, we'll discuss a number of information-gathering steps that will help you organize your thoughts during the early stages of the design process. The design scenarios that follow will help you test your understanding of the concepts presented throughout this text. Be forewarned that this chapter does contain advanced material; while this book provides most of the information required for the design scenarios, this chapter is intended for readers who have previous security experience.

Initial Design Steps

Before jumping headfirst into the design process, you may find it helpful to develop a plan of attack. A structured approach to the design process helps reduce the chances of inadvertently introducing security vulnerabilities and performance bottlenecks into your solution. The guidelines presented in this section will aid you in assessing the security requirements of your network, choosing the proper security services and mechanisms, and incorporating these services and mechanisms in your network. You may find the following nine-step procedure helpful during the initial stages of the design process:

- **Step 1** Identify your assets and assess their value.
- **Step 2** Identify as many threats as possible.
- **Step 3** Select the appropriate security services to protect your assets based on the threats identified in step 2.
- **Step 4** Choose the best security mechanisms (cryptographic or otherwise) to implement the services chosen in step 3. Your choice of mechanisms should take into account the characteristics of the communication protocols, devices, and applications within your network.
- **Step 5** Determine whether there is a need that the security services and mechanisms you have chosen remain bound to application data, or whether the data only needs to be secured while in transit. (The former situation requires Application layer security, while the latter may benefit from session-based security.)
- **Step 6** If session-based security is a better fit, identify the most appropriate network layer at which to place the security mechanisms.
- **Step 7** Determine whether security can and should be implemented within security gateways, or if it must be incorporated directly into end-user hosts and devices.
- **Step 8** Identify any existing security protocols that meet the requirements of your particular network environment or application(s).
- **Step 9** If one or more of your primary requirements cannot be satisfied using existing security protocols, assess the benefits, drawbacks, and level of difficulty associated with developing new security protocols.

We'll elaborate on each of these design steps in the following sections.

Step 1: Identify Your Assets and Assess Their Value

Before you can select the appropriate security services and mechanisms for your network, you must understand what it is that you are trying to protect. The average service provider and corporate networks contain many valuable resources:

- **Stored and transmitted data** Financial and medical records, proprietary design information, personnel records, confidential e-mail, and customer information
- **Transmitted media streams** Copyrighted audio and video content and voice/video communications

- **Network resources** Routing and provisioning equipment, application servers, and database servers
- **Various other services and applications** Internet access, IP telephony, voice- and video-conferencing, cable and satellite television, audio- and video-on-demand, home shopping and banking, electronic commerce, and online gaming

In addition to identifying your assets, you should assign each one a value. The value should reflect the cost of replacing the asset if destroyed; any lost revenue stemming from network or service downtime (denial of service); compensating others for negative publicity or lost revenue resulting from the disclosure of sensitive information; and the perceived monetary, personal, or professional value of the asset to attackers. The level of security employed to protect each asset should be proportional to the value assigned to the asset; keep in mind that professional thieves, corporate spies, and other intruders are more likely to focus their attacks on obtaining high-value items.

Step 2: Identifying the Threats

Before you can choose the necessary security services to protect your assets, you must determine the threats that exist on your network. This step involves identifying the individuals or groups most likely to benefit from attacking your network resources, data, services, and applications, and then anticipating the techniques they will use to do so.

In Chapter 3, we introduced a number of categories of attackers:

- Casual hackers and crackers
- Employees and insiders
- Thieves and career criminals
- Corporate spies and hired professionals
- Foreign governments and terrorist organizations

The motivations and skill levels of these groups vary greatly, as does their associated level of threat. In Chapter 3, we also identified many popular classes of attacks:

- Eavesdropping on network traffic and traffic patterns
- Impersonation of users and devices
- Denial of service (DoS)
- Data modification
- Packet replay

- Routing attacks
- Protocol-specific vulnerabilities

In addition to the attacks discussed in Chapter 3, you must also consider vulnerabilities in network software (host operating systems and applications) and firmware running on hardware devices. Such vulnerabilities often result from improper software development practices, such as incorporating secret backdoors, not performing adequate input validation to prevent buffer overflows, and allowing sensitive memory to be written to disk (virtual memory swap files). These vulnerabilities and other security-related news are often posted to mailing lists and websites, such as these:

- 2600 Magazine Online (http://www.2600.com)
- @stake (http://www.l0pht.com)
- BugNet (http://www.bugnet.com)
- Bug Traq Mailing List (see http://www.securityfocus.com/popups/forums/bugtraq/fag.shtml for more information)
- CERT Coordination Center (http://www.cert.org)
- Insecure.org (http://www.insecure.org)
- Nomad Mobile Research Center (http://www.nmrc.org)
- NT Bugtraq (http://www.ntbugtraq.com)
- Phrack Magazine Online (http://www.phrack.com)
- Security Focus (http://www.securityfocus.com)

The number and variety of attacks grows each day as new vulnerabilities are discovered. It is nearly impossible to anticipate every attack, so focus on those attacks that you feel will result in the greatest amount of damage.

Step 3: Selecting the Appropriate Security Services

Once you have identified the threats to your network, you can select the security services appropriate for countering those threats. In Chapter 2, we examined a number of security services, including identity and data origin authentication, authorization and access control, availability, confidentiality, integrity, and nonrepudiation. Table 12-1 lists the basic categories of attacks and the security services appropriate for defending against them. It also lists additional countermeasures, many of which were discussed in Chapter 3.

Category of Attack	Appropriate Security Service(s)	Other Countermeasures
Eavesdropping on user data and traffic patterns	Confidentiality	Subnet partitioning and the use of Layer 2 switches Use of tunneling security gateways to conceal traffic patterns Traffic padding and routing control
Impersonation	Strong authentication	Enforcing strong password selection policies Adequately protecting keying material (for instance, the private keys used to generate digital signatures) Not basing authentication on host addresses or other easily falsified information
Denial of service or availability	Authentication (both identity and data origin authentication), message integrity, and high availability	Rate-limiting traffic as it enters network boundaries Application of periodic security patches to correct protocol flaws Use of redundant systems for fail-over
Data modification	Message integrity (primary) access control and authorization (secondary), and confidentiality (secondary)	N/A
Packet replay	Time-varying authentication (both identity and data origin authentication)	Frequent password updates and key refreshes Use of sequence numbers and timestamping IPSec's anti-replay service
Denial of involvement in an electronic transaction	Nonrepudiation (including optional timestamping)	A binding legal framework for enforcing nonrepudiation
Routing attacks	Authentication (device and data origin authentication) and message integrity	Limiting the exchange of routing information to router-only subnets Monitoring the sources of routing information updates Disabling source routing
Protocol-specific vulnerabilities	All services	Use of security protocols, such as IPSec, that run at the lower layers of the protocol stack (below problematic Application and Transport layer protocols) Keeping up-to-date with security and operating system patches

Table 12-1 Network Security Threats and Countermeasures

Step 4: Choosing Suitable Security Mechanisms

The next step in designing a security infrastructure is choosing the security mechanisms (both cryptographic and noncryptographic) best suited to your particular network environment. Table 12-2 lists the cryptographic mechanisms for implementing the primary security services. (Refer to Chapter 2 for more information on these cryptographic security mechanisms.)

Before choosing security mechanisms, you must analyze the communication protocols, devices, and applications present in your network for any restrictions and limitations that will dictate your use of security. Table 12-3 lists a number of questions to ask before selecting security mechanisms.

Every network application has specific QoS requirements that both necessitate and limit the use of certain security mechanisms. In Table 12-4, we present a number of QoS-related concerns when choosing cryptographic security mechanisms. Refer to Chapter 4 for a detailed discussion of the QoS requirements for many popular network applications, and Chapter 8 for more information regarding the effects of cryptography and security protocols on QoS.

Security Service	Cryptographic Security Mechanism(s)
Access control	Encrypting data and supplying decryption keys to authorized individuals only.
Authentication (identity)	Proving possession of a private key: Based on a user's ability to decrypt data encrypted with his or her public key By generating a digital signature
Authentication (data origin)	Message authentication codes (for data in transit) and digital signatures (for stored data)
Confidentiality	Symmetric encryption and digital envelopes
Integrity	Message authentication codes (for data in transit) and digital signatures (for stored data)
Nonrepudiation	Digital signatures (used in conjunction with optional timestamping)[1]

[1] In addition to cryptographic mechanisms, nonrepudiation requires a legal framework for defining the relationship between communicating parties, enforcing the use of nonrepudiation services, and upholding these service in a court of law.

Table 12-2 Cryptographic Mechanisms for Implementing the Various Security Services

Network Component	Restrictions and Limitations
Communication protocol	**The type of communication protocols in use** Must the security mechanisms be resistant to packet loss or bit errors during transmission (in other words, are you using unreliable communication protocols or protocols that do not support error correction)? **The maximum payload size per packet or frame and existing header-to-payload ratios** Will security mechanisms consume excessive bandwidth when using small payload sizes (for instance, the bandwidth consumed when transmitting padding, IVs, MACs, large encryption keys, and digital certificates)?
Device	**The amount of processing power, RAM, and storage available for cryptographic and other security-related operations and code** Does each device contain enough resources (processing power, memory, and storage) to support all the necessary security mechanisms? Should cryptographic security mechanisms be implemented within software or dedicated hardware? *The number of simultaneous connections each device must maintain* Does the device have the processing power to support the required number of simultaneous sessions, and the available memory to maintain a context (cipher suite, keying material, and other security-related parameters) for each connection?
Application	**The QoS requirements associated with each application** How will the security mechanisms (especially those based on public-key cryptography) affect the overall delivery of application content in terms of bandwidth, latency, jitter, packet loss, and availability? **The sensitivity of the data being exchanged and the length of time the data must remain secure** What key sizes are appropriate for securing the data, and how do larger key sizes affect QoS within the application?

Table 12-3 General Concerns When Choosing Security Mechanisms

QoS Characteristic	Concerns
Bandwidth and throughput	Encryption and integrity mechanisms must not reduce throughput below the level required by the application content. Encryption and integrity mechanisms, such as block cipher padding, IVs, and MACs, must not consume large amounts of bandwidth in proportion to application data.
Latency (delay)	Cryptographic processing on routers and end-user devices should not contribute substantially to delay in real-time media streams. Computationally expensive public-key cryptographic operations should be limited to end-entity authentication and initial key exchange, and should not significantly delay session establishment. Key refreshes and renewals should be performed asynchronous to media stream delivery, and should employ symmetric key encryption keys (KEKs).
Jitter (delay variation)	Cryptographic processing on routers and end-user devices should not contribute significantly to jitter in real-time media streams. Key refreshes and renewals should be performed asynchronous to media stream delivery, and should employ KEKs. Buffering aids in smoothing out the jitter in a signal, but introduces additional signal delay.
Packet loss	Encryption and integrity mechanisms must be resistant to packet loss when not employing reliable communication protocols. To avoid packet loss, cryptographic processing should not significantly slow the rate at which routers and security gateways process incoming data.
Availability	Security mechanisms must not contribute to DoS attacks for services and applications with very high availability requirements.

Table 12-4 QoS Considerations When Choosing Cryptographic Security Mechanisms

Step 5: Identifying the Need for Persistent Security Services and Mechanisms

If there is a requirement in your application for binding security services and mechanisms directly to user data (data-oriented security), your application may call for the use of an Application layer security protocol. (We discussed such a scenario in "Binding Security Services and Mechanisms to Data" in Chapter 7.) Otherwise, you may be able to take advantage of a session-based approach that protects data only while in transit. For a more detailed handling of persistent and session-based security services, refer to Chapter 7, and to the sections in Chapter 2 entitled "Store-and-Forward vs. Session-Based Encryption" and "Using Public-Key Algorithms."

Step 6: Choosing a Network Layer

If session-based security meets your needs, you must determine the appropriate network layer(s) at which to place security services and mechanisms. Your choice of network layer depends on a number of factors, including

- The extent of security coverage required by each application (in other words, whether an application demands complete end-to-end security, or a point-to-point solution will suffice).
- The number of upper-layer applications requiring security.
- Whether each application dictates the use of unique security policy settings, or the same general policy settings apply to multiple applications.
- The level of transparency required by upper-layer applications in terms of integration effort and policy configuration.
- Whether security functionality can and should be included natively within a protocol stack.

We discussed each of these considerations in detail in Chapter 7.

Step 7: Choosing Between Host-Based Security and Security Gateways

In addition to selecting the appropriate network layer(s) in which to place security services and mechanisms, you must analyze the benefits and drawbacks of implementing security either within security gateways, or directly within end-user hosts and devices. As we discussed in Chapter 7, some situations facilitate the use of security gateways, while others require host-based security. Your decision should be based on a number of considerations, including the following:

- The extent of coverage required by each application. (End-to-end security requirements cannot be met using security gateways.)
- The number and location of hosts and devices requiring similar security services.

- The level of effort associated with the implementation, configuration, and maintenance of a host-based approach.
- Whether there is a need to distinguish between traffic flows (to avoid the overhead associated with applying security services to traffic that does not require security).
- Whether there is a requirement to maintain user contexts from one session to the next for purposes of authentication.
- The difficulties either approach presents when integrating with existing security components and policies.

Chapter 7 provides a more detailed discussion of the pros and cons of security gateways versus host-based security.

Step 8: Identifying Existing Security Protocols That Meet Your Needs

In many cases, you may find that existing network security protocols meet your needs. There is a wide variety of protocols from which to choose, and each protocol offers different characteristics that make it particularly suited to some environments, but not others. We compare the applicability of some of the more commonly used security protocols in Table 12-5.

Security Protocol	Network Layer	Applicability
S/MIME	Application	Store-and-forward applications requiring end-to-end security and data-oriented security services and mechanisms
SSL and TLS	Transport	Session-based applications providing end-to-end protection of transmitted data Simpler to implement than IPSec when only one or two applications require security Requires a connection-oriented Transport layer communication protocol (TCP, for example)
IPSec	Network	Session-based applications providing end-to-end *or* tunneling protection of transmitted data May be used to protect multiple upper-layer applications and protocols requiring the same security policy settings

Table 12-5 The Applicability of Existing Security Protocols

Security Protocol	Network Layer	Applicability
DOCSIS 1.1 BPI+	Link (or Data Link)	Session-based applications requiring *only* point-to-point security High level of transparency to upper-layer protocols operating at the Network, Transport, and Application layers

Table 12-5 The Applicability of Existing Security Protocols *(continued)*

We describe each of the protocols in Table 12-5 at length in Chapters 6, 7, and 9. We also provide a more detailed comparison of these protocols in the section entitled "Comparing Existing Security Protocols" in Chapter 7.

NOTE:
Even though high performance was a design criterion for the security protocols listed in Table 12-5, most of them do not account for QoS. As a result, many of these protocols work well for securing best-effort traffic, but may be unable to meet the needs of applications with strict QoS requirements (such as real-time, streaming multimedia).

Step 9: Designing a New Protocol

If no existing security protocols match the requirements of your environment, you may need to consider developing a proprietary solution. However, you should always consider this your last line of defense. Before choosing to design your own security protocols, you should assess the benefits, drawbacks, and level of difficulty associated with developing the protocols. Your considerations should include

- Whether network devices can be upgraded to support existing security protocols by simply adding faster processors, more memory, or greater storage resources.
- Whether slight modifications to the requirements of the applications bring existing security protocols within acceptable tolerances.
- The consequences of security vulnerabilities introduced by improper design decisions, and the level of damage that these weaknesses may cause.
- Any interoperability issues associated with integrating a proprietary protocol into an existing network environment.

- Whether communicating peers outside your organization or network will accept and interoperate with a proprietary protocol (if there is a need for outside communication).
- The costs associated with developing a proprietary solution—is it greater than the value of the asset you are trying to protect?
- How difficult a proprietary solution may be to maintain.

In Chapter 5, we discussed the role and importance of public scrutiny in the design process. While public scrutiny may work well for weeding out design flaws during the development of open and widely available standards, it may be impractical or undesirable for proprietary solutions. If public analysis and debate is not an option, consider hiring one or more security consultants to review the design under a *nondisclosure agreement (NDA)*. An NDA is a legally binding agreement between two parties that prevents one or both parties from sharing proprietary or otherwise sensitive information, gained from the relationship, with a third party.

Sample Design Scenarios

To demonstrate and reinforce the concepts presented in this book, the following sections present two hands-on design scenarios. The first scenario involves identifying flaws in an existing design, while the second scenario instructs you to build a security infrastructure from scratch following the nine-step procedure presented earlier in this chapter. You may not need any references aside from this book; however, feel free to use any resources at your disposal to complete the scenarios.

CAUTION:
The design scenarios presented in this section are meant to resemble real-life applications. However, they are intended solely for educational purposes—not as solutions to real life problems.

Scenario 1: A Flawed Design

Our first design scenario involves locating the mistakes that someone else has made in designing a security infrastructure. In this scenario, you are a security consultant tasked with the job of analyzing an existing design to identify any vulnerabilities and poor design choices. The security solution you will be analyzing is for a business-to-business (B2B) video conferencing application. The network environment (illustrated in Figure 12-1) consists of end-user devices, which we call video conferencing terminals (VCTs), communicating over an IP network. For purposes of this scenario, we will assume that the service provider's network includes the necessary IP integrated services to support the QoS requirements for video conferencing.

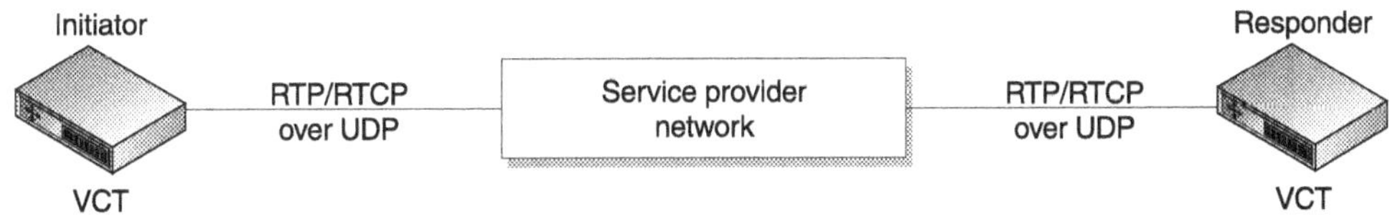

Figure 12-1 The network environment for a simple B2B video conferencing application

The video conferencing application has the following characteristics:

- It transmits real-time voice and video data. The combined bandwidth required for a single voice and video data stream is 500 Kbps, so bidirectional communication requires 1 Mbps.
- It uses IP as a network layer protocol.
- It uses a combination of RTP and RTCP for media stream transport. (Both audio and video data are transmitted in the same RTP message.) RTP and RTCP both run on top of UDP.
- Communication sessions involve *only* two parties: the initiating and responding VCTs. In other words, the system relies on no other devices for call establishment, termination, and signaling; all exchanges occur directly between the initiator and responder.

NOTE:
In practice, additional exchanges would be required between the VCTs and IntServ routers within the service provider's network to reserve the appropriate resources for supporting the application's QoS requirements. The VCTs would most likely communicate with the routers using RSVP (the Resource Reservation Protocol).

The VCTs are embedded devices with limited processing power, memory, and storage. They have the following features and characteristics:

- A 200MHz RISC processor
- Limited RAM for cryptographic operations
- Limited flash memory for storing cryptographic code
- Dedicated hardware MPEG encoders/decoders
- Dedicated cryptographic hardware for performing traffic encryption (if cryptographic operations cannot be implemented within software)
- Ethernet controller (capable of 10 Mbps throughput minimum)
- External input/output (I/O) connectors for audio and video equipment
- No native support for any security protocols within the protocol stack

At a minimum, the security design for the video conferencing application must account for

- Authentication of the communicating parties
- End-to-end privacy of transmitted video conferencing data
- The QoS requirements of the application and the limited resources available on the VCTs

A Flawed Solution

Peter, a hardworking, yet inexperienced security professional, has proposed a solution to this scenario. His solution consists of components for authentication, initial key exchange, traffic encryption, message integrity, nonrepudiation, and periodic key renewals. To meet the minimum requirements for security as outlined by the scenario, Peter has chosen cryptographic mechanisms for authentication and traffic encryption. He has also deemed message integrity and nonrepudiation crucial for business communication, and incorporated the corresponding cryptographic mechanisms. Through a detailed analysis of existing security protocols, Peter determined that a new protocol was necessary to meet the delay and jitter requirements of the video conferencing application. He has placed the new protocol within the Application layer, and implemented all security services and mechanisms directly within the end-user devices (the VCTs). Table 12-6 provides a quick summary of Peter's choice of cryptographic mechanisms for implementing the various components of his solution.

Component	Cryptographic Mechanism(s)
Authentication	RSA digital signatures using SHA-1 (with a 2048-bit modulus)
Initial key exchange	RSA encryption and decryption (with a 2048-bit modulus)
Traffic encryption	RC4 stream cipher using 128-bit encryption keys
Message integrity	SHA-1 message digest generated over the contents of various messages
Nonrepudiation	RSA digital signatures using SHA-1 (with a 2048-bit modulus)
Key renewal	RSA encryption and decryption (similar to that used for the initial key exchange)
Certificate usage	X.509v3 certificates for exchanging the public-key material necessary for the verification of digital signatures and for key exchange

Table 12-6 Peter's Choice of Cryptographic Security Mechanisms

The authentication and initial key exchange occur during a handshaking session between the initiating and responding VCTs. The handshake (illustrated in Figure 12-2) must be successfully completed before the communicating parties can begin exchanging voice and video data.

A handshaking session follows these steps:

1. The initiator sends a *ConnReq* (connection request) message to the responder containing its X.509 digital certificate and an RSA-SHA-1 digital signature generated over the contents of the message minus the signature. The initiator generates the signature using its RSA private key.
2. The responder extracts the public key from the initiator's certificate, and uses it to verify the signature attached to the message. If the signature verifies successfully, the initiator is authenticated.
3. Upon successfully authenticating the initiator, the responder replies with a *ConnRep* (connection reply) message containing its X.509 digital certificate and an RSA-SHA-1 digital signature generated over the contents of the message. The responder uses its RSA public key to generate the digital signature.
4. The initiator extracts the public key from the responder's certificate and uses it to verify the responder's signature. If the signature verifies correctly, the responder is authenticated.
5. The initiator generates a random nonce using a pseudorandom number generator (PRNG). It seeds the PRNG using a combination of the current date and time and the MAC address of its network adapter. It then encrypts the nonce using the public key of the responder, and sends the encrypted nonce to the responder in a *KeyExchange* message.

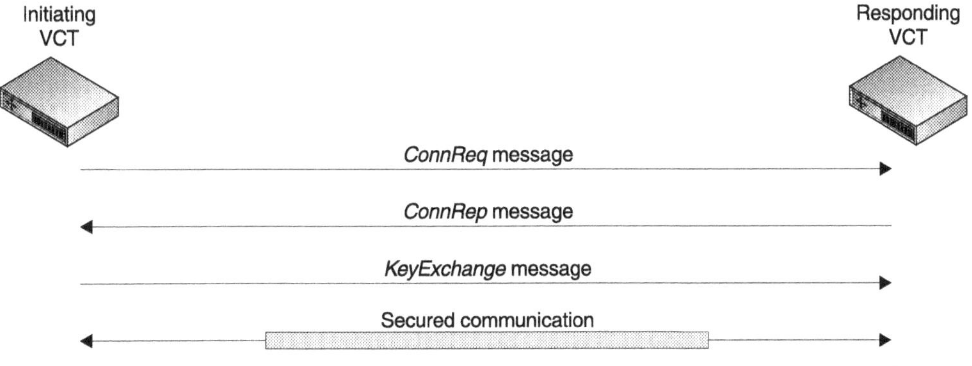

Figure 12-2 The handshaking sequence employed by Peter's security protocol to provide authentication and initial key exchange

6. The responder uses its RSA private key to decrypt the nonce contained within the *KeyExchange* message.
7. Both parties derive the RC4 encryption keys for RTP by supplying the nonce as input to an SHA-1 message digest. To derive the RC4 encryption key for RTCP, both parties take the resulting digest of the random nonce and pass it through an SHA-1 algorithm a second time.

Following the handshaking session, both parties immediately begin exchanging encrypted data.

For traffic encryption, Peter has chosen the RC4 stream cipher due to its high level of performance and relative simplicity. He feels that the use of a fast encryption algorithm might allow cryptographic operations to take place in software, as opposed to requiring dedicated hardware. (His rationale is that software-based solutions are less expensive to develop and manufacture than their hardware-based counterparts.) The use of RC4 has also allowed Peter to select a larger key size (128 bits) without affecting the throughput of the encryption algorithm.

Peter's protocol uses the RC4 algorithm to encrypt voice and video traffic as one contiguous stream of data. For example, when RTP messages are transmitted containing 1KB of voice and video data, the first kilobyte of the keystream will be used to encrypt the payload of the first RTP message. The second kilobyte of the keystream will then be used to encrypt the payload of the second RTP message, and so on, until the communication session terminates. The recipient of the data will generate an identical keystream to that of the sender, and use the keystream to decrypt the contents of each incoming RTP packet. The protocol uses a similar scheme for encrypting RTCP messages. However, it encrypts the entire contents of RTCP messages, and only the payload of RTP messages (as illustrated in Figure 12-3). Even when transmitting data at the rate required by the video conferencing application, the RC4 keystream should not repeat for the lifetime of a communication session.

For message integrity, Peter has decided to append an SHA-1 digest to the end of each encrypted RTP and RTCP message. The SHA-1 digest is intended to prevent malicious users from modifying the contents of the RTP and RTCP messages.

In addition to using RSA digital signatures for authentication, Peter also uses them for nonrepudiation. Both communicating parties generate digital signatures over the entire contents of the bidirectional communication session. The signature generation process is as follows:

- After outbound traffic leaves the MPEG encoder (in compressed format), it is passed through an SHA-1 message digest. This operation occurs before the VCT encrypts the traffic.

- The same operation occurs for inbound traffic. However, the digest operation follows decryption and precedes MPEG decoding.
- Once the conferencing session terminates, both parties append their inbound digest to their outbound digest, and pass the result through an SHA-1 message digest. They then encrypt the result with their corresponding RSA private keys.

Both parties store copies of their outbound and inbound media streams along with their respective digital signatures. The validity of a stored session may be checked at any time by verifying the attached digital signature. Peter intends that businesses use this nonrepudiation scheme to settle legal disputes by demonstrating involvement of both parties in a conference, and to ensure the integrity of verbal commitments made during a conference.

The final component of Peter's solution is periodic key renewal. Both the initiator and responder maintain a timer that indicates when the renewal must take place. For security reasons, a key renewal occurs once every 120 seconds, and it is the responsibility of the initiator to generate and transmit a new random value to the responder prior to expiration of the timer. To exchange keying material securely, the initiator generates a new random nonce, encrypts it with the public key of the responder, and sends the encrypted nonce to the responder in a *KeyRenew* message. The *KeyRenew* message is transmitted as part of the media stream. The responder extracts the message before passing the media stream to its MPEG decoder and decrypts the nonce using its RSA private key. Both parties generate new keying material by supplying the random value as input to an SHA-1 message digest. To begin using the newly negotiated keys, the initiator and responder exchange *BeginEncrypt* messages.

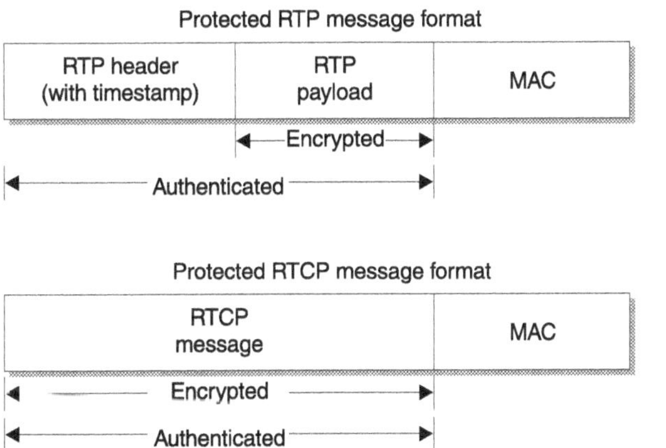

Figure 12-3 Peter's protocol provides message integrity by appending a MAC (SHA-1 digest) to the end of each RTP and RTCP message.

NOTE:
Since public-key cryptographic operations occur relatively infrequently (only during the initial handshake and for periodic key renewal), Peter has chosen to implement the RSA algorithm in software. This will help reduce the production costs of the VCTs.

What's Wrong with This Picture? Analysis of a Flawed Design

How many flaws did you identify in Peter's design? Here is a list of some of the more important ones (although more flaws may exist than are presented here):

- The use of 2048-bit RSA modulus sizes may be excessive for this particular application. A more appropriate value for balancing security and performance might be 1024-bit. The use of 786-bit modulus sizes may be too low, given the sensitive business-related nature of communication.
- The handshaking session does not account for authentication failures in steps 2 and 4. It should return an error code in the event of unsuccessful digital signature verification to inform the other end of the failure. This flaw could lead to a potential DoS attack.
- In step 3, the responder uses its public key to generate the signature, when it should be using its private key. An individual or device can be authenticated based on possession of its private key, since only that individual or device should have access to the key. (That was an easy one).
- In this case, end-entity authentication relies on digital certificates for the dissemination of public keys. However, the solution defines no hierarchy for establishing trust in the certificates. In other words, the initiator and responder do not share a common trusted third party. In addition, the design does not specify whether the end user or the device is being authenticated.
- In step 4, the initiator does not verify that the responder in step 3 matches the original responder from step 1. As a result, anyone can respond to the initiator's connection request with a falsified *ConnRep* message. As long as the message contains a valid digital signature, it would be accepted by the initiator. Once again, the initiator and responder should share a common trusted root in order to build trust in one another.
- The method used to seed the PRNG in step 5 is inadequate, because the initiator gathers entropy from very predictable sources. (Review Chapter 2 for a discussion of acceptable entropy-gathering techniques.)

- The handshaking session performs no checks to ensure that the security parameters (in this case, keying material) have been properly established by both communicating parties. Prior to encrypting RTP and RTCP messages, the initiator and responder should exchange some form of encrypted test message to verify that data encrypted by one party can be decrypted by the other, and vice versa. This is similar to the *Finished* message defined by SSL and TLS (see Chapter 6). If either party fails to decrypt the test message of its peer, it should immediately transmit an error code.
- The handshaking session is susceptible to replay attacks. By capturing and replaying the *ConnReq* and *KeyExchange* messages sent by the initiator in steps 1 and 5, an attacker could reauthenticate to the responder. To counter this type of attack, each side might include a timestamp value within its digital signature. Alternatively, both the initiator and responder could generate and exchange separate random nonces. The initiator would derive the security parameters for its outbound traffic based on the nonce provided by the responder, and would derive inbound parameters from its own nonce. The responder would do just the opposite, so that its inbound and outbound parameters would match the outbound and inbound parameters of the initiator, respectively. The random nonce would have to change from one connection to the next.
- The method of generating encryption keys using an SHA-1 digest operation produces RC4 keys that are longer than 128 bits. This is a minor issue, since the 160-bit output of the SHA-1 algorithm could easily be truncated to 128 bits. However, MD5 might have been a better choice given that the length of its output is 128 bits. (In other words, you avoid having to perform the truncation when using MD5.)
- The use of a stream cipher for encrypting real-time multimedia traffic exchanged via a connectionless and unreliable transport protocol, such as UDP, is a poor design choice. Just one lost packet will desynchronize the initiator and responder's keystreams, thus preventing communication until the keystreams are resynchronized. (For a more complete explanation of keystream desynchronization, refer to Chapter 8.)

  One solution to this problem would be the use of reliable Transport layer protocols that support retransmission. However, retransmission adversely affects the delivery of data in real-time multimedia applications. As a result, it is rarely used. Another solution would be to encrypt each individual RTP and RTCP message as an autonomous unit by restarting the keystream generator for each encryption operation. Restarting the keystream generator might prevent desynchronization, but it would require the use of a separate key for each encrypted message in order to prevent keystreams from being reused.

NOTE:
In real-time multimedia applications, lost packets are rarely retransmitted. Low levels of packet loss result in only momentary signal degradation. Retransmission, on the other hand, causes signal delay and jitter that hinders the timely delivery of real-time data.

- In the context of this solution, the SHA-1 digest provides no additional benefits in terms of integrity protection, and it consumes valuable bandwidth. To ensure the integrity of a message, you must either encrypt the SHA-1 digest (to produce a digital signature) or incorporate a shared secret key into the input of the message digest to produce an HMAC.
- Not only is nonrepudiation for real-time multimedia traffic impractical, but the mechanism presented in this scenario is fundamentally flawed. When digital signatures are used for authentication, integrity, and nonrepudiation, they must be generated by the true source of the data, and should always be transmitted and stored along with the original data. Peter's method only satisfies the first requirement. Since the two parties never share their digital signatures, it is relatively easy for one or both of them to modify the stored media streams and generate a new digital signature over the modified data. The other side will have no way of knowing that the data is not authentic.

  In addition, the idea of generating a digital signature over the entire bidirectional communication session is a bit absurd. Without a connection to an external storage device, the VCTs will be incapable of saving copies of the media stream. Even with access to external storage, a one-hour conference would require 3.6GB of storage space.

  Furthermore, even if this method were not flawed, it would be susceptible to packet loss. Losing a single packet during transmission will create a mismatch in the outbound traffic of one host and the inbound traffic of the other, and the digital signatures resulting from the data streams will not coincide.
- Without a trusted timestamping server, it is difficult to prove exactly when the communication session occurred for purposes of nonrepudiation.

- Unless the communication session is a matter of national security, a key renewal rate of once every 120 seconds is too frequent. Given the length of most conference calls, key renewal probably isn't necessary. The only situation requiring key renewal might be when the communication session is persistent (in other words, when the session never terminates unless by accident).
- When using a single traffic-encryption key, it may expire before the two communicating parties can successfully complete a key renewal. By maintaining a set of active keys with overlapping lifetimes, the initiator will always have valid keying material available. This technique is used by DOCSIS 1.1 BPI+ (see Chapter 9).
- The key renewal process requires an excessive amount of processing power on the part of the responder and may interfere with content delivery. For each key renewal, the responder must perform a decryption using its RSA private key. Since Peter has chosen to implement the RSA algorithm in software, the RSA decryption operation (using a 2048-bit modulus) will likely require at least 2.5–3 seconds on a 100MHz RISC processor. (Chapter 8 presents a number of performance metrics for the various cryptographic algorithms.)

  Peter should have considered the use of a multitier key exchange mechanism that employs a combination of public-key and symmetric cryptography. Following authentication, both parties could negotiate a key encryption key (KEK) for use in encrypting key renewal messages. (We discussed KEKs in Chapter 2.)

Scenario 2: Designing Security from the Ground Up

Now it's your turn to apply what you have learned. In this scenario, you are tasked with designing a security infrastructure from the ground up. For the purposes of this exercise, try not to burden yourself with the mundane details of handshaking protocols and message formats. Instead, concentrate on the high-level details of the infrastructure, using the nine initial design steps discussed at the beginning of this chapter. Later in this section, we'll examine a sample solution containing responses to the nine initial design steps; you can use these to check your design.

Background Information and Design Criteria

We base this scenario on the application defined in Scenario 1. However, we include a few twists that make the security design process a bit more interesting. Whereas Scenario 1 specifies only B2B communication, Scenario 2 expands the application to include home and small office subscribers. In addition, certain government agencies require that the communication system support electronic surveillance (wiretapping). The network environment (see Figure 12-4) consists of the initiating and responding VCTs, a call management server (CMS), a proxy server (PS), an electronic surveillance server (ESS), and a billing records server (BRS).

The video conferencing application now has the following characteristics:

- The application transmits real-time voice and video data. The combined bandwidth required for the voice and video data is 300 Kbps, so bidirectional communication requires 600 Kbps.
- Communication sessions between the initiator and responder also involve interaction with the CMS and PS.
- The CMS acts as a proxy for signaling messages related to call establishment and termination, and manages the initial establishment of keying material and periodic key renewals between the VCTs. When electronic surveillance is enabled, the CMS forwards all keying material to the ESS.
- The connection between the CMS and the VCTs uses TCP as a reliable and connection-oriented transport protocol.
- The CMS sends billing information messages to the BRS over UDP (a connectionless and unreliable transport protocol).

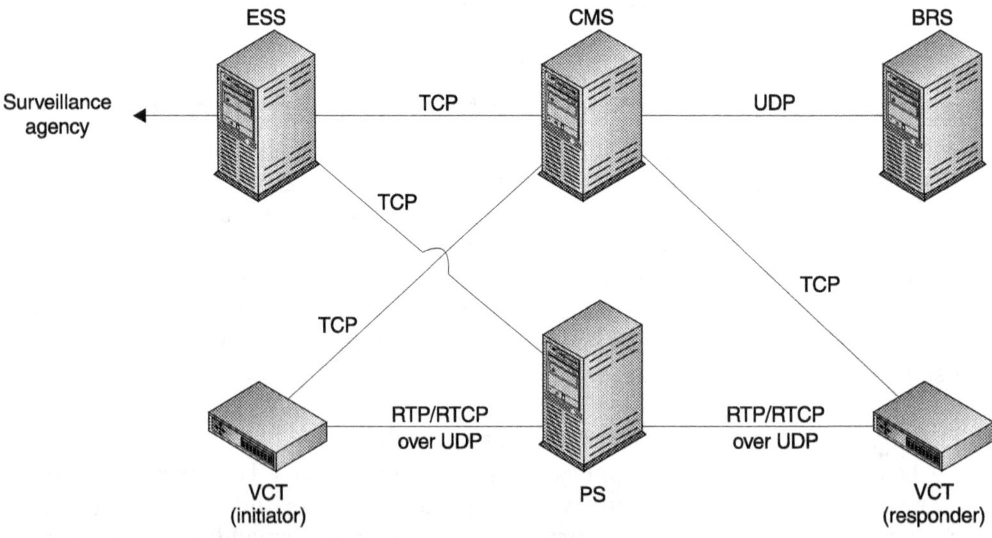

Figure 12-4 The network environment for a video conferencing application that supports electronic surveillance

- All traffic between the initiator and responder must flow through the PS.
- The PS typically forwards traffic between the initiator and responder only. However, when electronic surveillance is enabled, the PS also captures and forwards entire communication sessions to the ESS over TCP.
- All components can lie in different subnets.

NOTE:
TCP was chosen for the PS-ESS interface to prevent additional packet loss in the surveillance signal. Since the data will be stored and replayed at a later time, and the PS-ESS interface will only be used to capture a small number of simultaneous communication sessions, retransmission should not interfere with electronic surveillance.

- The connection between the PS and the VCTs uses UDP as a transport layer protocol.
- The video conferencing application uses IP as a network layer protocol.
- The application uses a combination of RTP and RTCP for media stream transport. RTP and RTCP both run on top of UDP.

Aside from the VCTs, we assume that all network components have ample resources for security-related operations, and that the service provider operates multiple CMSs, PSs, ESSs, and BRSs to accommodate the requests of a very large number of subscribers. The VCTs have the following features and characteristics:

- A 100MHz RISC processor
- Limited RAM for cryptographic operations
- Limited flash memory for storing cryptographic code
- Dedicated hardware MPEG encoders/decoders
- Dedicated cryptographic hardware for performing traffic encryption
- Ethernet controller (capable of 10 Mbps throughput minimum)
- External input/output (I/O) connectors for audio and video equipment
- No native support for any security protocols within the protocol stack
- Factory installed RSA key pairs and X.509v3 digital certificates

At a minimum, your security design must account for

- End-to-end privacy of transmitted video conferencing data
- Authentication of the initiating and responding VCTs to the CMS

- Protection by the service provider from theft of service
- The ability for authorized government agencies to monitor traffic between any two VCTs on the network

The QoS requirements of the video conferencing application and the limited resources available on the VCTs

A Possible Solution

This section provides a sample solution to Scenario 2 using the nine steps outlined earlier in this chapter. This is not the only solution, nor is it necessarily the best solution. However, it serves to demonstrate the concepts presented in this book by meeting all the security requirements outlined in the scenario and complying with the QoS requirements of the application.

Step 1: Determine the assets that require protection. For the purposes of this exercise, we'll limit the assets to voice and video content transmitted between the initiator and responder, keying material exchanged between the VCTs and the CMS, and surveillance information exchanged on the CMS-ESS and PS-ESS interfaces. We consider the traffic exchanged during a video conferencing session to be highly sensitive information. The assets also include the video conferencing service offered by the provider.

Step 2: Identify the possible threats to those assets. We will limit the possible threats to eavesdropping, data modification, impersonation, and theft of service/QoS for this scenario. Although they would be possible in this scenario, we will not burden ourselves with preventing DoS, replay, and routing attacks.

Step 3: Choose the appropriate security services to counter the threats. The CMS must authenticate both VCTs (the initiator and responder) prior to call establishment to prevent rogue devices from stealing network service. All initial key-exchange and key-renewal messages sent from the CMS to the initiating and responding VCTs must be encrypted and must include integrity protection to prevent leakage and modification of keying material. The CMS-VCT interface also requires a connection-oriented and reliable communication protocol, such as TCP, to prevent loss of keying information due to dropped packets. (Such a loss could be used as a DoS attack to halt an ongoing communication session.)

To prevent the disclosure of sensitive subscriber information, we must ensure the end-to-end confidentiality of RTP media streams and RTCP control messages exchanged between the initiating and responding VCTs. Notice that we are not requiring integrity protection for RTP/RTCP traffic. Encrypted voice and video signals are very difficult to modify without detection, and in many cases, message integrity is not required. Furthermore, integrity mechanisms, such as HMACs, introduce additional processing delay and consume network bandwidth.

The VCT-PS interface does not need to implement security services of its own. Since the PS simply forwards encrypted traffic from the initiating VCT to the responding VCT, there is no need for the PS to apply any security mechanisms. Instead, the virtual VCT-VCT interface is responsible for applying the necessary encryption and optional message-authentication and integrity mechanisms to the RTP and RTCP traffic prior to passing it to the PS. The same is true for the PS-ESS interface. However, the PS-ESS interface does require mutual authentication in order to prevent the PS from supplying an intruder with access to sensitive media streams and to keep the ESS from accepting modified media streams from rogue devices.

To prevent eavesdropping, all billing-related information sent from the CMS to the BRS must be encrypted. An attacker might otherwise use information obtained from billing records to determine the type of services to which a particular customer subscribes, as well as how frequently the user accesses the services. The attacker might also be able to glean valuable corporate information for purposes of competitive marketing. To prevent theft of service resulting from modified billing records, the billing information messages must include the appropriate integrity mechanisms. The CMS and BRS must mutually authenticate one another in order to prevent the BRS from accepting billing information messages from an unauthorized host or device, and to keep the CMS from supplying billing information to entities other than the BRS.

The CMS-ESS interface requires the same level of security as the VCT-CMS interface. Since the CMS and ESS use this interface to communicate security parameters for electronic surveillance, the interface must support authentication, end-to-end confidentiality, and message integrity. If there is no authentication mechanism in place, an attacker might be able to obtain the keying material required to decrypt the communication session of his or her choice. Table 12-7 summarizes the required security services for each network interface.

NOTE:

None of the interfaces in this scenario require nonrepudiation. In most cases, nonrepudiation does not apply to session-based communication, and is better suited for stored electronic documents.

Interface	Required Security Services	Optional Services
VCT-CMS	Client-side authentication, end-to-end confidentiality, and message integrity	N/A
VCT-VCT	End-to-end confidentiality	Message integrity
VCT-PS	End-to-end confidentiality	Message integrity
CMS-BRS	Mutual authentication, end-to-end confidentiality, and message integrity	N/A
CMS-ESS	Mutual authentication, end-to-end confidentiality, and message integrity	N/A
PS-ESS	Mutual authentication and end-to-end confidentiality	Message integrity

Table 12-7 Security Services Required for Each Interface in Scenario 2

Step 4: Select security mechanisms based on the characteristics of your network and applications. Before we begin choosing security mechanisms for implementing the security services identified in step 3, let's take a look at the QoS characteristics of the video conferencing application. Aside from the bandwidth necessary for video transmission, voice communication is the most restrictive component of the application. (Chapter 4 discusses the various QoS requirements for common network applications.) Consequently, the application requires 600 Kbps of bandwidth, an end-to-end delay of less than 300 milliseconds (100–150 ms is desirable), and jitter in the range of 10–20 milliseconds for the media streams. To meet these requirements, you should consider the following when selecting cryptographic security mechanisms:

- The throughput of all traffic encryption algorithms (whether implemented in software or hardware) should exceed the required throughput for the application. For the video conferencing application, which requires a throughput of 600 Kbps (for bidirectional communication), AES in CBC mode should suffice. Even when implemented in software on the 100MHz embedded RISC processor, AES is capable of encrypting 1–1.5MB of data or more per second. (We presented performance figures for the various cryptographic algorithms in Chapter 8.) AES is also a good choice from the point of view of interoperability; AES is standards based and widely available to all vendors.

  A software implementation of 3DES, on the other hand, may not suffice, as the algorithm is only capable of encrypting 0.2–0.3MB of data per second on a 100MHz embedded RISC processor (and that's with 100 percent processor utilization). While the resulting encryption rate of 2 Mbps still far exceeds the throughput requirement of the video conferencing application, we must take into account the actual processing power available for cryptographic operations. For example, if only 10 percent of the available processing power were set aside for traffic encryption, then a software-based solution would fail miserably. In fact, it would limit bidirectional throughput to 200 Kbps.

It's a good idea, whenever possible, to implement traffic encryption within hardware. Dedicated hardware implementations offload cryptographic operations from the main system processor, freeing up the processor for its primary tasks. The processing rates of cryptographic hardware are also much less likely to fluctuate during normal operation than a shared processor, resulting in less signal degradation.

NOTE:
For this solution, you may also consider AES in OCB mode. As discussed in Chapter 8, OCB mode combines encryption and data origin authentication into a single operation. This can improve performance substantially on embedded devices (especially when implementing traffic encryption algorithms in hardware), and consumes less bandwidth than a separate encryption algorithm and HMAC.

- Periodic key renewal should add minimal latency and jitter to the delivery of application content. It is best to implement these operations out of band (asynchronous to the transport of encrypted media streams). Public-key operations are too computationally expensive to use for encrypting key-renewal messages. The use of symmetric KEKs improves performance considerably.
- HMACs consume bandwidth by appending a set quantity of data to each authenticated message (16 bytes for HMAC-MD5 and 20 bytes for HMAC-SHA-1). In most cases, this only poses a problem when the size of the HMAC digest is greater than 5–10 percent of the overall message length.

NOTE:
We present many more cryptographic performance concerns in Chapter 8.

The only interface with these stringent QoS requirements runs between the initiating and responding VCTs (the virtual VCT-VCT interface, which includes the PS-VCT interfaces). The delay and jitter requirements do not apply to security operations that occur at the beginning of a call (authentication and initial key exchange) or out of band (asynchronous to the media stream transport). We assume that the PS contributes very little delay and jitter to the media stream when forwarding traffic between the initiator and the responder. Table 12-8 identifies the most appropriate security mechanisms for implementing the security services on each network interface.

Interface	Required Security Mechanisms
VCT-CMS	X.509 certificates and optional RSA-SHA-1 digital signatures for authentication (the authentication procedure depends on your choice of security protocol). RSA encryption and decryption for key exchange (because each VCT already contains a factory-installed RSA key pair). The CMS should use the public key obtained from a VCT's digital certificate to encrypt the session keys before distributing them to the VCT. The VCT then uses its private key to decrypt the keying material. Symmetric cryptography for traffic encryption. (The ciphers available will depend on your choice of security protocol.) SHA-1 HMACs for message authentication and integrity.
VCT-VCT	Symmetric cryptography for traffic encryption. Since we are transmitting real-time traffic over an unreliable Transport layer protocol, we will use a block cipher (AES in CBC or OCB mode with 128-bit keys). Optional SHA-1 HMAC for message integrity. We might also consider using a block cipher in OCB mode to combine encryption and message authentication into a single operation. Note that while MD5 is faster than SHA-1, the use of MD5 would increase code size, because each VCT would need to implement SHA-1 for digital signature generation/verification and MD5 for the creation of HMACs.
VCT-PS	Symmetric cryptography for traffic encryption and optional message integrity. (This is already provided by the VCT-VCT interface.)
CMS-BRS	Authentication and key exchange mechanisms will depend on your choice of security protocol. Symmetric cryptography for traffic encryption. (The ciphers available will depend on your choice of security protocol.) SHA-1 HMACs for message authentication and integrity.
CMS-ESS	Authentication and key exchange mechanisms will depend on your choice of security protocol. Symmetric cryptography for traffic encryption. (The ciphers available will depend on your choice of security protocol.) SHA-1 HMACs for message authentication and integrity.
PS-ESS	Symmetric cryptography for traffic encryption and optional message integrity. (This is already provided by the VCT-VCT interface.)

Table 12-8 Cryptographic Security Mechanisms Required for Each Interface in Scenario 2

NOTE:
Digital signatures are denoted as optional because you can also authenticate communicating parties based on their ability to decrypt random nonces exchanged during a handshaking session. (For more details, review the section entitled "Security Protocol Tuning" in Chapter 8.)

This solution employs a very simple certificate hierarchy, as illustrated in Figure 12-5. The hierarchy is very similar to the DOCSIS 1.1 BPI+ certificate hierarchy described in Chapter 9. At the top of the hierarchy sits the VCT Root CA, which is trusted by all CMSs. Just below the top level of the hierarchy lies subordinate manufacturer CAs. The last level of the hierarchy consists of video conferencing terminals. The VCT Root CA is a centrally managed entity responsible for issuing X.509v3 certificates to subordinate VCT manufacturer CAs. The manufacturers then issue a X.509v3 device certificate to each VCT they produce. Every VCT device certificate must be installed during manufacturing, along with the corresponding RSA public/private key pair.

Step 5: Determine whether there is a need to bind security services and mechanisms to application data. The two most common reasons for binding security services and mechanisms to application data are to secure data in a store-and-forward environment, and to ensure nonrepudiation. The only time in this scenario when security services and mechanisms must remain bound to data is when the RTP/RTCP messages are stored on the ESS (since we are including no additional security services or mechanisms on the PS-ESS interface). This single requirement, however, necessitates the use of an Application layer security protocol for protecting traffic on the VCT-VCT interface (the

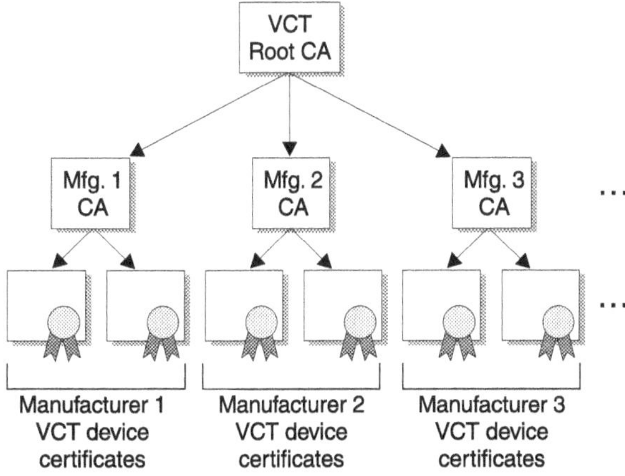

Figure 12-5 The video conferencing application employs a simple certificate hierarchy.

interface on which RTP and RTCP messages are exchanged). We opt for the use of session-based security protocols on all other network interfaces.

Step 6: Choose the appropriate network layer at which to place the security mechanisms. All interfaces identified in this scenario require routable, end-to-end security services. As a result, we must implement security at or above the Network layer. Since we must protect multiple Transport layer communication protocols (TCP and UDP), we may wish to use a Network layer security protocol for the CMS-BRS, VCT-CMS, and CMS-ESS interfaces. An added benefit of using a Network layer security protocol is that it will prevent us from having to implement more than one security protocol within the CMS (since the CMS uses multiple Transport layer communication protocols).

We have already established the need for an Application layer security protocol to secure the RTP and RTCP traffic.

Step 7: Determine whether security services can and should be implemented within a gateway, or if they must be implemented within the end-user hosts and devices. In a way, the CMS acts as a security gateway for authentication, initial key exchange, and key renewals between the initiator and responder. However, the CMS does not perform authentication and key exchange/renewal in lieu of the end-user devices, and each VCT must directly support the necessary security services and mechanisms. Accordingly, we choose to implement host-based security within the VCTs. All of the remaining network elements must directly support the necessary security services and mechanisms as well, so the solution contains no security gateways.

Step 8: Consider existing security protocols, and identify those that meet the needs of your environment. The most obvious choice for a Network layer security protocol in any IP-based network is IPSec. IPSec provides all the necessary end-to-end security services for the VCT-CMS, CMS-BRS, and CMS-ESS interfaces. We also choose IPSec for end-entity authentication on the PS-ESS interface (although we set the encryption and message authentication transforms to null). IPSec offers anti-replay protection, which will prevent attackers from capturing network traffic and later retransmitting the data to gain unauthorized access to network resources.

We choose IKE as the key-management protocol for IPSec. One main reason for choosing IKE is ease of integration; most IPSec implementations come with built-in support for IKE. Since all VCTs contain factory-installed digital certificates, the VCT-CMS interface uses IKE with digital certificates for authentication. Authentication on all the remaining IPSec-enabled interfaces employs IKE with preshared keys. Within this scenario, there are three primary reasons for choosing to use preshared keys:

- All components are on the same provider network.
- The number of systems using this type of authentication will be relatively small, and the communication endpoints will rarely change.
- No certificate hierarchy exists for distributing certificates to network devices and servers within the service provider's network.

NOTE:
The Transport layer security protocols we have discussed up to this point—SSL and TLS—require a reliable, connection-oriented Transport layer communication protocol, such as TCP. Some of the interfaces within the video conferencing application use UDP, which is neither connection-oriented nor reliable. If we were to use SSL or TLS to secure the interfaces that use TCP, we would have to employ a second protocol to support the interfaces running UDP.

Table 12-9 lists the appropriate security protocols for each network interface in the video conferencing application.

Again, since the PS and ESS apply no security mechanisms of their own, they rely on the encryption and message integrity mechanisms provided by the VCT-VCT interface.

Interface	Security Protocol/Authentication and Key Management Protocol
VCT-CMS	IPSec/IKE with digital certificates
VCT-VCT	RTP and RTCP Application layer security/network-based call signaling messages sent between a VCT and CMS
VCT-PS	Encryption and message integrity provided by RTP/RTCP Application layer security across the VCT-VCT interface
CMS-BRS	IPSec/IKE with preshared keys
CMS-ESS	IPSec/IKE with preshared keys
PS-ESS	IPSec with null encryption and authentication/IKE with preshared keys Encryption and message integrity provided by RTP/RTCP Application layer security across the VCT-VCT interface

Table 12-9 Security and Key Management Protocols Used by the Network Interfaces in Scenario 2

Step 9: Consider designing a new protocol only when your requirements cannot be met by existing security protocols. Since we have determined that the security requirements for media stream transport within the video-conferencing application cannot be met using existing security protocols, we *may* need to design our own. At this point, we must carefully consider the benefits and drawbacks of developing a proprietary solution. This process involves many variables that are unique to a particular organization, its network, and its relationship with other organizations. Without these variables, we have no way of determining whether a proprietary solution makes sound business sense, and so our solution ends here. In the real world, you *will* have access to these variables, and, hopefully, the information presented in this book will empower you to create an efficient and successful broadband network security infrastructure.

Appendix A

TCP/IP Primer

The TCP/IP suite provides a rich set of protocols allowing computers and devices of all types to communicate. Developed primarily during the 1970s, the TCP/IP suite is now the most widely used set of communication protocols in the world. The IETF formally adopted TCP/IP as a standard in 1980, and it has since become the basis for all Internet traffic. In addition to the TCP and IP protocols, UDP, ICMP, and ARP play an important role in the suite. We discuss each of these standards in this appendix. The protocols are outlined in the following IETF documents:

- RFC 768: User Datagram Protocol (Postel, 1980)
- RFC 791: Internet Protocol (University of Southern California, 1981a)
- RFC 792: Internet Control Message Protocol (Postel, 1981)
- RFC 793: Transmission Control Protocol (University of Southern California, 1981b)
- RFC 826: An Ethernet Address Resolution Protocol (Plummer, 1982)
- RFC 2460: Internet Protocol, Version 6 (IPv6) Specification (Deering and Hinden, 1998)

The TCP/IP suite is organized into layers, each layer being responsible for a particular subset of the functions necessary for network communication. The

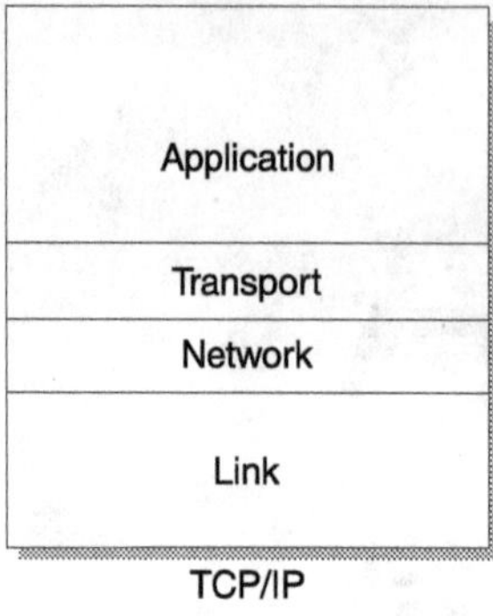

Figure A-1 The TCP/IP network model consists of four layers that correspond to layers of the OSI model.

traditional TCP/IP network model, illustrated in Figure A-1, consists of four layers: Link, Network, Transport, and Application. (Review Chapter 4 for a discussion of the functions performed at each layer.) Layering offers many benefits. For example, protocols operating at one layer can be implemented independently from those running at another layer; only the interfaces between the layers must be known. Furthermore, components at one layer can be replaced with no effect upon other layers, as long as the replacement affords similar functions and it interfaces correctly with existing components. Figure A-2 lists many of the protocols and applications contained within the TCP/IP suite.

Encapsulation

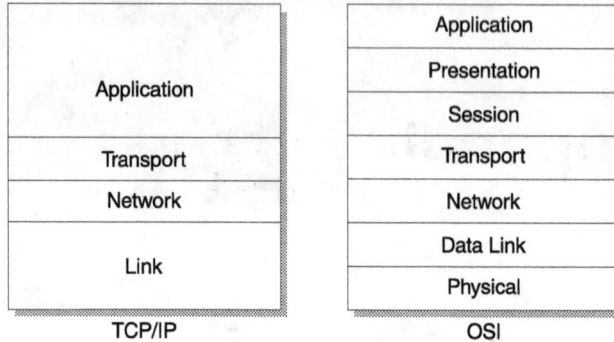

Each layer within the TCP/IP protocol suite encapsulates data from upper-layer protocols before passing it down to a lower layer. *Encapsulation* is the process of wrapping data within fields that identify the source, destination, and purpose of the data. These fields contain information necessary for the destination host to interpret the data upon receipt. Fields preceding the data are known as *headers,* and those following the data are *trailers.* A terminology exists for describing encapsulated data at each network layer. For TCP/IP, the encapsulation process begins with an application that passes user data to the TCP layer. TCP creates a

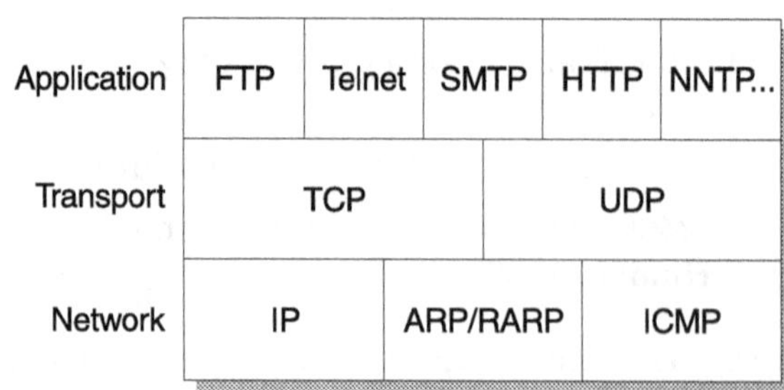

Figure A-2 The TCP/IP suite includes many protocols and applications.

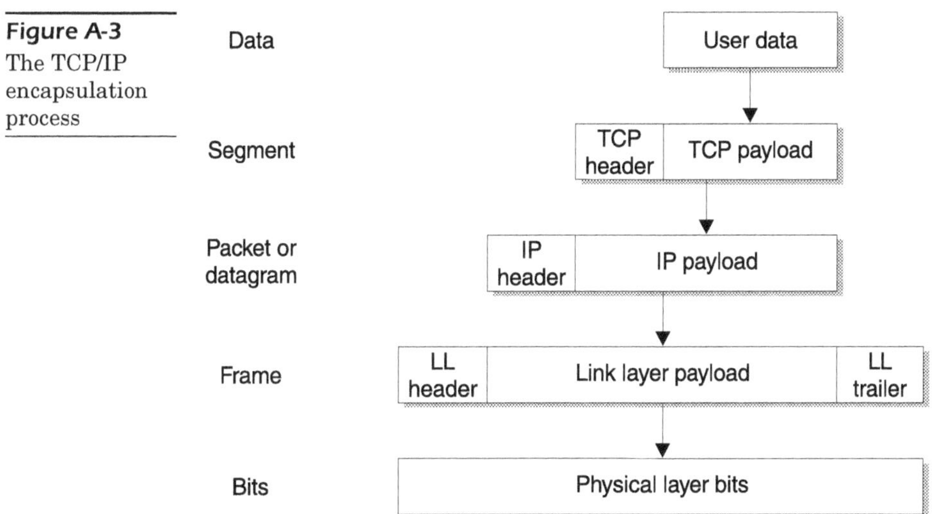

Figure A-3 The TCP/IP encapsulation process

segment, which includes a TCP header and the user data, and then passes the segment to the IP layer. IP adds its own header information to create a *datagram,* or *packet,* before handing the data off to the Link layer. The Link layer converts the datagram to a *frame* by inserting a header and a trailer. It then passes the data to a network interface card as a sequence of bits for transmission across a physical medium. Figure A-3 illustrates the encapsulation process for TCP/IP.

Internet Protocol

The *Internet Protocol (IP)* is a connectionless protocol that provides logical addressing but unreliable delivery of datagrams. As an unreliable communications protocol, IP offers no guarantee that a packet will reach its intended destination. Instead, IP relies on upper-layer protocols, such as TCP, for acknowledgment and retransmission of lost packets or packets containing errors. The connectionless nature of IP means that it maintains no state information from one datagram to the next, and that it does not formally establish a session prior to transmitting data. These activities are also responsibilities of an upper- layer protocol or application. (For a more detailed discussion of connection-oriented/connectionless and reliable/unreliable communication protocols, review the section entitled "Communication Protocol Characteristics" in Chapter 4.)

IP runs at the Network layer, and is responsible for attaching a logical address to every outbound datagram. (We'll discuss logical addressing in the following sections.) It individually transmits each datagram as an autonomous, self-contained unit; no relationship exists between the datagrams until they have been passed to an upper-layer protocol, unless the datagrams contain fragments of a larger IP packet. (We'll discuss fragmentation later in the chapter, in the "IP Fragmentation and Reassembly" section.) As a result, packets may arrive in a different order than that in which they were originally transmitted. This occurs when two packets follow different paths to their destination. Some routers may dynamically adjust forwarding paths if there is some change in network conditions (for example, an unexpected increase in network traffic or a route failure), so routes do not necessarily remain the same from one packet to the next. Figure A-4 illustrates why two packets may arrive out of order. If a host transmits Packet A at time *t* and Packet B at time $t + s$ seconds, then B will arrive before A if Packet B's total trip time is $\geq s$ seconds less than A's. In general, this will not create a problem, but it does require the destination host to reorder the incoming packets. Again, this is usually the responsibility of an upper-layer protocol, such as TCP.

The most current version of IP is IPv6. IPv6 offers a number of enhancements over its predecessor, IPv4, to security and to quality of service. Unlike IPv4, all IPv6 protocol implementations must support the IPSec Authentication Header and Encapsulating Security Payload protocols. (Review Chapter 6 for explanations of these protocols.) IPv6 also offers more flexible packet *classification* and *prioritization* than IPv4 for facilitating QoS over packet-based networks. (Review Chapter 4 for a discussion of packet classification/prioritization and QoS.) Finally, IPv6 increases the available address space of IPv4 from 2^{32} to 2^{128} hosts. Although IPv6 has been available as an IETF draft standard for some time (Deering and Hinden, 1998), most TCP/IP implementations still follow version 4.

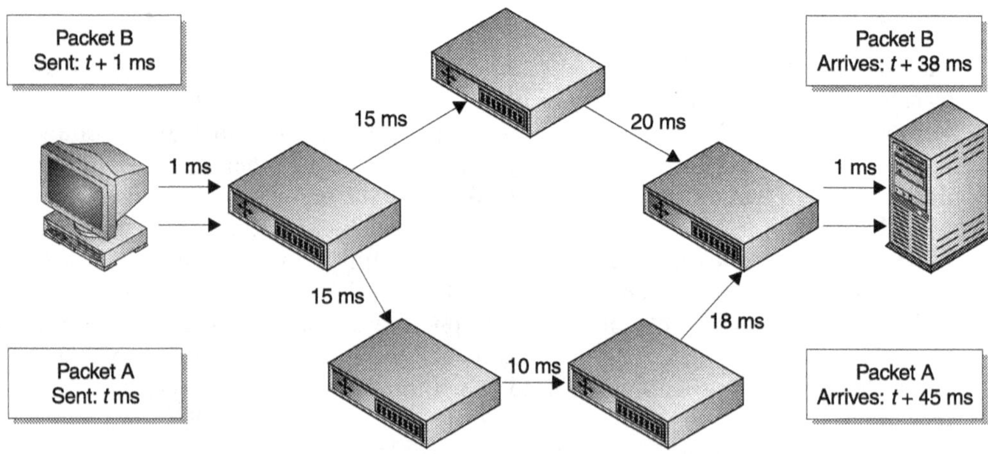

Figure A-4 IP packets may arrive out of order when following different routing paths.

NOTE:
Right about now, you may be asking yourself, "Why do we need to increase the number of available IP addresses? IPv4 already supports over four billion." Well, even though IPv4 supports up to 4,294,967,296 hosts, the Internet is running out of IP addresses, due to inefficient network classification schemes that result in many unused addresses.

IP Headers

An IP header is normally 20 bytes long, but it may contain an additional four bytes of options. The header precedes ICMP messages and upper-layer protocol information (for example, TCP and UDP segments).

Figure A-5 diagrams the format of an IPv4 header, which consists of the following fields:

- **Version** A 4-bit field indicating the protocol version.
- **Header Length** A 4-bit field containing the length of the IP header, including options, in words. (A word consists of 32 bits.) The maximum header length is 60 bytes.
- **Type of Service (TOS)** An 8-bit field specifying how routing devices and applications should handle the datagram. Bits 0–2 indicate the precedence of the datagram. Bits 3–7 enable different TOS values for minimizing delay, maximizing throughput, maximizing reliability, and minimizing monetary cost (Almquist, 1992). The final bit is unused, and must be set to 0.
- **Total Length** A 16-bit field indicating the length of the IP datagram (header and payload) in bytes. The maximum length of an IP packet is 65,535 bytes, which is the largest number that can be represented by a 16-bit field. When the Total Length field is coupled with the Header Length field, the recipient of a datagram can determine where the header ends, and where the data begins and ends.
- **Identifier** A 16-bit field providing a unique identifier for each datagram. In the event that a datagram becomes fragmented, the Identifier field provides a mechanism for reconstructing the original datagram from the fragments. All fragments belonging to the same datagram have identical Identifier fields.
- **Flags** A 3-bit field containing flags for describing fragmented packets. The *more fragment* bit indicates that the packet is a fragment. This flag is turned off in the final fragment of a datagram to indicate that the last fragment has been received and that the datagram can be reassembled. The *don't fragment* bit tells routing devices not to fragment the packet. Instead, a device will discard the packet and return an ICMP error to the sender. The remaining bit is unused.

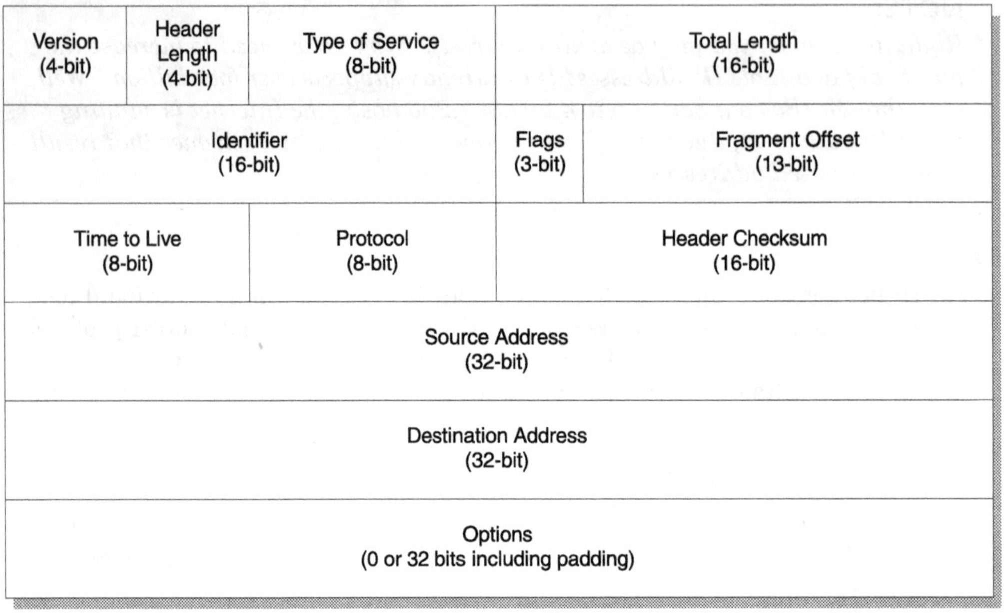

Version (4-bit)	Header Length (4-bit)	Type of Service (8-bit)	Total Length (16-bit)	
Identifier (16-bit)			Flags (3-bit)	Fragment Offset (13-bit)
Time to Live (8-bit)		Protocol (8-bit)	Header Checksum (16-bit)	
Source Address (32-bit)				
Destination Address (32-bit)				
Options (0 or 32 bits including padding)				

Figure A-5 The format of an IPv4 header

- **Fragment Offset** A 13-bit field used in reconstructing a datagram from its fragments. It identifies the location of a fragment within the original datagram.
- **Time to Live (TTL)** An 8-bit field defining the maximum lifetime of a datagram. Every router along the path between the sender and recipient decrements the TTL by one before forwarding the packet. A datagram containing a TTL of 0 will be discarded, and an ICMP message will be returned to the sender. If caught in a routing loop, the TTL will eventually expire, thus preventing the datagram from circulating around the network forever.
- **Protocol** An 8-bit field identifying the upper-layer protocol encapsulated within the IP packet. See RFC 1700 (Reynolds and Postel, 1994) for a list of available values.
- **Header Checksum** A 16-bit field containing a checksum over the entire IP header. IP generates the checksum by taking the ones's complement of the header.
- **Source Address** A 32-bit field indicating the source host address of the datagram.

- **Destination Address** A 32-bit field containing the destination host address of the datagram.
- **Options** If present, this is a 4-byte field containing padding so that it ends on a 32-bit boundary. In most cases, the options are unnecessary and are rarely used.

IPv4 defines the following options:

- **Security (11 bits)** Contains Department of Defense codes pertaining to security and handling restrictions for the packet.
- **Loose Source Routing (variable-length)** Allows the sender to dictate the route taken by the datagram. The packet must traverse the list of addresses supplied by the sender, but can pass through additional addresses as well.
- **Strict Source Routing (variable-length)** Similar to the Loose Source Routing option, except that the datagram can traverse *only* the IP addresses specified by the source.
- **Record Route (variable-length)** Causes each router to record its IP address as a means of tracking the path taken by the datagram.
- **Timestamp (variable-length)** Similar to the Record Route option. Causes each router to also record the time the datagram passed through the router.
- **Stream ID (4 bits)** Allows an IP datagram to carry a stream identifier along networks that do not support the use of stream identifiers. (A discussion of stream identifiers is beyond the scope of this text; however, Delgrossi and Berger, 1995, provides a description of how they may be used.)

CAUTION:
For security purposes, most routers ignore loose and strict source routing. These options allow attackers to capture network traffic by rerouting it through the network segment of their choice.

IPv6 headers differ considerably from those of IPv4. Figure A-6 diagrams the format of an IPv6 header, which consists of the following fields:

- **Version** A 4-bit field identifying the IP version.
- **Traffic Class** An 8-bit field used to distinguish between low-priority and high-priority packets, or between packets that require varying levels of quality of service (QoS). For example, the traffic class for streaming multimedia might indicate a much higher priority than that of best-effort Internet access services.

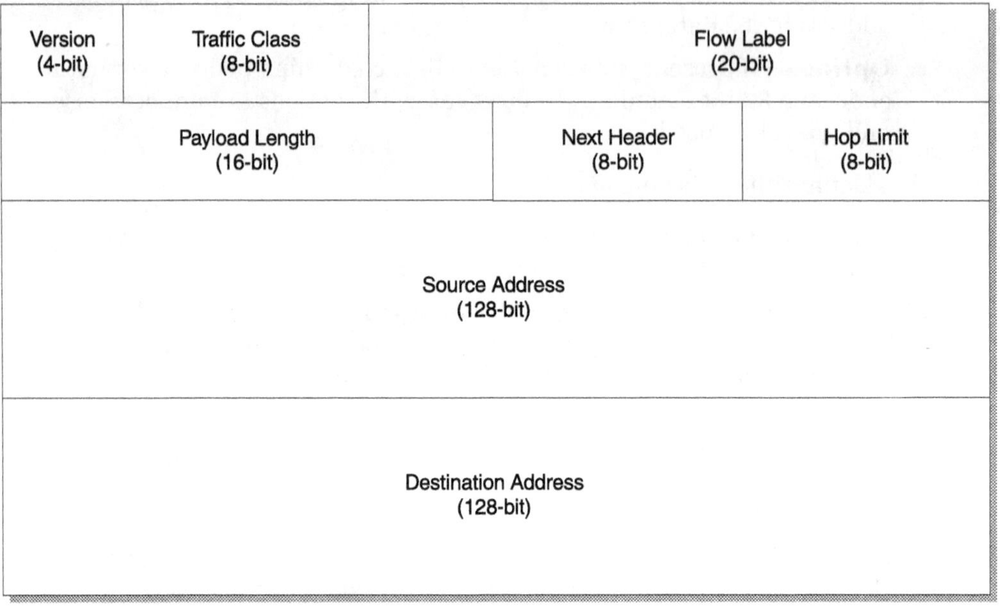

Figure A-6 The format of an IPv6 header

- **Flow Label** A 20-bit field used to label sequences of packets that require special handling (enhanced QoS, for instance) by routers.
- **Payload Length** A 16-bit field indicating the length in bytes of the upper-layer payload that follows the IP header.
- **Next Header** An 8-bit field identifying the upper-layer protocol header that follows the IP header. See RFC 1700 (Reynolds and Postel, 1994) for a list of available values.
- **Hop Limit** An 8-bit field defining the maximum lifetime of a datagram. Every router along the path between the sender and recipient decrements the Hop Limit by one before forwarding the packet. A datagram containing a Hop Limit of 0 will be discarded, and an ICMP message will be returned to the sender. If caught in a routing loop, the Hop Limit will eventually expire, thus preventing the datagram from circulating around the network forever.
- **Source Address** A 128-bit field containing the source host address of the datagram.
- **Destination Address** A 128-bit field containing the host address of the destination.

IPv6 Extension Headers

IPv6 inserts optional information between the IP header and upper-layer protocol headers. The options are encoded within their own headers, which are known as *extension headers*. An IPv6 datagram can contain zero or more extension headers, with each header type indicated by the Next Header field of the preceding header. Intermediate routing devices process only IPv6 headers, ignoring all extension headers. Only the final destination processes the extension headers. The contents of each extension header determine whether subsequent headers should be processed, so all extension headers must be handled in the order in which they appear in the IP datagram. IPv6 currently includes six extension headers:

- **Hop-by-Hop Options** Contains optional information that must be inspected by each node on the path between the source and the destination.
- **Routing** Similar to the Loose Source Routing option of IPv4. Allows the sender of a datagram to specify IP addresses that the packet must traverse on the way to its destination.
- **Fragment** Identifies a datagram fragment. This header includes an identifier and a fragment offset.
- **Destination Options** Contains optional information that is examined only by the destination host.
- **Authentication** Used by the IPSec Authentication Header (AH) protocol. (Refer to Chapter 6 for a discussion of the AH protocol.)
- **Encapsulating Security Payload** Used by the IPSec Encapsulating Security Payload (ESP) protocol. (Refer to Chapter 6 for a discussion of the ESP protocol.)

IP Routing

When two hosts sit on the same network segment, or *subnet*, they may exchange IP packets directly without first contacting a router. However, when the source address and the destination address of an IP packet lie on different subnets, the packet must pass through one or more routing devices in order to reach the destination network. If the sender determines that the destination host is not on the local subnet, it passes the packet to its default gateway. A *default gateway* is a router or multihomed host capable of determining the best forwarding path for delivering the packet.

NOTE:

A multihomed *host is a computer containing two or more* network interface cards (NICs), *each connected to a different subnet.*

The default gateway processes the IP header of the packet to determine the destination network. It is important to remember that routers make forwarding decisions based on the destination network, not on the entire address of the destination host. To separate the network portion of the address from the host portion, the default gateway uses a subnet mask. A *subnet mask* consists of a series of 1's and 0's that, when logically ANDed with an IP address in binary format, separates the network and host portions of the address; the network portion corresponds to the bits set to 1 in the subnet mask, and the host portion corresponds to the 0 bits.

Some simple examples may help clarify the use of subnet masks:

- For an IP address of 192.168.0.1 and a subnet mask of 255.255.255.0, the network portion of the address would be 192.168.0, and the host portion would be 1.
- For an IP address of 10.1.1.10 and a subnet mask of 255.0.0.0, the network portion of the address would be 10, and the host portion would be 1.1.10.

After processing the IP header, the default gateway checks its *routing tables* to determine the interface on which to forward the packet to the destination network. This process occurs at each routing device along the forwarding path until the packet is received by a router attached to the destination network. (For a complete discussion of IP routing, refer to Stevens, 1994.)

IP Fragmentation and Reassembly

As an IP datagram traverses the path between its source and destination, it may cross network segments that have smaller *maximum transmission units (MTUs)* than its source network. An MTU is the maximum-size *protocol data unit (PDU)* supported by the network. For Ethernet, the MTU is 1500 bytes, which means that an Ethernet frame can contain only 1500 bytes of upper-layer data. The difference in MTU sizes arises from the variety of physical layer network technologies employed by each segment. Fragmentation allows an IP datagram to traverse networks that have an MTU smaller than the size of the datagram.

When IP receives a datagram from an upper-layer protocol, it determines the MTU for the network by querying the appropriate network interface. If the size of the datagram is larger than the MTU for the network segment, IP fragments the datagram. Each fragment arrives independently, and the fragments are then reassembled by the destination host. In order for the destination host to reassemble the packet, each fragment must contain a fragment identifier and an offset value. All fragments from the same original IP datagram are assigned identical *fragment identifiers*. The *offset value* indicates the fragment's position in the datagram. Table A-1 lists MTUs for many popular network architectures.

Table A-1 Maximum Transmission Units (MTUs) for Various Network Mediums

Media Type	MTU (in Bytes)
Ethernet	1500
IEEE 802.3	1492
Token Ring	4464 (4 Mbps), 17,914 (16 Mbps)
X.25	576
FDDI	4352
Hyperchannel	65,535
PPP	1500
SLIP	1006 (logical restriction)

Address Resolution Protocol and Reverse Address Resolution Protocol

Link layer protocols, such as Ethernet, use a different addressing scheme than IP. Each interface on an Ethernet network is uniquely identified by a 48-bit Ethernet address (also known as a *MAC* or *hardware* address). The Link layer uses this address to deliver frames from one network node to another. Link layer addressing schemes are flat (that is, they do not group addresses together into subnets). Consequently, they do not scale well in large routed network environments without a Network layer protocol to perform logical addressing. Network layer protocols, such as IP, add scalability by dividing the address space in such a way that hosts can be grouped and identified by the network segments to which they belong. This grouping aids in routing packets from one network to another. IP uses either a 32-bit or a 128-bit address to identify hosts, depending on whether you are using IPv4 or IPv6. This presents a challenge, since the IP address must be converted to a Link layer address before transmitting the packet. The *Address Resolution Protocol* (*ARP*) performs this conversion.

To discover the hardware address of the destination, the source host broadcasts an ARP message containing the destination's IP address. All hosts on the subnet read the ARP message, but only the host with the corresponding IP address responds with its hardware address. If the destination host lies within another subnet, the source host instead asks for the hardware address of the router that is acting as its default gateway. The router then sends a similar ARP message to discover the hardware address of the destination host, or of the next router in the forwarding path. This process repeats until the packet reaches its destination.

Reverse ARP (*RARP*) maps hardware addresses to logical addresses. RARP is used primarily by diskless workstations to determine their IP addresses during initialization. Since a diskless workstation contains no permanent storage, it has no way of remembering its IP address. However, it can always read the hardware address contained within its NIC. To obtain an IP address, the workstation sends an RARP message containing its hardware address to a host responsible for distributing and managing IP addresses for the network. The host selects an IP address from a list of available addresses and sends a response to the workstation.

Internet Control Message Protocol

The *Internet Control Message Protocol* (*ICMP*) communicates error codes and other status information between two hosts. A number of common network utilities employ ICMP for obtaining status information during network troubleshooting. Two of the most commonly used utilities are Ping and Traceroute. The *Ping* utility uses ICMP echo messages to determine the operational status of a remote host. A Ping client sends an *echo request* to a Ping server, and the server responds with an *echo reply*. If the server replies, the client knows that the server is running and listening on its network interface.

The *Traceroute* program determines the route taken by an IP datagram on its way from the source to the destination. Traceroute begins by transmitting an IP datagram with the TTL field in the IP header set to 1. At the next hop, the TTL is decremented to 0. Instead of forwarding the packet, the first router in the path between the source and destination drops the packet, and responds with an ICMP *time exceeded* message. The sender records the address of the router and sends a second datagram with a TTL of 2. This time, the first router strips the IP header from the datagram, creates a new header with a TTL of 1, and forwards the datagram to the next router on the path between the source and destination. The second router discards the packet and sends another *time exceeded* message to the source host. Again, the host records the address of the router and sends another IP datagram with a TTL of 3. This process continues until a datagram reaches the destination host. Unlike a router, the destination host will process a datagram with a TTL of 1 without generating an ICMP response. To elicit a response, the Traceroute program sends each datagram to a port that the destination host is not likely to be listening on. This causes the destination to respond with an ICMP *port unreachable* message. This indicates to the sender that it has reached the final destination.

Although ICMP messages are encapsulated within IP datagrams, they differ from upper-layer protocol messages in that they do not insert their own protocol headers. ICMP encapsulation is illustrated here:

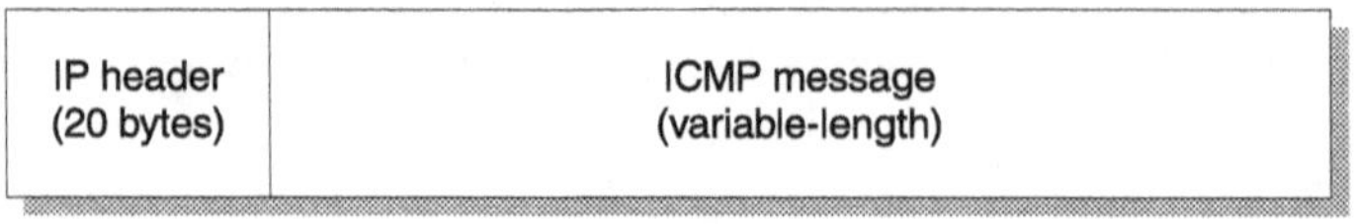

IP header (20 bytes)	ICMP message (variable-length)

Notice that the ICMP message directly follows the IP header.

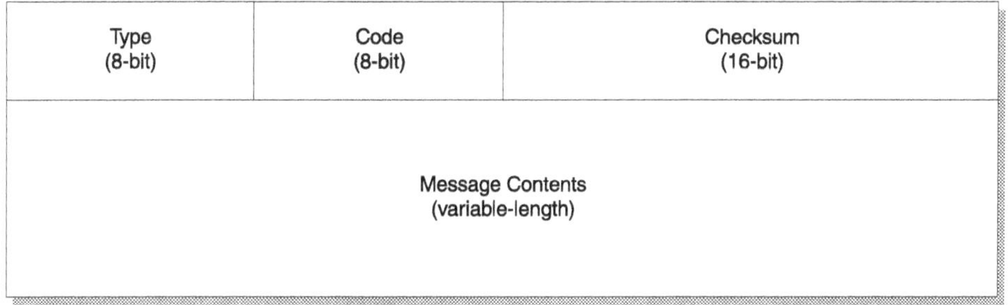

Figure A-7 The format of an ICMP message

Figure A-7 diagrams the format of an ICMP message, which consists of the following fields:

- **Type** An 8-bit field containing the ICMP message type (see Table A-2).
- **Code** An 8-bit field identifying additional information about the ICMP message. Many message types include their own unique codes (see Tables A-3 through A-6).
- **Checksum** A 16-bit field containing a checksum over the entire ICMP message.
- **Message Contents** A variable-length field containing the contents of the ICMP message.

Table A-2 ICMP Message Types

Type	Description
0	Echo reply
3	Destination unreachable
4	Source quench
5	Redirect
8	Echo request
9	Router advertisement
10	Route solicitation
11	Time exceeded
12	Parameter problem
13	Timestamp request
14	Timestamp reply
15	Information request
16	Information reply
17	Address mask request
18	Address mask reply

Table A-3 Destination Unreachable (Type 0) Codes

Code	Description
0	Network unreachable
1	Host unreachable
2	Protocol unreachable
3	Port unreachable
4	Fragmentation needed and DF bit set
5	Source route failed
6	Destination network unknown
7	Destination host unknown
8	Source host isolated
9	Destination network administratively prohibited
10	Destination host administratively prohibited
11	Network unreachable for TOS
12	Host unreachable for TOS
13	Communication administratively prohibited by filtering
14	Host precedence violation
15	Precedence cutoff in effect

Table A-4 Redirect (Type 5) Codes

Code	Description
0	Redirect for network
1	Redirect for host
2	Redirect for type-of-service and network
3	Redirect for type-of-service and host

Table A-5 Time Exceeded (Type 11) Codes

Code	Description
0	Time-to-live equals 0 during transit.
1	Time-to-live equals 0 during reassembly.

Table A-6 Parameter Problem (Type 12) Codes

Code	Description
0	IP header bad
1	Required option missing

Transmission Control Protocol

The *Transmission Control Protocol (TCP)* is a connection-oriented communications protocol running at the Transport layer. Hosts communicating via TCP formally establish a logical connection—a *virtual circuit (VC)*—prior to data exchange. They maintain the VC and related session-state information until they have finished communicating. Creation and termination of a VC requires the exchange of a series of TCP messages. To establish a virtual circuit, TCP peers use a three-way handshake routine (see Figure A-8). The three-way TCP handshake follows these steps:

1. The source host (Host A) sends an SYN request to the destination host (Host B).
2. Host B responds by acknowledging the SYN request, and transmits a SYN request of its own.
3. Host A completes the connection by acknowledging Host B's SYN request.

Graceful session termination requires the exchange of four messages, as illustrated in Figure A-9:

1. The host wishing to terminate the session—in this case, Host A—transmits a TCP FIN request.
2. Host B responds by acknowledging the FIN request.
3. Host B then sends an FIN request of its own to Host A.
4. Host A acknowledges the FIN request from Host B, and the connection is terminated.

TCP provides for reliable transmission of data via a combination of sequence numbers, timeouts, checksums, acknowledgments, and retransmissions. The sender assigns each outgoing packet a sequence number, which the host uses to detect lost packets and reassemble packets that arrive out of order. The sequence number for the first transmitted packet is called the *initial sequence number (ISN)*. The sender increments the sequence number by one for each byte

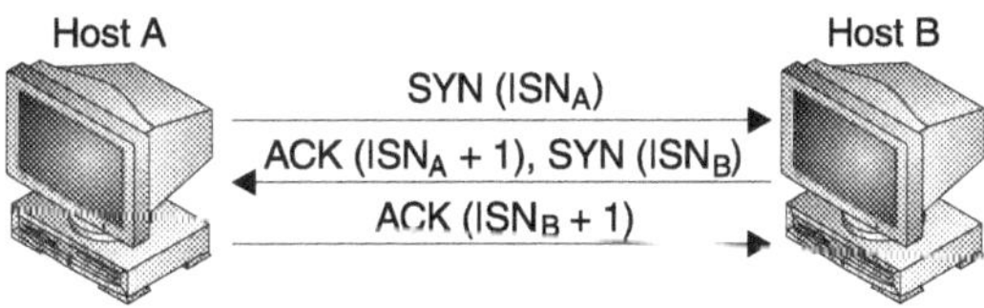

Figure A-8 TCP uses a three-way handshake to establish a connection.

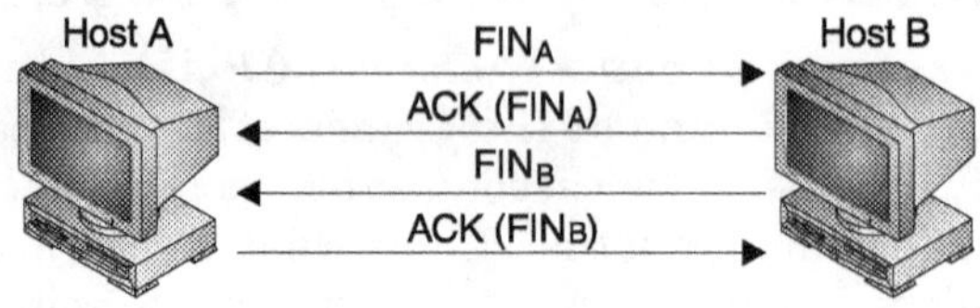

Figure A-9 TCP connection termination requires that communicating parties exchange four messages.

sent. The receiving host acknowledges successful receipt of each packet by sending an ACK message, which contains the next sequence number it expects to receive, back to the sender. To be received successfully, the packet must not contain bit errors. Prior to acknowledgment, the recipient verifies that the TCP segment (header and payload) contained within the packet matches the precomputed checksum in the TCP header. The sender also maintains a timeout counter for transmitted packets; if the sender does not receive an acknowledgment prior to expiration of the counter, it retransmits the packet. As we will soon discuss, the number of packets that the sender can transmit prior to receiving an acknowledgment is controlled by means of a technique known as *windowing.*

TCP uses a combination of the source address, destination address, source port, and destination port to identify a connection. The combination of an address and port number is known as a *socket*. Every TCP session can be identified uniquely via a pair of sockets—one for the source host and another for the destination. The destination uses the port numbers to demultiplex IP datagrams into their respective Transport layer sessions.

TCP Headers

A TCP header sits between an IP header and the upper-layer data. TCP encapsulation is illustrated here:

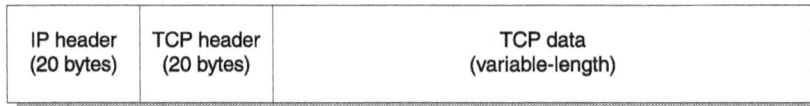

IP header (20 bytes)	TCP header (20 bytes)	TCP data (variable-length)

The length of a TCP header without options is 20 bytes. Since the maximum length of a TCP header is 60 bytes, the Options field may contain up to an additional 40 bytes of information.

Figure A-10 diagrams the format of a TCP header, which consists of the following fields:

- **Source Port** A 16-bit field indicating the source port for the communication session.

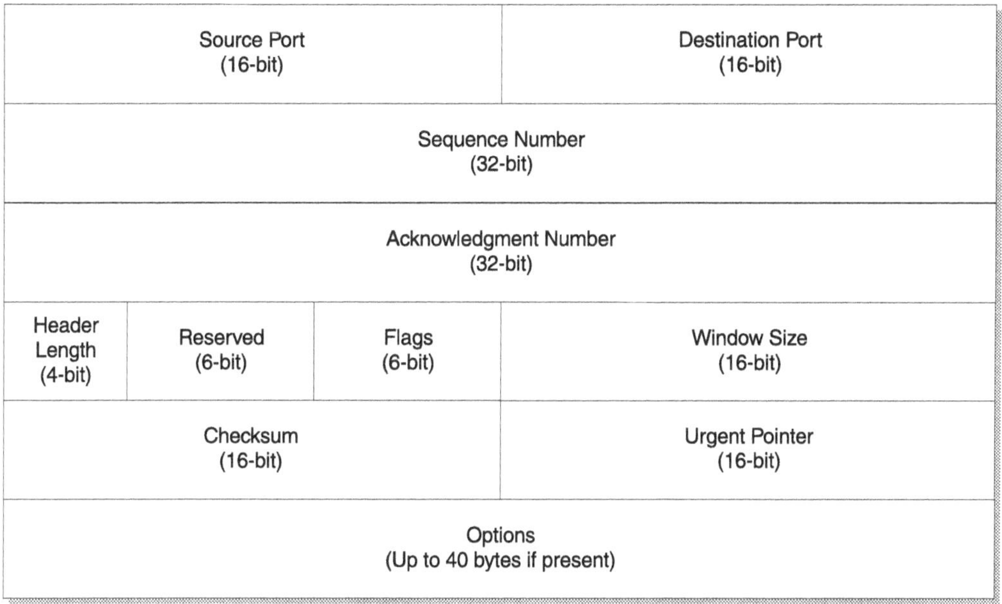

Figure A-10 The format of a TCP header

- **Destination Port** A 16-bit field identifying the destination port for the session.
- **Sequence Number** A 32-bit field containing the sequence number of the packet. The sequence number identifies a packet's position in the bitstream between the sender and recipient.
- **Acknowledgment Number** A 32-bit field indicating the sequence number of the next packet that the sender expects to receive.
- **Header Length** A 4-bit field containing the length of the TCP header in 32-bit words.
- **Reserved** A 6-bit field reserved for future use. Currently, all bits in this field must be set to 0.
- **Flags** A 6-bit field consisting of six 1-bit flags that are used for indicating the purpose of the TCP segment.
- **Window Size** A 16-bit field indicating the maximum number of bytes that the sender is allowed to send before receiving an acknowledgment. The recipient uses this field for flow control.
- **Checksum** A 16-bit field containing a checksum, calculated by the sender, over the entire TCP segment (header and payload).

- **Urgent Pointer** A 16-bit field used for transmitting emergency information. The value contained within this field is a positive offset from the packet's sequence number to the last byte of urgent data.
- **Options** A variable-length field containing optional TCP information. The length of this field must be a multiple of 8 bits.

The 6 bits within the Flags field correspond to the following flags:

- **URG** Indicates that the offset contained within the Urgent Pointer field is valid.
- **ACK** Indicates that the sequence number contained within the Acknowledgment field is valid.
- **PSH** Indicates that the recipient of the packet should pass the data to the Application layer process immediately.
- **RST** Indicates that the sender of the packet wishes to reset the connection.
- **SYN** Indicates that the sender of the packet is requesting to synchronize sequence numbers for establishing a connection.
- **FIN** Indicates that the sender of the packet has finished sending data and wishes to terminate the session.

Windowing

Even a host with modest computational resources can often process incoming packets more quickly than the average network can transmit them. Occasionally, busy routing devices and network servers may become congested trying to process a large number of incoming requests from multiple sources. *Congestion* occurs when the source host transmits packets faster than the destination device can process them. If there is no method in place for controlling the rate of transmission, the incoming data can overflow the receive buffers on the destination host, thus causing packet loss. To limit packet loss, TCP uses a technique known as *windowing*. Windowing provides *flow control* by limiting the number of packets that can be transmitted by the sender prior to receiving a TCP acknowledgment from the recipient. The *sliding window protocol* allows a recipient to dynamically adjust the window size according to the current network traffic conditions. By expanding or reducing the window size, the recipient can increase or decrease the frequency of incoming packets. The 16-bit Window Size field within a TCP header limits the size of a window to 65,535 bytes. However, there are scaling options for expanding the window size beyond this value.

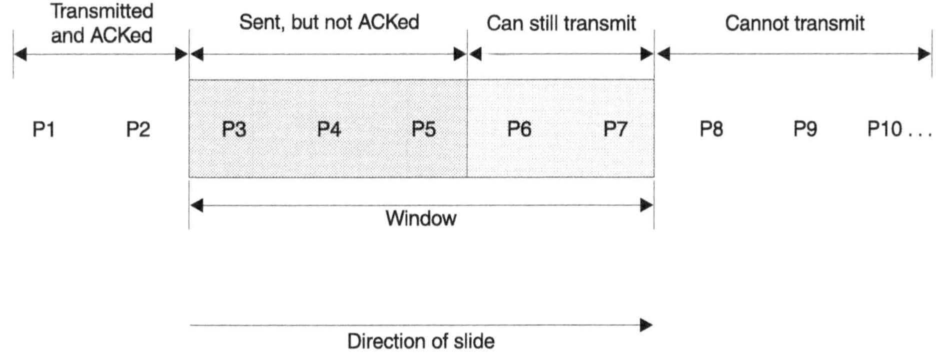

Figure A-11 Windowing allows the recipient of a TCP data stream to prevent packet loss by limiting the rate at which the sender transmits packets.

Figure A-11 (adapted from Stevens, 1994) illustrates the windowing scheme used by the sliding window protocol. The current window size is five packets. Packets P1 and P2 have already been transmitted and acknowledged. Packets in the left-hand side of the window (P3–P5) have been transmitted, but not acknowledged. The window indicates that Packets P6 and P7 can still be transmitted without waiting for an acknowledgment from the recipient. The sender cannot transmit packets beyond P7 until receiving further acknowledgments. The window slides from left to right—in other words, it opens to the right and closes to the left as the sender receives acknowledgments from the recipient.

User Datagram Protocol

Unlike TCP, the *User Datagram Protocol (UDP)* is an unreliable, connectionless communication protocol. UDP offers no mechanism for establishing virtual circuits, and the sending host does not verify the availability of the recipient before it transmits data. It simply addresses and forwards each packet, and then forgets about it. UDP offers no guarantee that the packet will reach the intended recipient.

While UDP includes a Checksum field for detecting errors, no mechanism exists for recovering from transmission errors. UDP simply discards packets containing errors without generating an error message. It is left to the upper-layer application to reassemble the segments in the proper order and retransmit lost or corrupted packets. Despite its unreliable behavior, UDP is not as useless as it may at first seem. The computer networks of today are much more reliable than those of past decades, and the possibility of bit errors and lost packets is very small. The lack of error recovery and retransmission mechanisms and the absence of session state information result in less protocol overhead than is necessary for TCP.

UDP Headers

IP header (20 bytes)	UDP header (8 bytes)	UDP data (variable-length)

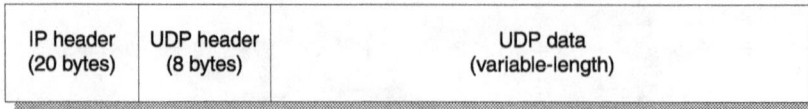

UDP headers are very simple in comparison to TCP headers, and consist of only eight bytes. The smaller header size results in a lower header-to-payload ratio than exists in TCP. Low header-to-payload ratios become important when transmitting a large number of small packets, because they allow the header information to consume less bandwidth.

Figure A-12 diagrams the format of a UDP header, which contains the following fields:

- **Source Port** A 16-bit field identifying the port on the source host.
- **Destination Port** A 16-bit field indicating the port on the destination host.
- **UDP Length** A 16-bit field containing the length of the UDP header and encapsulated data.
- **Checksum** An optional 16-bit field containing a checksum over the UDP header and data.

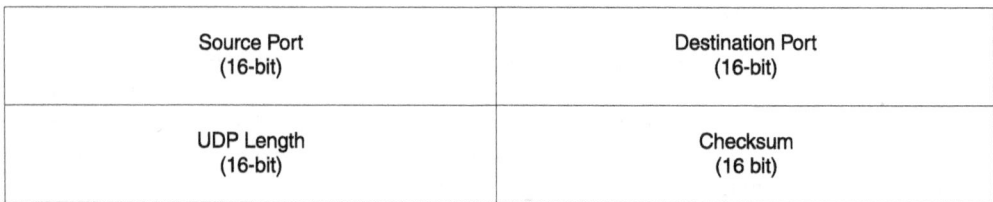

Source Port (16-bit)	Destination Port (16-bit)
UDP Length (16-bit)	Checksum (16 bit)

Figure A-12 The format of a UDP header

Resources

Almquist, P. July, 1992. *RFC 1349: Type of Service in the Internet Protocol*. The Internet Engineering Task Force.

Deering, S., and R. Hinden. December, 1998. *RFC 2460: Internet Protocol, Version 6 (IPv6) Specification*. The Internet Engineering Task Force.Plummer, David. November, 1982. *RFC 826: An Ethernet Address Resolution Protocol*. The Internet Engineering Task Force.

Postel, J. August, 1980. *RFC 768: User Datagram Protocol*. The Internet Engineering Task Force.

—September, 1981. *RFC 792: Internet Control Message Protocol*. The Internet Engineering Task Force.

Reynolds, J., and J. Postel. October, 1994. *RFC 1700: Assigned Numbers*. The Internet Engineering Task Force.

Stevens, W. Richard. 1994. *TCP/IP Illustrated Volume 1: The Protocols*. Reading, MA: Addison-Wesley.

Delgrossi, L., and L. Berger, Editors. August 1995. *RFC 1819: Internet Stream Protocol Version 2 (ST2) Protocol Specification—Version ST2+*. The Internet Engineering Task Force.

University of Southern California. September, 1981. *RFC 791: Internet Protocol*. The Internet Engineering Task Force.

—September, 1981. *RFC 793: Transmission Control Protocol*. The Internet Engineering Task Force.

Appendix B

Digital Certificates and Public-Key Infrastructure

This appendix provides supplementary material for those readers new to digital certificates and public-key infrastructure. Some of the discussions in Chapters 9–11 require a basic understanding of these topics. In this appendix, we will discuss the usage and contents of digital certificates and certificate revocation lists. We'll also introduce many fundamental public-key infrastructure concepts and associated terminology.

Digital Certificates

In Chapter 2, we discussed the difficulties associated with key exchange when symmetric encryption is used. We introduced public-key cryptography, and explained how it provides a more efficient and secure mechanism for key exchange than symmetric encryption. We also described how to use public-key cryptography for authentication and nonrepudiation. Before users can reap the benefits of public-key cryptography, however, they must first exchange their public keys. The public keys need not be encrypted during the exchange, as the security of public-key cryptosystems rests in the secrecy of the private key. While this lack of encryption poses no security threat, the exchange may still be subject to a man-in-the-middle or impersonation attack.

Chapter 3 describes a man-in-the-middle attack in which an intruder, Mallory, intercepts the initial public-key exchange between two communicating parties, Alice and Bob. Mallory substitutes his public key for those of Alice and Bob, and forwards the altered key exchange messages to their final destinations. Following the exchange, Alice and Bob believe they have each other's public keys, when, in fact, they both have Mallory's public key. Any messages they encrypt with Mallory's public key can be decrypted with the corresponding private key, which Mallory possesses. This allows Mallory to decrypt all future communication between Alice and Bob. (Refer to the section entitled "Man-in-the-Middle Attacks" in Chapter 3 for further discussion of this example.)

A *digital certificate* greatly reduces the effectiveness of man-in-the-middle attacks by cryptographically binding a user's public key to his or her identity. This prevents an intruder from replacing public keys without detection. By exchanging digital certificates instead of public keys, communicating parties can ensure that any public key they receive belongs to the individual or organization whose name appears on the certificate. In addition to preventing man-in-the-middle attacks, digital certificates also provide for scaleable key management and distribution in environments with many users.

Before a user can trust the public key contained in a certificate, however, a user must also trust the source of the certificate. This requires the use of a *trusted third party (TTP).* The TTPs responsible for issuing, managing, and revoking digital certificates are known as *certification authorities (CAs).* Before a CA issues a certificate to a user, it first authenticates the user's identity. When implemented properly, this process guarantees that a malicious user cannot request a certificate under another user's identity. To prevent an attacker from altering the contents of a certificate, a CA digitally signs each certificate, using its private key. The signature encompasses the contents of all fields in the certificate. We discuss the format of X.509 certificates later in this appendix, in the section entitled "Contents of a Digital Certificate."

NOTE:

The credibility of all certificates issued by a CA depends on the secrecy of the CA's private key. If a CA's key is ever compromised, then all certificates issued by that CA must be revoked or considered invalid. Most public CAs, as well as many private CAs (those internal to specific organizations), generate and store private keys within hardware security modules (HSMs). Since the private key never leaves the HSM, the HSM must perform all signing operations during certificate generation.

Digital certificates facilitate the exchange of signed and encrypted e-mail, the establishment of SSL sessions, the verification of digital signatures attached to signed applications and software upgrades, and much more. Some European countries have begun instituting digital signature laws for facilitating legally binding electronic transactions. These transactions will undoubtedly involve digital certificates.

Certificate Types and Classes

Two major categories of certificates exist today: end-entity certificates and CA certificates. A CA issues *end-entity certificates* to subjects who will not be issuing digital certificates to others. Examples include the users of secure e-mail systems, SSL-enabled Web servers, and software vendors (who would use the certificates for code signing). *CA certificates* are issued exclusively to certification authorities who, in turn, issue their own certificates. A certificate can be further classified by the type of public key that it contains (RSA, Diffie-Hellman, or ECC, for example) or the protocol with which it is associated (WTLS, PGP, and so on).

Public CAs, such as Verisign, that issue digital certificates as a service often classify them according to the level of authentication required to obtain a certificate. Here is a sample classification scheme:

- **Class 1** A basic certificate usually requires only a valid e-mail address as a suitable form of authentication. These certificates are issued to individuals for encrypting personal e-mail and for other low-security tasks; they offer no identity assurance.
- **Class 2** Authentication might involve a phone call and detailed information about the subject (in other words, more than just an e-mail address). These certificates may be issued to individuals for personal and business communication that requires a moderate level of security. Uses for Class 2 certificates might include encrypting and signing personal or business e-mail, as well as SSL-client side authentication.
- **Class 3** Authentication might involve an onsite visit and valid photo ID, such as a drivers license. Class 3 certificates may be issued to individuals and organizations for personal and business communication requiring a high level of security. Uses for Class 3 certificates might include high-dollar financial transactions, mission-critical B2B operations, and access to SSL-enabled Web servers.
- **Class 4** Authentication might require government-level approval, such as that used to obtain a security clearance, or a stringent background check during an onsite visit. Class 4 certificates would probably be intended for individuals or organizations—including subordinate CAs—requiring the highest level of security.

NOTE:

The class of a certificate may also dictate certain requirements for private-key storage. In general, the higher the certificate class, the more secure the private key must be kept. For instance, an issuing CA may require that private keys for all Class 4 and some Class 3 certificates be stored in HSMs.

Contents of a Digital Certificate

By far, the most common format for digital certificates is X.509. The X.509 standard, originally proposed by the Telecom Standardization group of the ITU, defines the fields within a digital certificate and describes the values that those fields can contain (International Telecommunication Union , 1997). Three versions of X.509 certificates exist today. Versions 2 and 3 include additional fields and functionality not present in Version 1 certificates. Version 2 includes two additional fields for the reuse of subject and issuer names, and Version 3 certificates incorporate extension fields for transporting supplementary information within an X.509 certificate. X.509 certificates are frequently encoded with *Abstract Syntax Notation 1 (ASN.1),* according to *Distinguished Encoded Rules (DER).* ASN.1 and DER encoding provide a common format for transferring data across multiple systems that have different internal data representation formats. (For more information on ASN.1 and DER encoding, refer to Kaliski, 1993).

Figure B-1 diagrams the format of an X.509 certificate, which has the following valid fields:

- **Version** Designates the version number associated with a certificate. The version number indicates the format of a certificate in terms of the attributes and values that may appear in the certificate. The most common version in use today is version 3.
- **Serial Number** Contains the serial number of the certificate. All certificates issued by a single CA must contain unique serial numbers.
- **Signature Algorithm ID** Indicates the public-key algorithm and message digest combination used to generate the signature attached to the certificate.
- **Issuer** Identifies the CA that has issued the certificate.
- **Validity Period** Specifies the start and end dates for the period during which the certificate is valid. A certificate may be invalidated prior to the end date of the certificate. (See the section entitled "Certificate Revocation," later in this appendix.)
- **Subject** Identifies the owner of the public key contained within the certificate. The subject is the end entity that requested the certificate.
- **Subject Public Key Information** Contains the public key of the subject, along with information designating which algorithm (RSA or ECC, for example) the key can be used with.
- **Issuer Unique Identifier** Uniquely identifies the issuer of the certificate in the event that the issuer name contained within the Issuer field is being reused. This field is optional and is present only in Version 2 and 3 certificates.

Figure B-1 The format of an X.509 digital certificate

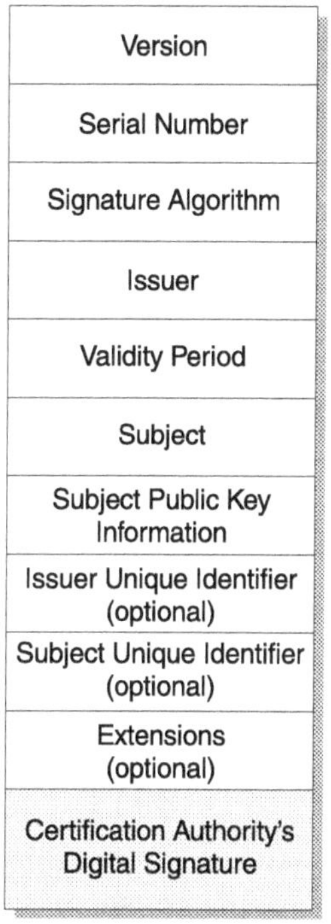

- **Subject Unique Identifier** Uniquely identifies the subject of the certificate in the event that the subject name contained within the Subject field is being reused. This field is optional and is present only in Version 2 and 3 certificates.
- **Extensions** Allows the issuer to embed additional information within the certificate while remaining compatible with the format specified by the X.509 standard. Extensions are optional, and can only appear in Version 3 certificates.
- **Certification Authority's Digital Signature** Contains a digital signature generated with the private key of the issuing CA. The signature encompasses all other fields within the certificate.

Certificate Extensions

There are occasions when the issuer of a certificate may wish to include fields in addition to those specified by X.509 versions 1 and 2. For instance, by including key usage restrictions, the issuer can limit use of the public key contained within a certificate to encryption, signing, key exchange, or any combination of these operations. This is not possible with X.509v1 or v2 certificates. X.509v3, on the other hand, includes support for certificate extensions. Extensions allow the issuer of a certificate to specify arbitrary fields for inclusion within the certificate. The Public-Key Infrastructure X.509 (PKIX) Working Group within the IETF has outlined numerous extensions for use in Internet-based PKIs. For a complete list of these extensions, see RFC 2459 (Housley et al., 1999). Here are some of the more popular certificate extensions:

- **Basic Constraints** Indicates whether the subject identified within the certificate is a CA; and, if so, lists the maximum depth for a certification path passing through that CA. (We'll discuss certification paths later in this appendix.)
- **Certificate Policy** Specifies the policy under which the certificate has been issued and contains identifiers describing the purpose of the certificate.
- **Key Usage** Restricts the usage of the public key contained within the certificate.
- **Extended Key Usage** Indicates one or more uses for the public key in addition to those defined within the key usage extension.

RFC 2459 defines the following key usage settings:

- **Digital Signature** The public key can be used for generating digital signatures in support of security services (for example, authentication and message integrity) other than nonrepudiation and certificate/CRL signing.
- **Nonrepudiation** The public key can be used to verify a digital signature for purposes of nonrepudiation.
- **Key Encipherment** The public key can be used to encrypt other keys for secure transport.
- **Data Encipherment** The public key can be used to encrypt user data.
- **Key Agreement** The public key can be used for key agreement.
- **Encipherment Only** The public key can only be used for encryption during key agreement. This flag must be used in conjunction with the Key Agreement flag.
- **Decipherment Only** The public key can only be used for decryption during key agreement. This flag must be used in conjunction with the Key Agreement flag.
- **Certificate Signing** The public key can be used to verify signatures contained within digital certificates.
- **CRL Signing** The public key can be used to verify signatures attached to certificate revocation lists. (See the section entitled "Certificate Revocation," later in this appendix.)

CAUTION:

If the key usage extension is present in a TLS server-side certificate, the digital signature and key encipherment bits must be set to true. *If the certificate contains a Diffie-Hellman public key, the key agreement bit must be set to* true *(Dierks and Allen, 1999).*

The following key usage extension values are defined by RFC 2459:

- **Server Authentication** The public key can be used for server-side TLS authentication.
- **Client Authentication** The public key can be used for client-side TLS authentication.
- **Code Signing** The public key can be used to verify signatures attached to downloadable executable code. This public key corresponds to the private key used to sign the code.
- **E-mail Protection** The public key can be used to protect electronic mail messages.
- **Timestamping** The public key can be used to bind the current time with the digest of an object.

In addition to these extensions, there are various extensions relating to the IPSec operations (IPSec tunnel, IPSec end system, and IPSec user).

Each extension includes a flag, known as a *criticality indicator*, that determines how an application should handle unrecognized extension types. If a user or application comes across an unrecognized extension marked as *noncritical*, the extension can be ignored, and processing of the certificate can continue. Conversely, an unrecognized *critical* extension will result in immediate rejection of the certificate. While the use of critical extensions ensures proper handling of crucial certificate fields, the presence of numerous critical extensions may reduce interoperability when certificates are deployed outside of the organization responsible for defining the extensions.

Many Microsoft Windows–based operating systems include an ASN.1 parser for processing X.509 digital certificates and certificate revocation lists. Double-clicking a certificate file icon, shown at left, opens a window displaying the contents of the certificate. Figure B-2 shows the window that appears when the certificate file for the CableLabs DOCSIS Root CA is double-clicked.

Validating a Digital Certificate

When an application receives a certificate, it performs a number of steps to verify the integrity and validity of the certificate. It must also identify a relationship between the user of the application and the issuer of the certificate before it

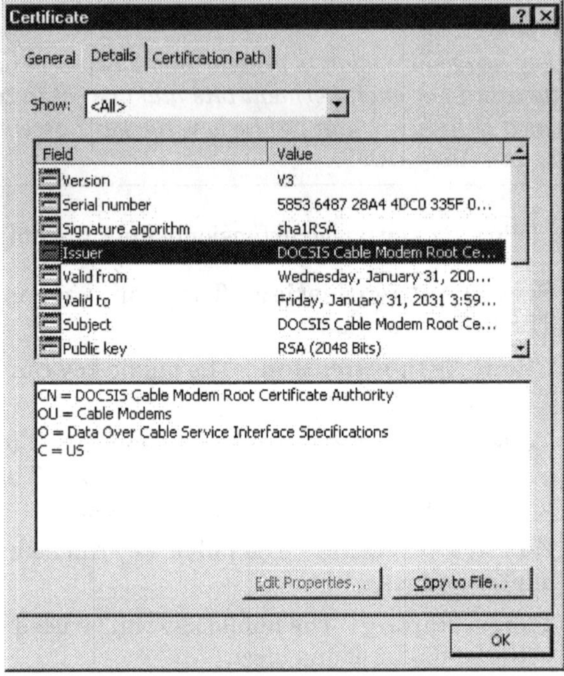

Figure B-2 When using Microsoft Windows, you can view the contents of a certificate by double-clicking the certificate's icon.

will trust the public key contained within the certificate. Certificate processing and validation follows these steps:

1. The application first checks the certificate signature using the public key of the issuer and the algorithms specified within the Signature Algorithm ID field of the certificate.
2. If the signature verifies correctly, the application checks the validity period of the certificate.
3. If the current date and time are within the validity period indicated by the certificate, the application then parses any certificate extensions.
4. If the application recognizes all critical extensions within the certificate, the application looks for policy or usage constraints.
5. If the certificate is being used in accordance with the purpose specified by any policy or usage extensions (for example, certificate policy, key usage, or extended key usage), the application verifies that the certificate has not been revoked. (See the upcoming section entitled "Certificate Revocation" for a description of this process.)

If all the preceding steps are successfully completed and the certificate has not been revoked, the application attempts to determine a trust relationship between the user of the application and the issuer of the certificate. Communicating applications may establish trust by exchanging certificate chains, or an application may establish a trust path on its own by using a process known as *path discovery*. (We describe path discovery in our discussion of public-key infrastructure in later sections.)

NOTE:
A certificate chain is a list of certificates where the issuer of each certificate is the subject contained within the previous certificate in the chain.

Certificate Revocation

The validity period associated with a certificate provides a built-in mechanism for limiting the lifetime of the certificate. Depending on usage patterns and the required level of security, certificate lifetimes may range from a single communication session to many years. Some network devices contain X.509 certificates whose validity periods span the expected lifetime of the device (often 20 or 30 years). Once the validity period has expired, the certificate will no longer be considered valid. This allows the certificate issuer to phase out old certificates and enforce certificate renewal policies. However, there are a number of instances when a certificate should be considered invalid prior to its expiration. For example, a compromised private key, a change in employment status, or the issuance of a new certificate might all prematurely invalidate an existing certificate. The process of invalidating a certificate prior to its expiration is known as *revocation*.

Once a CA determines that a certificate should be revoked, it places the certificate in a *certificate revocation list (CRL)*. Applications use CRLs to check the status of a certificate during certificate validation. CAs periodical distribute updated CRLs containing lists of all nonexpired certificates that have been revoked by the CAs. Each CA determines the frequency of distribution based on the level of security required by its users. The less often a CA posts updates, the greater the possibility that a revoked certificate can be exploited by a malicious user. The difference in time between revocation and distribution of an updated CRL containing the revoked certificate is known as the *time granularity* problem.

NOTE:
If the private key belonging to a CA is ever compromised, no certificate signed using that key can be trusted. As a result, the CA certificate should be revoked; this will, in effect, revoke all certificates issued by that CA.

Contents of a CRL

A typical CRL consists of an issuer name, the time and data when the CRL was issued, the time and date when the next update will occur, a list of revoked certificates, and extensions. Each CRL contains a signature generated by the issuing CA to prevent tampering. Figure B-3 diagrams the format of an X.509 CRL, which contains the following fields:

- **Version** Contains the version number for the CRL; the most recent version is X.509v2.
- **Signature Algorithm ID** Indicates the public-key algorithm and message digest combination used to generate the signature attached to the CRL.
- **Issuer** Identifies the entity that issued the CRL.
- **This Update** Indicates the date and time the CRL was issued.
- **Next Update** Indicates the latest date and time of issuance for the next CRL.
- **User Certificate** Contains the serial number of a revoked certificate.
- **Revocation Date** Designates the date and time the certificate was revoked.

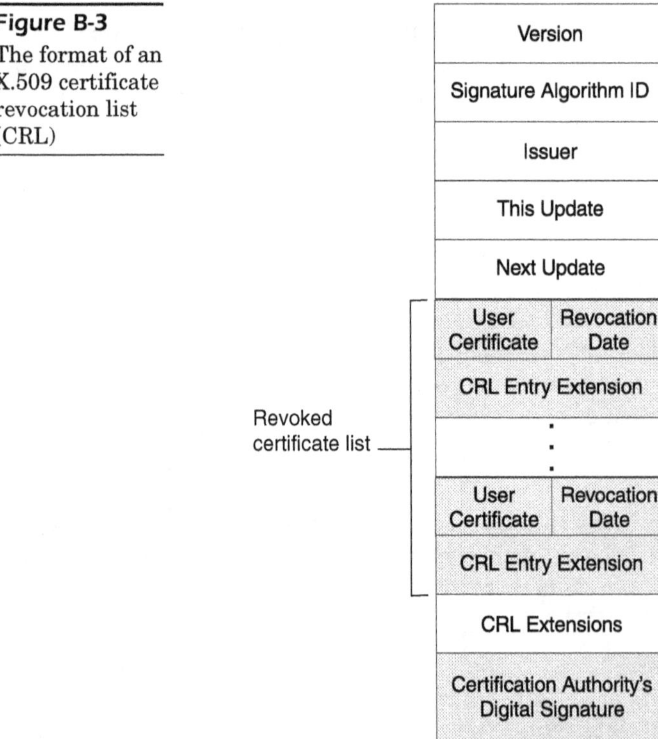

Figure B-3 The format of an X.509 certificate revocation list (CRL)

- **CRL Extensions** Allows the issuer to embed additional information within the CRL while remaining compatible with the X.509 standard. These extensions apply to the CRL as a whole.
- **CRL Entry Extension** Extensions applying to a particular entry in the CRL. All CRL entry extensions should be marked as noncritical.
- **Certification Authority's Digital Signature** Contains a digital signature generated using the private key of the issuing CA. The signature encompasses all other fields within the CRL.

The PKIX Working Group has defined a number of CRL and CRL entry extensions. RFC 2459 contains a complete list of defined extensions for X.509v2 CRLs. In this appendix, we discuss one important CRL entry extension: reason code. The reason code indicates the reason for revoking the certificate, and is always marked as noncritical. RFC 2459 defines the following reason codes:

- **Key Compromise** The private key that belongs to the subject of the end-entity certificate is suspected of being compromised.
- **CA Compromise** The private key belonging to the issuing CA is suspected of being compromised.
- **Affiliation Change** Information within the certificate, such as the subject's name, employment status, or title, has changed.
- **Superceded** The certificate has been replaced by another certificate.
- **Cessation of Operation** The certificate is no longer valid for the purpose it was originally issued.
- **Certificate Hold** The certificate has been suspended, and may subsequently be revoked or removed from the CRL.
- **Remove from CRL** This is used by delta CRLs to identify certificates that should be removed from the most recent base CRL.

Status Checking

The CRLs discussed in the previous sections are known as *static* CRLs. Once issued, the certificate list contained within a static CRL does not change until the next update. To avoid stale revocation information, a CA must use the smallest update interval possible. However, frequently downloading large CRLs can clog a slow network connection. *Delta* CRLs were introduced as an alternative to distributing entire CRLs. A delta CRL contains the list of certificates revoked since the dissemination of the last base CRL. (A *base* CRL contains a complete list of all revoked certificates at the time of creation.) To determine the status of a certificate, an application checks the base CRL and each delta CRL issued since the most recent base CRL. A CA can specify a CRL *distribution point* within each end entity certificate that it generates. The CRL distribution point indicates the network location where an application can download the most current CRLs and delta CRLs.

Online status checking provides a viable alternative to static CRLs. Using the *Online Certificate Status Protocol (OCSP),* an application can obtain the true status of a certificate in real time. Instead of downloading a CRL, an OCSP *requestor* submits a status request to an OSCP *responder*. The responder looks up the status of the certificate in an LDAP directory or other repository, and transmits a response message back to the requestor. The response contains the exact status of the certificate at the time the response was created. Online status checking greatly reduces the overhead associated with downloading large CRLs and eliminates the time granularity problem. However, OCSP must have network access to centrally managed OCSP responders, which are responsible for replying to certificate status requests. This prevents an application from performing status checking when it is not connected to a network.

NOTE:

Some OCSP implementations reference static CRLs during certificate status checking. These implementations do not provide the true status of the certificate.

Public-Key Infrastructure

Wide-scale deployment of public-key cryptography requires a system for maintaining keys and digital certificates, and establishing trust between communicating parties. A *public-key infrastructure (PKI)* provides the necessary protocols, tools, and associated policies and procedures for accomplishing this task. Feghhi et al., in their book *Digital Certificates: Applied Internet Security,* offer the following detailed definition of public-key infrastructure (Fegghi, Fegghi, and Williams, 1999):

> . . . [a PKI] defines a set of agreed-upon standards, certification authorities, structures between multiple certification authorities, methods to discover and validate certification paths, operational protocols, management protocols, interoperable tools, and supporting legislation.

In the remainder of this section, we'll discuss various PKI trust models and the role of certification authorities in a public-key infrastructure.

CA Operations

Certification authorities are central to any PKI. They act as trusted third parties, allowing users to build confidence in one another's identity through mutual trust in the CA. A CA's operational responsibilities include certificate enrollment, distribution, maintenance, and revocation. Collectively, these actions are referred to as *certificate lifecycle management*. CAs often publish *certificate practices statements (CPSs),* which outline policies and procedures for responding to end-entity certificate requests, authenticating subjects prior to certificate issuance, generating and storing keys and certificates, revoking certificates, and other operations related to the management and maintenance of public keys and digital certificates. In addition to lifecycle management, a CPS may also impose limitations on the legal liability accepted by the CA, and define policies and procedures that must be followed by the operators of the CA.

The first responsibility of a CA is certificate enrollment, which follows these steps:

1. A user generates a public/private key pair on his or her computer, and safely stores the private key. (The PKCS #8 and PKCS #12 standards address safe storage of private keys.)
2. The user submits a certificate request to the CA, usually in PKCS #10 format, that contains the user's public key and identifying information (for example, the user's name, affiliation, and title).
3. The CA authenticates the user prior to generating and issuing the certificate. The authentication mechanism determines the level of trust that can be placed in the certificate.
4. The CA transmits the certificate to the user. Certificate responses often take the form of PKCS #7 messages.

Geographically disperse organizations may make use of local *registration authorities (RAs)* to offload registration and authentication duties to a regional office. This allows the CA to concentrate on core certificate management operations, such as generation, distribution, and revocation. Local RAs have the added benefit of reducing long-distance travel when authentication requires an on-site visit.

NOTE:
When a certificate request is made through a web browser, the browser handles key generation and storage of the user's private key.

In the preceding scenario, no key pair generation occurrs at the CA. In the event that an organization wishes to use *dual key pairs*—separate keys for signing and encryption—the CA may generate and store the key pair used for encryption. End entities generate their own key pairs for signing, and maintain exclusive control of the private key for purposes of nonrepudiation. Why would you use dual key pairs?

Consider, for instance, a corporation that has mandated the use of dual key pairs for its employees. If an employee were to leave the company and destroy both key pairs, the employer could recover the keys used for encryption. This would allow the employer to decrypt any message encrypted by the terminated employee. Without this ability, the employer could lose important financial records, intellectual property, and other business documents. By requiring users to generate their own key pairs for signing, the employer can enforce nonrepudiation. If this were not the case, a user would be able to deny ever signing a message or participating in an electronic transaction, because others also possessed the private key. Loss of the private key for signing would pose no problem to the employer, since signature verification requires only the corresponding public key.

Certificate storage and distribution are also the responsibilities of the CA. Following the generation of a certificate, a CA usually publishes the certificate to a *repository*. Repositories take many forms, including directories, relational databases, and files within a local or network file system. To obtain public keys for encryption and signature verification, users and certificate-enabled applications periodically download certificates from these repositories. CAs commonly employ *X.500* or *Lightweight Directory Access Protocol (LDAP)* directories for the purpose of certificate distribution.

The final operation performed by a CA is revocation. We discussed certificate revocation at length in the earlier section entitled “Certificate Revocation.”

Trust Models

The relationship between certification authorities in a PKI defines the trust model. Three primary models exist: strict hierarchy, mesh, and hybrid. A *strict hierarchy* consists of a *root* CA and any number of subordinate CAs and end entities (see Figure B-4).

At the top level of the hierarchy (level 0) sits the root CA. The root CA issues its own *self-signed* certificate. Just below the top lies the first level of subordinate CAs, whose certificates have been issued by the root CA. The hierarchy can include any number of levels containing subordinate CAs, but each CA at level n of the hierarchy must be certified by one, and only one, CA at level n minus 1. The end entities, or subjects, reside at the bottom of each branch in the hierarchy. The use of a single root CA and a strict top-down hierarchical structure provides simplicity, scalability, and centralized control. However, it offers little flexibility for large PKI deployments. Take for instance, the following scenarios:

- Within a large Internet-based PKI, organizational relationships may not follow a hierarchical structure. As a result, it may be impractical for multiple organizations to agree upon a single root CA.
- To avoid issuing a large number of overlapping certificates, two companies not sharing a common root CA may wish to establish a trust relationship between their respective PKIs by directly certifying each other’s CAs.

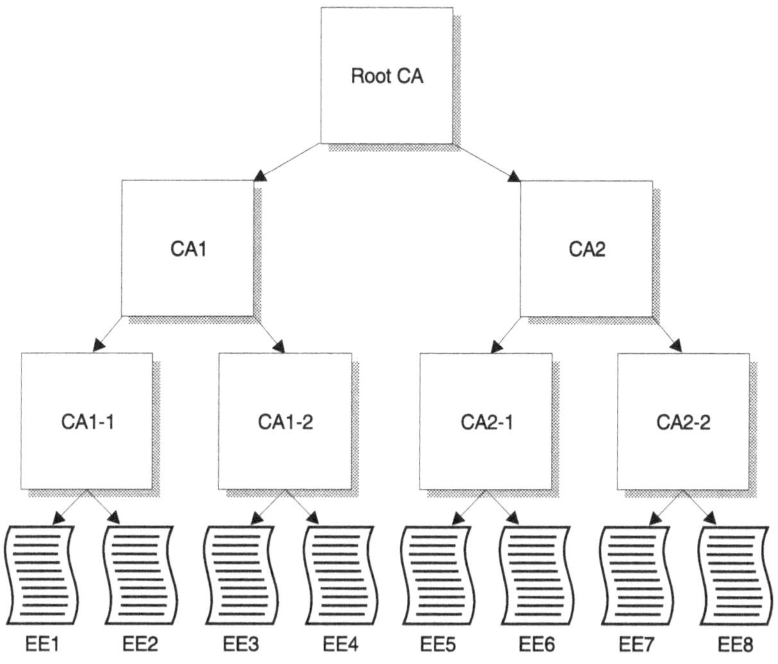

Figure B-4 A strict top-down PKI hierarchy.

The latter case, known as *cross-certification,* establishes a one- or two-way peer-to-peer trust relationship between two CAs (see Figure B-5). Cross-certification is an important characteristic of *mesh* models. It greatly improves flexibility by allowing a PKI to exactly match the internal structure and trust relationships within an organization.

Mesh models come in two primary flavors: complete and partial. In a *complete mesh*, every CA cross-certifies with all other CAs in the mesh, as illustrated in Figure B-6. While this model keeps certification paths extremely short, it

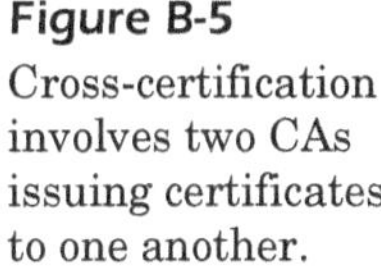

Figure B-5 Cross-certification involves two CAs issuing certificates to one another.

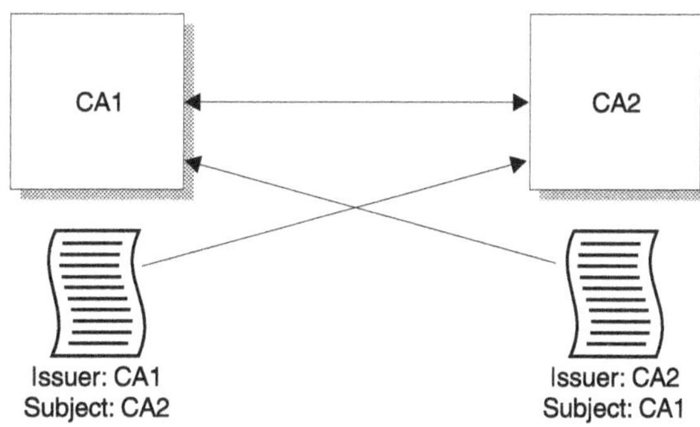

Figure B-6
An example of a complete mesh

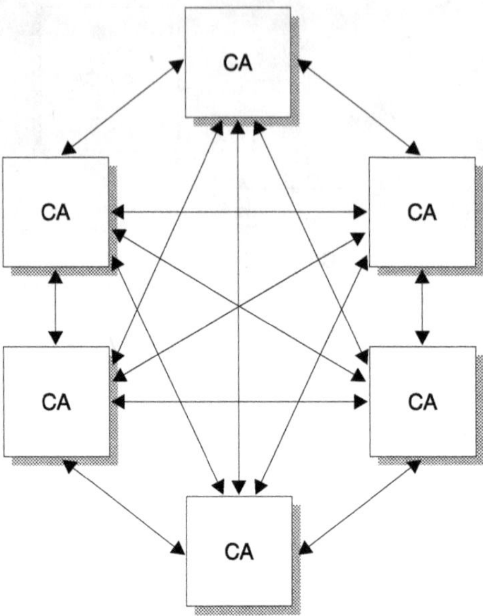

severely lacks scalability. For *n* CAs, the number of cross-certifications required is $n \times$ (n minus 1). This number grows very quickly as the number of CAs increases.

NOTE:
A certification path *defines the trust relationship between two entities in terms of the individual relationships between various members of a PKI. These "individual relationships" may exist between an end entity and a CA, a subordinate CA and the root CA, or two cross-certified CAs. Communicating parties establish trust by traversing a certification path until they determine a common TTP.*

A *partial mesh* consists of an arbitrary number of relationships between CAs; each CA can cross-certify with as many or as few CAs as it wishes (see Figure B-7). In general, this improves scalability by reducing the number of cross-certifications, but increases the length of certification paths. In Figure B-7, it's easy to see that the shortest certification path from CA3 to CA5 travels through CAs 1 and 2 or CAs 2 and 4. Mesh models lack centralized management, as there is no root CA. In addition, it is often difficult to control and validate certification paths traversing many cross-certified CAs.

The final organizational structure is the *hybrid* model. This model combines the flexibility of a partial mesh with the scalability and centralized management of a strict top-down hierarchy. The hybrid model is very similar to a strict hierarchy (see Figure B-8), but allows peer-to-peer relationships (that is, relationships in the horizontal direction) between CAs. Subordinate CAs can directly cross-certify to shorten certification paths, and root CAs can cross-certify for

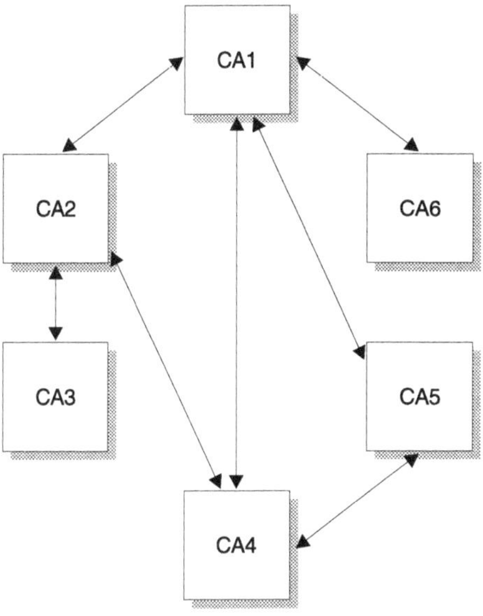

Figure B-7 An example of a partial mesh

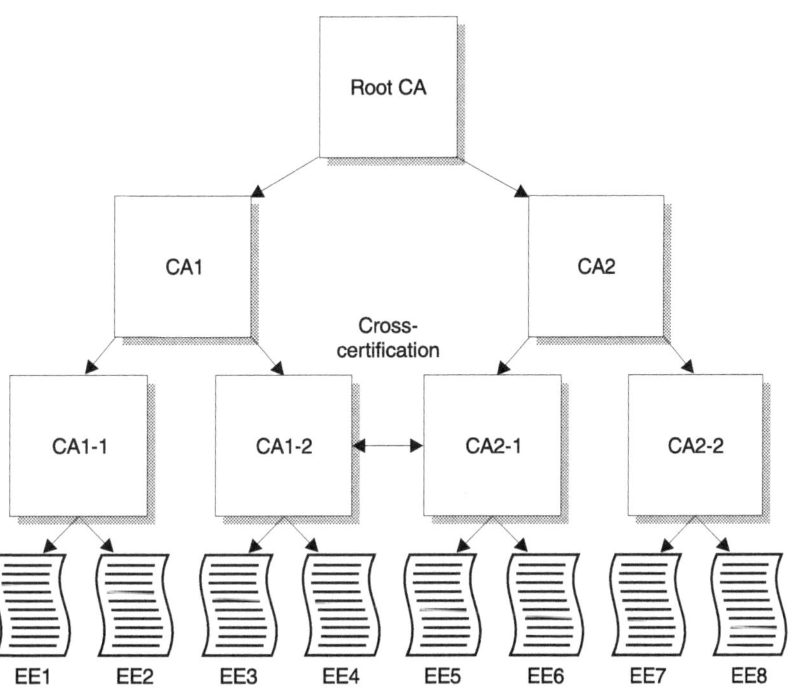

Figure B-8 The hybrid trust model combines the simplicity of the strict hierarchical model with the flexibility of the mesh model.

PKI interoperability (in other words, to ensure that certificates issued in one hierarchy are trusted by users in another).

Notice the cross-certification between CA1-2 and CA2-1 in Figure B-8; the certification path between any end entity issued a certificate by CA1-2 and another end entity issued a certificate by CA2-1 need only traverse the path between CA1-2 and CA2-1 (as opposed to traversing all the way to the root CA). Since the root CA sets policy for the entire hierarchy, subordinate CAs should not be allowed to cross-certify without the consent of the root CA. Some may view this restriction as a disadvantage of the hybrid model. The hybrid model has the following characteristics and advantages (Nash et al., 2001):

- Low administrative overhead
- High scalability
- Flexibility to match any organizational structure
- Simplified certificate distribution
- A fixed relationship between end entities and the root CA, allowing for transmission of certification paths

Path Discovery and Validation

Within a PKI, two communicating parties must identify a common TTP before they can establish confidence in one another's certificates. *Path discovery,* also called *path construction*, is the process of determining a trust relationship between two communicating parties by systematically traversing a certification path. The certification path should begin with an end entity and terminate at a shared CA or pair of cross-certified CAs, and can contain any number of subordinate CAs.

For a strict hierarchy, path discovery is very simple. Referring back to Figure B-4, two entities establish trust by traveling up the hierarchy until their paths cross at a shared CA or until they reach the root CA. Users and certificate-enabled applications within a strict hierarchy typically send a *certificate chain* along with every message and transaction. The chains consist of every certificate required to establish a certification path between communicating parties.

Well-constructed mesh models can simplify path processing by significantly reducing the length of certification paths. On the other hand, a poorly designed mesh, lacking well-defined and documented trust relationships, can greatly increase the overhead associated with path discovery and validation. Path discovery times for hybrid models usually lie somewhere in between those for strict hierarchies and properly designed partial meshes.

Verification of all certificates in the certification path between two entities is known as *certificate path validation*. Path validation ensures that a definable and trusted relationship exists between all entities in the certification path, or all entities whose certificates appear in a certificate chain. The process for validating a certificate chain is similar to that for validating a single certificate, but includes the following additional steps (Nash et al., 2001):

- The subject and issuer names within successive certificates in the chain must match.
- The number of certificates following each CA certificate in the chain must comply with path length constraints.
- Every certificate in the chain, except for that of the end entity, must be marked as a CA certificate.

References

Dierks, T., and C. Allen. January, 1999. *RFC 2246: The TLS Protocol Version 1*. The Internet Engineering Task Force.

Feghhi, Jalal, Jalil Feghhi, and Peter Williams. 1999. *Digital Certificates: Applied Internet Security*. Reading, Massachusetts: Addison-Wesley.

Housley, R., W. Ford, W. Polk, and D. Solo. January, 1999. *RFC 2459: Internet X.509 Public Key Infrastructure Certificate and CRL Profile*. The Internet Engineering Task Force.

International Telecommunication Union. June, 1997. *ITU-T Recommendation X.509: Information Technology—Open Systems Interconnect—The Directory: Authentication Framework*.

Kaliski, Burton S., Jr. November, 1993. *A Layman's Guide to a Subset of ASN.1, BER, and DER*. RSA Laboratories.

Nash, Andrew, Bill Duane, Celia Joseph, and Derek Brink. 2001. *PKI: Implementing and Managing E-Security*. Berkeley, California: Osborne/McGraw-Hill.

Index

NUMBERS

3DES, 290
802.11 standard
  WEP, advantages/disadvantages, 234
  WEP case study, 165-167
  wireless networks and, 165

A

Abstract Syntax Notation One (ASN.1)
  object identifiers and, 301
  Presentation layer and, 107
access control, 20-21, 388
access control lists (ACLs), 20
access networks, 7-8, 115
access points (APs), 167
ACKs (acknowledgments), 114
ACLs (access control lists), 20
active attacks, 69
adaptive-chosen-cipher text attack, 97
adaptive-chosen-plaintext attack, 97
Address Resolution Protocol (ARP), 415, 425-426
address spoofing attacks, 83-85
  combined with DoS techniques, 84-85
  countermeasures, 89-90
  how they work, 83-84
administration, security, 266-268
ADSL (asymmetric DSL), 9, 122
Advanced Research Projects Agency Network (ARPANET), 4
AES (Advanced Encryption Standard), 34
agents, DDoS attacks, 76
AH (Authentication Header), 181-186
  anti-replay service and, 183
  compared with ESP, 172
  format, 183-184
  inbound/outbound packets, 185-186
  IPv4/IPv6 mutable and immutable fields, 181-182
  modes of operation, 185
  PacketCable networks and, 323
Alert protocol
  SSL, 210-211
  TLS, 211-212
algorithms. *See also* by type
  comparing block ciphers, 34-35
  comparing message digests, 38
  comparing public-key algorithms, 47-48
  defined, 22
  guidelines for choosing, 251-253
  types of, 23
American National Standards (ANSI), 146
anti-replay service
  AH and, 183
  ESP and, 186
  PacketCable and, 326
antivirus programs, 14
AP Request/AP Reply messages, Kerberos, 332-334, 337
API (application programming interface), iTV, 360
Application layer (layer 7)
  Kerberos and, 216
  OSI Reference Model, 107
  PacketCable and, 319
  securing stored data, 231
  transparency and, 226
application programming interface (API), iTV, 360
application security policy, DVB MHP, 375-378
applications
  network applications, 134-135
  real-time vs. elastic applications, 129-130
  security mechanisms and, 389
  transparency and, 225-228
  video conferencing, 395
APs (access points), 165
ARP (Address Resolution Protocol), 415, 425-426
ARPANET (Advanced Research Projects Agency Network), 4
AS (authentication server), 217, 218, 220
ASN.1 (Abstract Syntax Notation One)
  object identifiers and, 301
  Presentation layer and, 107
ASNI (American National Standards), 146
asset identification, design scenarios, 384-385, 406
asymmetric cryptography. *See* public-key cryptography
asymmetric DSL (ADSL), 9, 122

asynchronous bulk traffic, 129-130
ATM (Asynchronous Transfer Mode)
  capacity of, 114
  compared with IP, 139
  QoS over, 136-139
attackers, 62-66
  categories of, 65, 385
  Hacker Ethic and, 63
  motivations of, 64
attacks
  address spoofing, 83-85
  attackers and motivations, 62-66
  categories of, 68-69, 385-386
  cryptography attacks, 95-100
  data modification attacks, 77-79
  denial-of-service attacks, 75-77
  eavesdropping attacks, 69-73
  impersonation attacks, 73-75
  known vulnerabilities and, 386
  origination of, 67-68
  packet-replay attacks, 79-80
  passive vs. active, 69
  routing attacks, 80-83
  session hijacking attacks, 86-90
  social engineering, 100
  TCP/IP denial-of-service attacks, 90-95
  TCP/IP-specific attacks, 83
  TCP sequence number prediction, 85-86
  times of greatest vulnerability, 66-67
attenuation, 11
authentication
  cross-realm authentication, 219-220, 337-338
  cryptographic mechanisms for, 388
  DOCSIS 1.1 BPI+, 289-291
  DVB MHP, 369-371
  Kerberos, 217-219
  overview of, 19
  separating from encryption in SSL Handshake protocol, 206-207
  traffic encryption, 396
Authentication Header. *See* AH (Authentication Header)
authentication messages, DVB MHP, 363-369
  certificate files, 368-369
  hash files, 365-366
  overview of, 363-365
  signature files, 367-368
authentication server (AS), 217, 218, 220
Authorization Request/Response messages, CMTS, 292
authorization services, 20-21
Authorization state machine, BPKM, 280-285
  cable modem authorization, steps in, 282-285
  events, 281-282
  messages, 281
  operational matrix of, 285
  parameters, 283
  states of, 280
availability
  QoS and, 134-136, 390
  services, 21

B

backbone networks, 7-8, 114
bandwidth
  digital certificates and public keys and, 259
  fiber optics and, 11-12
  HMACs and, 261
  QoS and, 130-131, 390
  security services use of, 248
baseband, vs. broadband, 104
baseCRLs, 447
Baseline Privacy Interface (BPI), 154. *See also* DOCSIS 1.0 BPI; DOCSIS 1.1 BPI+
Baseline Privacy Key Management Protocol. *See* BPKM (Baseline Privacy Key Management Protocol)
Bellovin, Steven, 89
best effort QoS, 136
bidirectional transmission, DOCSIS, 273
binding services, to data, 223-224, 412
biometric authentication devices, 19
BITS (bump-in-the-stack) implementation, security protocols, 234-235
BITW (bump-in-the-wire) implementation, security protocols, 235-237
BlackICE Defender, 14-15
block ciphers
  comparing, 34-35
  as cryptographic tool, 54-56
  encryption speed of, 256
  modes of operation, 29-33, 54-55
  overview, 27-29
  performance and, 253-255
block replay attacks, 97-98
block size, 28
blocks, 27-28
BPKM (Baseline Privacy Key Management Protocol), 279-289
  Authorization state machine, 280-285
  encapsulation and, 276
  overview of, 279
  TEK state machine, 285-289
broadband technologies
  cable, 8-9, 115-120
  communication protocols, 113-114
  data encapsulation and, 111-112
  defined, 7
  DSL, 9, 120-123
  fiber optics, 11-12
  fixed wireless, 9-10, 123-126

future of, 10-11
hackers exploitation of, 67-68
history of telecommunication and, 4-7
origins of, 104-105
OSI Reference Model, 106-110
overview of, 3-4
security, common users, 12-15
security, network infrastructure, 15-16
service providers, 114-115
TCP/IP model, 110
two-way satellite, 10, 126-128
broadband technologies, QoS and, 129-143
availability, 134-136
bandwidth, 130-131
cell relay vs. packet switching, 136-139
IP networks and, 139-143
jitter, 132-133
latency, 132
packet loss, 133
parameters, 130-131
real-time vs. elastic applications, 129-130
Broadband Wireless Internet Forum (BWIF), 147-148
brute force attacks, 74
buffering, reducing jitter, 133, 265
bump-in-the-stack security protocol implementation, 234-235
bump-in-the-wire security protocol implementation, 235-237
BWIF (Broadband Wireless Internet Forum), 147-148

C

cable modem termination system. *See* CMTS (cable modem termination system)
cable modems
authorization, 282-285
cable networks and, 118-119
certificate formats, 297-301
certificate hierarchy, 292-297
certificate validation, 301-302
limited security resources of, 249
strong authentication of, 275
use of BPKM by, 276
cable networks, 115-120
cable modems, 118-119
data privacy, 275
DOCSIS and PacketCable and, 119-120
history of CATV, 115
mixing fiber optics and coaxial cabling, 116
transmission, 117-118
tree architecture of, 115-116
cable services, 8-9
cable television, 4
Cable Television Laboratories, 147
cable transmission, 117-118
CableLabs, 119-120, 311
Call Management Servers, PacketCable networks, 316
CAs (Certification Authorities)
certificate enrollment by, 449
certificate management by, 438-439
local system format, 353-354
manufacturer CAs, 293-297
MTA formats, 347-350
operations of, 448-450
service provider formats, 351-353
CATV (Community Antenna Television), 7, 115
CBC (cipher block-chaining) mode, 30-31, 56
cell relay vs. packet switching, 136-139, 312
certificates. *See* digital certificates; X.509 certificates
certification revocation lists. *See* CRLs (certification revocation lists)
CFB (cipher feedback) mode, 31-32, 56
chaining, 30, 99
ChangeCipherSpec protocol, SSL, 209-210
chosen-ciphertext attack, 97
cipher block-chaining (CBC) mode, 30-31, 56
cipher feedback (CFB) mode, 31-32, 56
cipher suites, 210-216
defined, 212
encryption algorithms, 215-216
key-exchange algorithms, 215
list of, 213-214
overview of, 210, 212, 214
ciphers. *See* algorithms
ciphers, block. *See* block ciphers
ciphers, stream. *See* stream ciphers
ciphertext algorithms, 22
ciphertext-only attacks, 96
cleartext algorithms, 22
CLECs (competitive local exchange carriers), 6
client-side authentication, SSL Handshake protocol, 207-208
CMS (Cryptographic Message Syntax), 232
CMTS (cable modem termination system)
authorization and, 276
certificate validation, 301-302
DOCSIS and, 119, 273
KEKs and TEKs and, 290
message integrity checks (MICs), 303
PacketCable and, 119
coaxial cable, 116
code footprint
algorithm selection and, 252
message digests, 261
performance and, 256
public-key algorithms, 259
code signing, 47
code verification certificates (CVCs), 301

Code Verification Signature (CVS), 305
collision-free output, message digests, 37
communication protocols, 113-114, 389
Community Antenna Television (CATV), 7, 115
compact security code, 250
competitive local exchange carriers (CLECs), 6
complete mesh, PKI trust model, 451-452
compression, encryption and, 263-264
confidentiality services, 18, 72, 388
congestion, TCP and, 432
connection-oriented protocols, 113, 429
connection states, SSL, 199-201
connectionless protocols
  IP as, 417
  overview of, 113
  UDP as, 433
controlled-load service, 140
Copy Protection System (CPS). *See* OpenCable Copy Protection System (CPS)
costs, of security, 52-53
countermeasures
  address spoofing attacks, 89-90
  cryptography attacks, 98-100
  data modification attacks, 78-79
  eavesdropping attacks, 71-73
  impersonation attacks, 75
  packet-replay attacks, 79-80
  routing attacks, 82-83
  session hijacking attacks, 89-90
  threats and, 387
CPE (customer premise equipment), 273
CPS (Copy Protection System). *See* OpenCable Copy Protection System (CPS)
CRCs (cyclic redundancy checksums), 108
credentials, false, 73-74
criticality indicators, X.509 certificates, 443
CRLs (certification revocation lists), 445-448
  CAs and, 445
  contents of, 445-446
  DOCSIS 1.1 BPI+ and, 303
  fields, 446-447
  MHP and, 374
  overview of, 374-375
  types of, 447
cross certification, 451
cross-realm authentication, 219-220, 337-338
cryptanalysis, 23
cryptographic hardware, 262
Cryptographic Message Syntax (CMS), 232
cryptographic service providers (CSPs), 22
cryptography. *See also* encryption; public-key cryptography; symmetric-key cryptography
  FIPS standards, 51-52
  interoperability, 57
  overview of, 21-23
  random number generation and, 23-25
  store-and-forward vs. session-based, 52-53
  tools, block ciphers, 54-56
  tools, message digests, 56
  tools, public-key algorithms, 56
  tools, stream ciphers, 53-54
cryptography attacks, 95-100
  block replay, 97-98
  countermeasures, 98-100
  cryptanalytic, 95-97
  list of, 95
  weak keys, 97
CSPs (cryptographic service providers), 22
customer premise equipment (CPE), 273
CVCs (code verification certificates), 301
CVS (Code Verification Signature), 305
cyclic redundancy checksums (CRCs), 108

D

DAC (discretionary access control), 20
data encapsulation. *See* encapsulation
Data Link layer
  DOCSIS operation at, 277
  logical link control (LLC) and, 109
data modification attacks
  countermeasures, 78-79
  how they work, 77-78
  security services for, 387
Data-Over-Cable Service Interface Specifications. *See* DOCSIS (Data-Over-Cable Service Interface Specifications)
data privacy
  BPI and, 154
  BPI+ and, 275
datagrams. *See also* UDP (User Datagram Protocol)
  IP fragmentation/reassembly and, 424-425
  TCP/IP encapsulation and, 417
DBS (digital broadcast satellite), 126
DDoS (distributed denial of service) attacks, 76
default gateways, 423-424
delays, QoS and, 132, 390
DeltaCRLs, 447
denial of service (DoS) attacks. *See* DoS (denial of service) attacks
DER (Distinguished Encoding Rules), 305
design, initial steps
  asset identification, 384-385
  host-based security vs. security gateways, 391-392
  network layer, selecting, 391
  overview of, 383-384
  security mechanisms, persistent, 390-391
  security mechanisms, selecting, 388-390
  security protocols, designing new, 393-394

security protocols, selecting, 392-393
security services, persistent, 390-391
security services, selecting, 386-387
threat identification, 385-386
design scenario 1, video conferencing, 394-403
analysis of flaws in, 400-403
characteristics of, 395-396
handshaking exchanges and, 397-398
security mechanisms, selecting, 396
signature generation, 398-399
design scenario 2, video conferencing
asset identification, 406
background information and design criteria, 403-406
binding services and mechanisms to data, 412
host-based security vs. security gateways, 412
network layer, selecting, 412
security mechanisms, selecting, 408-411
security protocols, designing new, 414
security protocols, selecting existing, 412-414
security services, selecting, 406-408
threat identification, 406
device drivers, NICs and, 229
DF certificate formats, 355-356
dictionary attacks, 74
Diff Serv (Differentiated Services Architecture), 142, 245
differentiated/priority-based QoS, 136
Diffie-Hellman algorithm, 42
digital broadcast satellite (DBS), 126
digital certificate formats, 297-301
DF format, 355-356
KDC format, 354-355
local system format, 353-354
MHP format, 372
MTA device format, 349-350
MTA manufacturer format, 348-349
MTA root format, 347-348
overview of, 441
service provider format, 352-353
service provider root format, 351-352
digital certificates, 437-448
as attack countermeasure, 99
bandwidth and, 259
classification of, 439
DOCSIS 1.1 BPI+ and, 152
DVB MHP and, 163
enrollment, 449
extensions, 346-347, 441-443
fields, 440-441
files, 368-369
H.235, 160
hot lists, 303
OpenCable CPS and, 165
overview of, 437-438
PacketCable and, 157
path validation, 454-455
PKI and, 41
revoking, 445-448
validation, 301-302, 357-358, 443-445
Digital Certificates:Applied Internet Security (Fegghi, Fegghi, and Williams), 448
digital envelopes
creating, 43-44
function of, 43
digital signatures, 44-47
code signing and, 47
creating and verifying, 45-46
nonrepudiation services, 224
PKCS#7 and, 305-307
time requirements of algorithms supporting, 257
Digital Subscriber Line. *See* DSL (Digital Subscriber Line)
digital subscriber line access multiplexer (DSLAM), 121
digital television (DTV), 117
Digital Video Broadcasting. *See* DVB Multimedia Home Platform
Direct Broadcast Satellite (DBS), 10
directory authentication, 370-371
discrete multitone modulation (DMT), 123
discretionary access control (DAC), 20
Distinguished Encoding Rules (DER), 305
distributed denial of service (DDoS) attacks, 76
distribution networks, 7-8, 114
distribution point, CRLs, 447
DMT (discrete multitone modulation), 123
DOCSIS 1.0 BPI
list of characteristics, 150-151
overview of, 154
PacketCable interface with, 157
DOCSIS 1.1 BPI+
applicability of, 393
certificate formats, 297-301
certificate hierarchy, 292-297
certificate validation, 301-302
certification revocation and hot lists, 303
encryption and authentication mechanisms, 289-291
list of characteristics, 151-153
MAC layer frame formats, 277-279
overview of, 154-155, 157, 275-277
signed software upgrade verification, 304-308
TFTP configuration files, 303-304
DOCSIS (Data-Over-Cable Service Interface Specifications)
compared with PacketCable, 313
overview of, 273-275
PacketCable interaction with, 311
standards, 119
document type definition (DTD), 375

DoS (denial of service) attacks, 75-77
- availability and, 21
- distributed, 76-77
- how they work, 75-76
- network security and, 15
- security services for, 387
- TCP/IP denial-of-service attacks, 90-95

DSL (Digital Subscriber Line), 120-123
- existing technologies, 9
- overview of, 120-122
- telephone and, 121
- varieties of, 122-123

DSL Forum, 147
DSLAM (digital subscriber line access multiplexer), 121
DTD (document type definition), 375
DTV (digital television), 117
dumpster diving, 100
DVB (Digital Video Broadcasting) Project, 148, 166
DVB-J environments, 361, 362
DVB Multimedia Home Platform
- application security policy, 375-378
- authentication messages, 363-369
- certificate revocation, 374-375
- hash files, 365-366
- Java security classes, 380
- list of characteristics, 161-163
- MHP platform, 360-362
- MHP security, 362-363
- object authentication process, 369-371
- overview of, 166, 359-360
- permission request file, 375-378
- return channel security, 379-380
- root certificates, 373-374
- signature files, 367-368, 368-369
- X.509 certificates and PKI hierarchy and, 372-375

dvb.certificates, 368
dvb.crl, 374-375
dvb.hashfile, 365
dvb.signaturefile, 367
dynamic policy managers, 20

E

eavesdropping attacks, 69-73
- countermeasures, 71-73
- as passive attack, 69
- security services for, 387
- use of packet sniffers for, 70-71

ECB (Electronic Code Book), 29-30, 55, 263
ECC algorithm, 48
echo request/reply, ICMP, 426
EDE (encrypt-decrypt-encrypt), 162
EH (Extended Header), DOCSIS BPI+, 278
elastic applications, real-time vs., 129-130
Electronic Code Book (ECB), 29-30, 55, 263
electronic surveillance, 404
EM emissions, capturing, 73
embedded devices, security resources of, 249-250
employee attacks, 65-66
Encapsulating Security Payload. *See* ESP (Encapsulating Security Payload)
encapsulation
- IPSec tunnel mode and, 176-177
- OSI Reference Model and, 111-112
- SSL Record layer, 202
- TCP/IP and, 416-417

encrypt-decrypt-encrypt (EDE), 152
encryption. *See also* cryptography
- benefits of transport mode for, 245
- compression and, 263-264
- as confidentiality mechanism, 18
- defined, 22
- DOCSIS 1.1 BPI+, 289-291
- separating from authentication, 206-207

end-entity certificates, 439
end-to-end protection
- compared with tunnel mode, 245
- host-based security and, 238
- security gateways and, 241
- security protocols and, 230-231

entropy gathering techniques, 400
error propagation, CBC mode, 31
ESP (Encapsulating Security Payload), 186-191
- compared with AH, 172
- header and packet format, 187-188
- inbound/outbound packets, 189-191
- modes of operation, 189
- overview of, 186-187
- PacketCable and, 323, 326

Ethernet networks, 105
ETSI (European Telecommunications Standards Institute), 147
exhaustive key searches, 23
Extended Header (EH), DOCSIS BPI+, 278
Extensible Markup Language (XML), 375
extension headers, IPv6, 423

F

FB (output feedback) mode, 32-33, 56
FCS (frame check sequence), 108
FDM (frequency-division multiplexing), 105
Federal Information Processing Standards (FIPS), 51-52
fiber optics
- limiting capture of EM emissions, 73
- overview of, 11-12

fiber rings, 116
fields

AH header, 184
cable modem end-entity, 300
digital certificates, 440-441
DOCSIS BPI+, 277-278
DOCSIS root CA, 298
DOCSIS subordinate manufacturer CA, 299
ESP header, 187-188
IPv4 header, 418-421
IPv6 header, 421-422
ISAKMP messages, 191-195
RTCP messages, 341
SAD, 180-181
SSL Handshake protocol, 203
SSL Record protocol, 201
TCP header, 430-432
UDP header, 434
X.509 certificates, 440-441
file authentication, DVB MHP, 369-370
FIPS (Federal Information Processing Standards), 51-52
fixed wireless transmission, 123-126
advantages over traditional wiring, 125
overview of, 9-10
point-to-multipoint vs. point-to-point, 123-124
flow control, TCP, 432
formats
AH, 183-184
Alert messages, 210
ChangeCipherSpec messages, 209-210
digital certificates, 441
DOCSIS BPI+, 277, 297-301
ESP, 187-188
ICMP messages, 427
IPv4, 420
IPv6, 422
SSL Handshake protocol, 203
TCP, 431
UDP, 434
X.509 certificates, 440
Fortezza cipher suite, 212
fragmentation, 424-425
fragmentation attacks, 91-92, 95
frame check sequence (FCS), 108
frames, TCP/IP encapsulation and, 417
frequency-division multiplexing (FDM), 105
full-duplex mode, Session layer, 107

G

gaming consoles, 249
gateways. *See* security gateways
geosynchronous earth orbit (GEO), 10
guaranteed/reservation-based QoS, 136
guaranteed services, 140-141

H

H-Series terminals, 158, 160
H.225, 157
H.235, 158, 165
H.245, 157
H.323 networks, 154
Hacker Ethic, 63
hackers, 63. *See also* attackers
Hackers: Heroes of the Computer Revolution (Levy), 63
half-duplex mode, Session layer, 107
handlers, DDoS attacks, 76
Handshake protocol, SSL, 202-209
client-side authentication, 207-208
fields, 203
format, 203
message types and parameters, 203-204
separating authentication from encryption, 206-207
session reuse, 208-209
state processing, 204
steps in SSL handshake, 205
handshaking exchanges
authentication failures and, 400
design scenarios and, 397-398
public keys and, 264
TCP handshake, 429
hard problems, 40-41
hardware accelerators, 262
hardware addresses, 425
hardware security modules (HSMs), 296-297
hash-based message authentication code (HMACs)
message integrity and, 39-40
security side effects of, 261-262
hash files, DVB MHP, 365-366
hashing functions, message digests, 36
HDTV (high-definition television), 117
headers
AH, 184
data encapsulation, 111
ESP, 187-188
IP, 419-423
ISAKMP, 192
TCP, 430-432
TCP/IP encapsulation and, 416
UDP, 434
HFC (hybrid fiber-coax), 273
high-definition television (HDTV), 117
high-speed Internet connections, 118
HMACs (hash-based message authentication code)
message integrity and, 39-40
security side effects of, 261-262
host-based security, 237-244

design scenarios and, 391-392, 412
extent of coverage, 238-239
implementation and maintenance, 241-242
integration with existing policies, 243
list of characteristics, 244
vs. security gateways, 391-392
traffic security and, 242-243
user contexts, 243
WAP Gap and, 239-241

HSMs (hardware security modules), 296-297
HTTP (Hypertext Transfer Protocol), 197
hybrid fiber-coax (HFC), 116, 273
hybrid, PKI trust model, 452-453
Hypertext Transfer Protocol (HTTP), 197

I

ICMP floods, 92
ICMP (Internet Control Message Protocol), 426-428
message codes, 428
messages, 427
overview of, 426
RFCs, 415

ICV (integrity check value), 184, 188
IEEE (Institute of Electrical and Electronic Engineers), 148
IETF (Internet Engineering Task Force), 148, 172
IKE (Internet Key Exchange), 191-197
ISAKMP messages, 191-195
ISAKMP phases and exchanges, 195-197
key management with, 413-414
overview of, 191
PacketCable and, 326-327

immutable/mutable fields, IPv4/IPv6, 181-182
iMode phones, 241
impersonation attacks, 73-75
countermeasures, 75
list of characteristics, 73-74
security services for, 387

inbound processing
AH, 186
ESP, 190-191

initial sequence numbers (ISNs), 85-86, 429
initialization vector (IV)
CBC mode and, 30-31
performance and, 255

insider attacks, 65-66
Institute of Electrical and Electronic Engineers (IEEE), 149
Integrated Services (IntServ), 140-141
integrity algorithms
DOCSIS 1.0 BPI and, 150
DOCSIS 1.1 BPI+ and, 152
DVB MHP and, 163
H.235, 154
OpenCable CPS and, 165
PacketCable and, 156

integrity check value (ICV), 184, 188
integrity services
cryptographic mechanisms for, 388
HMACs and, 39-40
message digests and, 36-39
messages, 396, 398
overview of, 19

interactive bulk traffic, 129-130
interactive burst-traffic, 129-130
interactive television (iTV), 360. *See also* DVB Multimedia Home Platform
internal attacks, 65-66
International Standards Organization (ISO), 106, 148
International Telecommunication Union (ITU), 148, 158
Internet access, security of, 273-275. *See also* DOCSIS 1.1 BPI+
Internet Control Message Protocol. *See* ICMP (Internet Control Message Protocol)
Internet Engineering Task Force (IEFT), 148, 172
Internet Key Exchange. *See* IKE (Internet Key Exchange)
Internet Security Associations and Key Management Protocol. *See* ISAKMP (Internet Security Associations and Key Management Protocol)
Internet Security Protocol. *See* IPSec (Internet Security Protocol)
Internet service providers (ISPs), 68, 121
interoperability, cryptography, 57
IntServ (Integrated Services), 140-141
IP fragmentation/reassembly, 424-425
IP headers, 419-423
IP (Internet Protocol). *See also* IPv4; IPv6
compared with ATM, 139
headers, 419-423
overview, 417-419
RFCs, 415
routing, 423-425

IP networks, QoS over, 139-143
IP telephony, PacketCable and, 158
IPSec (Internet Security Protocol)
advantages/disadvantages, 234
AH, 181-186
applicability of, 392
configuration settings, 230
DoS attacks and, 95
ESP, 186-191
IKE, 191-197
integration of, 229
overview of, 172-173
PacketCable and, 323, 326-327
SAD, 180-181

SAs, 177-179
SPD, 179-180
transparency and, 226
transport and tunnel modes, 174-177
X.509 certificates, 443
IPSec, key management, 332-337
AP Request/AP Reply messages, 332-334, 337
Rekey messages, 334, 335-336
Security Parameter Recovered messages, 334
Wake Up messages, 334
IPSec Working Group, 172
IPv4
fields and format, 418-421
IPv6 vs., 418
mutable and immutable fields, 181-182
options of, 421
IPv6
extension headers, 423
fields, 421-422
IPv4 vs., 418
mutable and immutable fields, 181-182
RFCs, 415
ISAKMP (Internet Security Associations and Key Management Protocol)
exchange types, 193, 195-197
header, 191-195
messages, 191-195
payload types, 193-195
phases, 195-197
ISNs (initial sequence numbers), 85-86, 429
ISO (International Standards Organization), 106, 148
ISPs (Internet service providers), 68, 121
ITU (International Telecommunication Union), 149, 161
iTV (interactive television), 360. *See also* DVB Multimedia Home Platform
IV (initialization vector)
CBC mode and, 30-31
performance and, 255

J

Java Secure Sockets Extension (JSSE), 380
Java security classes, DVB MHP, 380
jitter
buffering and, 265
QoS and, 132-133, 390
as side effect of security measures, 248-249
JSSE (Java Secure Sockets Extension), 380

K

KDC (key distribution center)
certificate formats, 354-355
cross-realm authentication and, 219
Kerberos and, 216-217
KEA (Key Exchange Algorithm), 212
KEKs (key-encryption keys), 35-36, 290
Kent, Stephen, 69
Kerberos, 216-220
authentication phases, 328
cross-realm authentication, 219-220
Kerberos authentication, 217-219
overview of, 216-217
PacketCable and, 328-331
public-key authentication, 220
Kerberos key management, 332-337
AP Request/AP Reply messages, 332-334, 337
Rekey messages, 334, 335-336
Security Parameter Recovered messages, 334
using with IPSec, 336-337
Wake Up messages, 334
key distribution
public-key cryptography, 42
symmetric-key cryptography, 35-36
key distribution center. *See* KDC (key distribution center)
key-encryption keys (KEKs), 35-36, 290
key escrow, NSA (National Security Association), 214
Key Exchange Algorithm (KEA), 212
key generation
impact of modulus size on, 258
RNGs and, 23-25
time required for, 251
key management. *See also* Kerberos key management
asynchronous, 265-266
with IKE, 413-414
RTP and RTCP, 343-345
security administration and, 267-268
Key Renew messages, 399
key requests/replies, DOCSIS BPI+, 289
key size, 258-260
keyed message digest. *See* HMACs (hash-based message authentication code)
keys, renewing, 396, 399, 403
keyspace, algorithms, 23
keystreams
defined, 25
generation of, 25-27
known-plaintext attacks, 96

L

Land.c, 91
last mile, 8, 115
latency
QoS and, 132, 390
as side effect of security measures, 248
Layer 2 switching, eavesdropping attacks and, 73

LDAP (Lightweight Directory Access Protocol), 450
Levy, Steven, 63
Lightweight Directory Access Protocol (LDAP), 450
Link layer security protocols, 229
LLC (logical link control), 109
LMDS (Local Multipoint Distribution Services), 9, 125
local loop, 8, 115
local system formats, CA certificates, 353-354
logical link control (LLC), 109

M

MAC addresses, 425
MAC layer frame formats, 277-279
MAC layer protocols, 153
MAC (media access control), 20, 109
man-in-the-middle attacks, 97-98, 438
manageability, security side effects and, 266-268
maximum transmission units (MTUs), 424-425
MD5SHA-1, 402
media access control (MAC), 20, 109
media streams, securing, 384
media terminal adapters. *See* MTAs (media terminal adapters)
message digests, 36-39
  applying, 56
  characteristics of, 36-37
  comparing, 38
  as integrity mechanism, 19
  preventing routing attacks, 82-83
  security side effects of, 261-262
message integrity checks (MICs), 303
MGs, MTA-to-MG communication, 340, 343
MHP. *See* DVB Multimedia Home Platform
MICs (message integrity checks), 303
MIME messages, 232
Mitnik attack, 88-89
Mitnik, Kevin, 88
MMH (Multilinear Modular Hash), 340
modems
  analog, 6
  FDM and, 105
  telephone and, 120
modulation techniques, 117
modulus size, 258-260, 400
Morris, Robert T., Jr., 76
MPEG compression, 135
MPLS (multiprotocol label switching), 132
MSOs (multiple system operators), 119, 157
MTA certificate formats
  device, 349-350
  manufacturer CA, 348-349
  root CA, 347-348
MTAs (media terminal adapters)
  communication, 340, 343-345
  PacketCable and, 158, 161
  physical protection of keying materials, 358
  PKINIT and, 329-330
  RTP messages and, 339-340
MTUs (maximum transmission units), 424-425
Multichannel Multipoint Distribution Services (MMDS), 9, 125
multifactor authentication, 19
Multilinear Modular Hash (MMH), 340
Multimedia Home Platform (MHP). *See* DVB Multimedia Home Platform
multiple system operators (MSOs), 119, 157
MultiPrime RSA, 260
multiprotocol label switching (MPLS), 132
mutable/immutable fields, IPv4/IPv6, 181-182

N

National Institute for Standards and Technology (NIST), 51
National Security Association (NSA), 212, 214
native implementation, security protocols, 234
network applications, QoS requirements of, 134-135
network devices, 109-110, 389
network infrastructure security, 15-16
network interface cards (NICs), 229
Network layer (layer 3)
  OSI Reference Model, 108-109
  PacketCable operation at, 157, 319
network layers
  security protocols and mechanisms, 172
  selecting in design process, 391, 412
network performance, QoS, 248-249
network resources, securing, 385
network security protocols, host-based vs. security gateways, 237-244
  extent of coverage, 238-239
  implementation, configuration, and maintenance, 241-242
  integration with existing policies, 243
  list of characteristics, 244
  traffic security and, 242-243
  user contexts and, 243
  WAP Gap and, 239-241
network security protocols, implementation
  application transparency and, 225-228
  binding to data, 223-224
  BITS implementation, 234-235
  BITW implementation, 235-237
  comparing protocols, 233-234

configuration and policy settings, 230
ease of integration, 228-229
extent of coverage, 230-231
native implementation, 234
performance and, 232-233
network security protocols, IPSec. *See* IPSec (Internet Security Protocol)
network security protocols, Kerberos. *See* Kerberos
network security protocols, SSL. *See* SSL (Secure Sockets Layer)
network security protocols, standards
DOCSIS 1.1 BPI+, 152
DVB MHP, 161
H.235, 158
OpenCable CPS, 165
PacketCable, 156
network service providers (NSPs), 121
NICs (network interface cards), 229
NIST (National Institute for Standards and Technology), 51
nonrepudiation services
cryptographic mechanisms and, 388
digital signatures and, 224
overview of, 19-20
RSA algorithm, 396
NSA (National Security Association), 212, 214
NSPs (network service providers), 121

O

Oakley protocol, 191
object identifier (OID), 301
OCSP (Online Certificate Status Protocol), 448
OID (object identifier), 301
one-way downstream transmission, 8
Online Certificate Status Protocol (OCSP), 448
OpenCable Copy Protection System (CPS)
list of characteristics, 164-165
overview of, 166-167
operations support system (OSS), 274
OSI Reference Model, 106-110
Application layer (layer 7), 107
chart of, 106
Network layer (layer 3), 108-109
Physical layer (layer 1), 109-110
Presentation layer (layer 6), 107
Session layer (layer 5), 107
Transport layer (layer 4), 107-108
OSS (operations support system), 274
outbound processing
AH, 185-186
ESP, 189-190
output feedback (FB) mode, 32-33, 56

P

packet encapsulation, TCP/IP, 417
packet interceptors, 235
packet loss, 133, 249, 390
Packet PDU, DOCSIS BPI+, 278-279
packet processing, inbound/outbound
AH, 185-186
ESP, 189-191
packet-replay attacks
countermeasures, 79-80
handshaking exchanges and, 401
how they work, 79
security services for, 387
packet sniffers, 70-71
packet-switching protocols, vs. cell relay, 136, 312
PacketCable, 120
Call Management Servers and, 316
cross-realm operation, 337-338
elements within, 314-316
IKE and, 326-327
interfaces and associated threats, 320-322
IPSec and, 323, 326
Kerberos and, 328-331
key management, 332-337
physical protection of keying materials, 358
protocols employed by, 317-318
secure software upgrades, 359
security services, 319, 323-325
SNMPv3 and, 327-328
threats to, 318-319
PacketCable, certificates, 345-358
certificate extensions, 346-347
certificate validation, 357-358
DF format, 355-356
KDC format, 354-355
local system formats, 353-354
MTA device formats, 349-350
MTA manufacturer formats, 348-349
MTA root format, 347-348
service provider formats, 352-353
service provider root format, 351-352
X.509 certificates and, 345
PacketCable, RTP/RTCP, 338-345
authentication, 340, 342
encrypted messages, 341
encryption transforms, 339, 342
functions of, 341-342
key management and, 343-345
overview, 339-341
voice telephony and, 338-339
PacketCable Security Specification
architecture of, 313-314
compared with DOCSIS, 313
list of characteristics, 155-157

list of technologies covered by, 312-313
overview of, 157-158
padding. *See also* keystreams
block ciphers and, 29
one-time, 26
performance and, 255
partial mesh, PKI trust model, 452-453
passive attacks, 69
password guessing attacks, 74
path discovery
PKI, 454-455
trusted paths and, 445
pay-per-view (PPV) television, 359-360
payload types, 265
PDAs, 249
PDUs (protocol data units), 112, 424
performance
algorithm selection and, 250-253
block ciphers and, 253-255
code footprint and, 256
IVs and padding and, 255
network security protocols, 232-233
stream ciphers and, 253
permanent virtual circuits (PVCs), 113
permission request file, DVB MHP, 375-378
persistent security services, 390-391
personal firewalls, 14
Physical layer (layer 1)
OSI Reference Model, 109-110
security of, 154
physical protection, PacketCable keying materials, 358
ping floods, 92
Ping utility, 426
PKCROSS, 338
PKCROSS (Hur), 220
PKCS (Public-Key Cryptography Standards), 48-51
PKCS#7, 305-307
PKI (public-key infrastructure), 448-455
CA operations, 448-450
digital certificates and, 41
DOCSIS 1.1 BPI+ and, 152
DVB MHP and, 161
H.235, 154
overview of, 448
PacketCable Security Specification, 157
path discovery and validation, 454-455
security administration and, 268
top-down hierarchy, 451
PKI trust models, 450-454
complete mesh, 451-452
hybrid, 452-453
partial mesh, 452-453
PKINIT, 329-331
PKINIT (Neuman), 220
plaintext, cryptographic algorithms, 22
POD (point-of deployment) modules, 161-162
point-to-multipoint, fixed wireless technologies, 123
point-to-point
fixed wireless technologies, 123
security protocols, 230
PPV (pay-per-view) television, 359-360
Presentation layer (layer 6), OSI Reference Model, 107
principals, Kerberos, 217
principle of least privilege, access control and, 21
private keys
decryption with, 40
public-key encryption and, 23
time requirements of, 257
PRNGs (pseudorandom number generators), 23-24
processors, algorithms and, 252
promiscuous mode, network adapters, 70-71
protocol architecture, SSL, 198
protocol data units (PDUs), 112, 424
protocol headers, 245
protocol-specific vulnerabilities, 387
protocol tuning, 264-265
pseudorandom number generators (PRNGs), 23-24
PSTN (Public Switched Telephone Network), 4, 120
public-key algorithms
applying, 56
comparing, 47-48
DOCSIS 1.0 BPI and, 151
DOCSIS 1.1 BPI+ and, 152
DVB MHP and, 162
encryption and, 23
H.235, 159
OpenCable CPS and, 165
PacketCable and, 156
security side effects of, 257-260
public-key authentication, Kerberos, 220
public-key cryptography, 40-51
comparing algorithms for, 47-48
digital envelopes, 43-44
digital signatures, 44-47
key distribution, exchange, and agreement, 42
overview of, 40-41
Public-Key Cryptography Standards (PKCS), 48-51
public keys
bandwidth and, 259
encryption with, 23, 40
operations, 264

Public Switched Telephone Network (PSTN), 4, 120
PVCs (permanent virtual circuits), 113

Q

QAM (quadrature amplitude modulation), 117
QPSK (quadrature phase shift keying), 117
quality of service (QoS)
  availability, 134-136
  bandwidth, 130-131
  cell relay vs. packet switching, 136-139
  cryptographic mechanisms and, 390
  degrees of, 136
  DOCSIS 1.0 BPI, 150
  DOCSIS 1.1 BPI+, 152
  DoS attacks and, 15
  DVB MHP, 161
  H.235, 158
  impact of security services on, 248-249
  IP networks, 139-143
  jitter, 132-133
  latency, 132
  OpenCable CPS, 164
  packet loss and, 133
  PacketCable, 155
  tunnel mode and, 245

R

radio frequency (RF) signals, 273
RAM, algorithm selection and, 252
random number generators (RNGs), 23-25
RARP (Reverse Address Resolution Protocol), 425-426
rate-adaptive, real-time applications, 129-130
RC4, 27, 168-169, 396, 398
RC6, 34
RCMM (Route Certificate Management Message), 374
Real-Time Control Protocol. *See* RTCP (Real-Time Control Protocol)
real-time multimedia. *See also* PacketCable
  jitter and, 248-249
  security side effects of, 265-266
Real-Time Transport Protocol. *See* RTP (Real-Time Transport Protocol)
real-time vs. elastic applications, 129-130
realms, Kerberos, 216
Record protocol, SSL, 199, 201-202
redundancy, availability and, 21
Rekey messages, Kerberos key management, 334, 335-336
reliable protocols, communication, 113
repeaters, Physical layer and, 109
replay attacks. *See* packet-replay attacks
repositories, digital certificates, 450
Resource Reservation Protocol (RSVP), 141
return channels
  MHP and, 360-361
  security of, 379-380
RF (radio frequency) signals, 273
RFCs
  IPSec, 172-173
  S/MIME, 232
  TCP/IP, 415
rigid, real-time applications, 129-130
Rivest, Ron, 27
RNGs (random number generators), 23-25
root CAs
  DOCSIS BPI+ and, 297
  DVB MHP, 373-374
  PKI trust model, 450
route authentication, 82-83
Route Certificate Management Message (RCMM), 374
routing
  high-performance security and, 250
  IP (Internet Protocol), 423-425
  Network layer and, 108
routing attacks, 80-83
  countermeasures, 82-83
  methods, 80
  security services for, 387
  use of routing protocols for, 80-82
routing loops, 80
routing protocols, 80-82
routing tables, 424
RSA algorithm, 48, 260, 396
RSVP (Resource Reservation Protocol), 141
RTCP (Real-Time Control Protocol)
  authentication algorithms, 342
  encrypted message format, 341
  encryption transforms, 342
  H.323 networks and, 157
  key management, 343-345
  overview of, 341-342
  packet loss and, 401-402
  QoS over IP, 142
RTP (Real-Time Transport Protocol)
  authentication algorithms, 340
  encryption transforms, 339
  H.323 networks and, 157
  key management, 343-345
  packet loss and, 401-402
  PacketCable's use of, 339-341
  QoS over IP, 142

S

S/MIME (Secure Multipurpose Internet Mail Extensions)
  advantages/disadvantages, 233

applicability of, 392
end-to-end protection with, 231
FIPS standards and, 52
overview of, 232
security integration and, 228
SAD (Security Associations Database), IPSec, 180-181
SAID (security association identifier), 276
SAs (security associations)
BPKM and, 276
bundles, 178
detection of, 179
function of, 177-178
identification parameters, 178
list of inbound/outbound in SAD, 180
SDKs (software development kits), 226-227
secret-key cryptography. *See* symmetric-key cryptography
Secure Multipurpose Internet Mail Extensions. *See* S/MIME (Secure Multipurpose Internet Mail Extensions)
Secure Sockets Layer. *See* SSL (Secure Sockets Layer)
security
balancing cost against performance, 57-58
common users, 12-15
factors in design out, 16
network infrastructure, 15-16
security administration
factors impacting, 266
infrastructure design and, 266-267
key management and, 267-268
PKI and, 268
security associations. *See* SAs (security associations)
Security Associations Database (SAD), IPSec, 180-181
security gateways, 237-244
defined, 237
design scenarios and, 391-392, 412
extent of coverage, 238-239
implementation and maintenance, 241-242
integration with existing policies, 243
list of characteristics, 244
traffic security and, 242-243
user contexts and, 243
WAP Gap and, 239-241
security mechanisms
binding to data, 412
overview of, 18
persistent, 390-391
selecting in design process, 388-390, 396, 408-411
security parameter index (SPI), 178
Security Parameter Recovered messages, Kerberos key management, 334
security protocols
designing new, 393-394, 414
selecting existing, 412-414
selecting in design process, 392-393
security services
authentication, 19
authorization, 20-21
availability, 21
binding to data, 412
confidentiality, 18
DOCSIS 1.0 BPI and, 150
DOCSIS 1.1 BPI+ and, 152
DVB MHP and, 162, 362-363
H.235, 159
integrity, 19
mechanisms for implementing, 388
nonrepudiation, 19-20
objectives of, 18
OpenCable CPS and, 164
PacketCable and, 155, 319, 323-325
persistent, 390-391
selecting in design process, 386-387, 406-408
security, side effects
algorithm selection and, 251-253
dedicated cryptographic hardware, 262
embedded devices and, 249-250
encryption and compression and, 263-264
manageability and, 266-268
message digests and HMACs, 261-262
network performance and QoS and, 248-249
overview of, 247
protocol tuning and, 264-265
public-key algorithms, 257-260
real-time multimedia and, 265-266
symmetric algorithms, 253-257
segments, TCP/IP, 417
selectors, SPD, 179-180
self signed certificates, 450
self-synchronous stream ciphers, 27
sequence numbering, 79-80
service provider formats, PacketCable
CA, 352-353
overview of, 350
root CA, 351-352
service providers, 15, 114-115
service theft, 15
session-based security, 52-53, 224
session hijacking attacks
countermeasures, 89-90
example scenarios, 86-87
how they work, 86
Mitnik attack, 88-89
session keys, Kerberos, 216
Session layer (layer 5), OSI Reference Model, 107
session reuse, SSL, 208-209
session states, SSL, 199-201, 265

SHA-1, 396, 398, 401
Shimomura, Tsutomu, 88-89
signature files, DVB MHP, 367-368
signature generation, 398-399
simplex mode, Session layer, 107
SKEME protocol, 191
Skipjack cipher suite, 212
sliding window protocol, 432
smart cards, 19, 249
Smurf, 91-92
SNMPv3
  Kerberized key management for, 332-337
  PacketCable networks and, 327-328
social engineering, 100
sockets, TCP, 430
software development kits (SDKs), 226-227
software upgrades
  DOCSIS 1.1 BPI+, 304-308
  PacketCable, 359
SONET (Synchronous Optical Network), 114
source routing, 81
SPD (Security Policy Database), 179-180
SPI (security parameter index), 178
SSL Handshake protocol. *See* Handshake protocol, SSL
SSL Record protocol, 201-202
SSL (Secure Sockets Layer)
  advantages/disadvantages, 233
  Alert protocol, 210
  applicability of, 392
  architecture of, 198
  ChangeCipherSpec protocol, 209-210
  ciphers supported, 210-216
  configuration settings, 230
  Handshake protocol, 202-209
  history of, 198
  online transaction protection, 12-13
  overview of, 197
  Record protocol, 201-202
  session and connection states, 199-201
  transparency and, 226
standards
  802.11 WEP encryption case study, 167-169
  DOCSIS, 119
  DOCSIS 1.0 BPI, 150-151, 154
  DOCSIS 1.1 BPI+, 151-155, 157
  DVB MHP, 161-163, 166
  H.235, 158-161
  OpenCabe CPS, 164-167
  PacketCable, 155-158
  PKCS, 48-51
  role of standardization, 147
  standards organization, 147-149
state
  connection states, 199-201
  session states, 199-201
  SSL Handshake protocol and, 204
state machines, BPKM, 279
  Authorization state machine, 280-285
  TEK state machine, 285-289
static CRLs, 447
statistical analysis, cryptography, 96
store-and-forward security, 52-53
stored data, securing, 231, 384
stream ciphers
  applying, 53-54
  keystream generation, 25-27
  performance and, 253
  RC4, 27
  real-time multimedia traffic and, 401
  synchronous, 27
subnet masks, IP Routing, 424
subordinate CAs, PKI trust model, 450
surge protection, 21
symmetric DSL (SDSL), 9
symmetric-key algorithms
  DOCSIS 1.0 BPI and, 151
  DOCSIS 1.1 BPI+ and, 152
  DVB MHP and, 162
  encryption and, 23
  H.235, 160
  OpenCable CPS and, 165
  PacketCable and, 156
symmetric-key cryptography, 25-40
  block ciphers, comparing, 34-35
  block ciphers, modes of operation, 29-33
  block ciphers, overview of, 27-29
  encryption speed of, 257
  HMACs, 39-40
  key distribution and management, 35-36
  message digests, 36-39
  security side effects of, 253-257
  stream ciphers, 25-27
  types of algorithms used for, 25
SYN floods, 91, 94
Synchronous Optical Network (SONET), 114
synchronous stream ciphers, 27
system performance, balancing security with, 57-58

T

TCP handshake, 429
TCP headers, 430-432
TCP/IP denial-of-service attacks, 90-95
  countermeasures, 94-95
  Land.c, 91
  list of, 90
  Smurf, 91-92
  SYN floods, 91
  Teardrop.c, 93

TCP/IP Reference Model, 110
TCP/IP Reference model, 416
TCP/IP-specific attacks, 83
TCP/IP suite, 415-435
  ARP/RARP, 425-426
  encapsulation in, 416-417
  ICMP, 426-428
  IP headers, 419-423
  IP overview, 417-419
  IP Routing, 423-425
  overview of, 415-416
  RFCs for, 415
  TCP headers, 430-432
  TCP overview, 429-430
  TCP windowing, 432-433
  UDP headers, 434
  UDP overview, 433
TCP (Transmission Control Protocol)
  headers, 430-432
  overview, 429-430
  RFCs, 415
  sequence number prediction, 85-86
  three-way handshake, 113
  windowing, 432-433
TDM (time-division multiplexing), 104
TEK state machine, BPKM, 285-289
  events, 288
  messages, 288
  operational matrix of, 287
  parameters, 289
  states of, 287
TEKs (traffic encryption keys)
  CMTS and, 290
  expiration of, 403
  session-based encryption and, 53
Telecommunications Act of 1996, 6
telecommunications, history of, 4-7
telephony
  DOCSIS and, 273-274
  history of telecommunications and, 4
  modems and, 120
  voice telephony, 338-339
teletext, 359
television
  cable television, 4
  CATV, 7, 115
  DTV, 117
  HDTV, 117
  history of telecommunications and, 4
  iTV, 360
  PPV, 359-360
TFN (Tribe FloodNet), 76
TFTP configuration files, DOCSIS 1.1 BPI+, 303-304
TGCP (Trunking Gateway Control Protocol), 343
TGS (ticket granting server), 217, 219-220
TGT (ticket-granting ticket), 218, 220
threats. *See also* attacks
  attack categories and, 387
  identifying in design process, 385-386, 406
  PacketCable, 318-322
throughput
  QoS and, 390
  slowdown, as side effect of security measures, 248
  traffic encryption algorithms and, 408-409
ticket granting server (TGS), 217, 219-220
ticket-granting ticket (TGT), 218, 220
time-division multiplexing (TDM), 104
time stamps
  countering packet-replay attacks with, 79-80
  as nonrepudiation mechanism, 20
time-synchronous tokens, 19
TLS (Transport Layer Security). *See also* SSL (Secure Sockets Layer)
  advantages/disadvantages, 233
  Alert protocol, 211-212
  applicability of, 392
  MHP return channels and, 379-380
  overview of, 198
  transparency and, 226
Token Ring networks, 104
total internal reflection, fiber optics, 11
Traceroute utility, 426
traffic encryption algorithms, 262, 396, 398
traffic encryption key. *See* TEKs (traffic encryption keys)
traffic flow analysis, 18
traffic padding, 18
traffic patterns, analysis by attackers, 72-73
traffic security, 242-243
trailers, encapsulation and, 111, 416
transmitted data, securing, 384
transparency, 225-228
  IPSec and, 226
  levels of, 228
  security gateways and, 241
  SSL/TLS and, 226-227
  Transport and Application layers and, 225-226
transport adjacency, 175
Transport layer (layer 4)
  OSI Reference Model, 107-108
  transparency and, 225-226
Transport Layer Security. *See* TLS (Transport Layer Security)
transport mode, IPSec. *See also* end-to-end protection

compared with tunnel mode, 178, 245
overview of, 174-175
Tribe FloodNet (TFN), 76
Trunking Gateway Control Protocol (TGCP), 343
trunks, of backbone networks, 114
trust models, PKI, 450-454
TTP (trusted third party), 400, 438
tunnel mode, IPSec
compared with transport mode, 178
overview of, 175-177
two-way satellite, 126-128
GEO satellite technology, 126-127
history of satellite technology, 126
overview of, 10
signal quality of, 127-128
two-way upstream transmission, cable services, 8

U

UDP (User Datagram Protocol)
headers, 434
overview, 433
RFCs, 415
unbiased output, RNGs, 23-24
uninterruptible power supply (UPSs), 21
unreliable protocols, 114
unshielded twisted pair (UTP) cable, 121
UPSs (uninterruptible power supplies), 21
user contexts, 243
UTP (unshielded twisted pair) cable, 121

V

validation, PKI, 454-455
VCs (virtual circuits), 107, 113, 429
VCTs (video conferencing terminals)
characteristics of, 395-396, 405-406
as end-user device, 394
video conferencing applications, 395
video conferencing terminals. *See* VCTs (video conferencing terminals)
voice packets, RTP, 339
voice telephony, PacketCable, 338-339
VPNs (virtual private networks)
high-performance security and, 250
IPSec tunnel mode and, 175-176
remote access security and, 13-14
vulnerabilities. *See* attacks

W

Wake Up messages, Kerberos, 334
WAP Forum, 240
WAP (Wireless Application Protocol) Gap, 240-241
weak keys, testing, 97, 99
WEP (Wired Equivalent Privacy)
advantages/disadvantages, 234
weaknesses of, 168-169
WLAN security and, 167
windowing, TCP, 430, 432-433
Wired Equivalent Privacy. *See* WEP (Wired Equivalent Privacy)
wireless communication, IEEE 802.11 standard, 167
Wireless LANs (WLANs), 167
Wireless Markup Language (WML), 240
Wireless Transport Layer Security (WTLS), 240
WLANs (Wireless LANs), 165
WML (Wireless Markup Language), 240
World Wide Web (WWW), 6
worms, Morris worm, 76
WTLS (Wireless Transport Layer Security), 240
WWW (World Wide Web), 6

X

X.500 directory, 450
X.509 certificates, 372-375
DOCSIS BPI+ and, 292
exchanging public keys, 396
extensions, 441-443
fields, 440-441
formats, 297, 440
overview of, 372-373, 440
PacketCable, 345
revoking, 374-375, 445-447
root certificates, 373-374
xDSL, 122
XML (Extensible Markup Language), 375

INTERNATIONAL CONTACT INFORMATION

AUSTRALIA
McGraw-Hill Book Company Australia Pty. Ltd.
TEL +61-2-9417-9899
FAX +61-2-9417-5687
http://www.mcgraw-hill.com.au
books-it_sydney@mcgraw-hill.com

CANADA
McGraw-Hill Ryerson Ltd.
TEL +905-430-5000
FAX +905-430-5020
http://www.mcgrawhill.ca

GREECE, MIDDLE EAST, NORTHERN AFRICA
McGraw-Hill Hellas
TEL +30-1-656-0990-3-4
FAX +30-1-654-5525

MEXICO (Also serving Latin America)
McGraw-Hill Interamericana Editores S.A. de C.V.
TEL +525-117-1583
FAX +525-117-1589
http://www.mcgraw-hill.com.mx
fernando_castellanos@mcgraw-hill.com

SINGAPORE (Serving Asia)
McGraw-Hill Book Company
TEL +65-863-1580
FAX +65-862-3354
http://www.mcgraw-hill.com.sg
mghasia@mcgraw-hill.com

SOUTH AFRICA
McGraw-Hill South Africa
TEL +27-11-622-7512
FAX +27-11-622-9045
robyn_swanepoel@mcgraw-hill.com

UNITED KINGDOM & EUROPE (Excluding Southern Europe)
McGraw-Hill Education Europe
TEL +44-1-628-502500
FAX +44-1-628-770224
http://www.mcgraw-hill.co.uk
computing_neurope@mcgraw-hill.com

ALL OTHER INQUIRIES Contact:
Osborne/McGraw-Hill
TEL +1-510-549-6600
FAX +1-510-883-7600
http://www.osborne.com
omg_international@mcgraw-hill.com

The Most Trusted Name in e-Security®

The Company

RSA Security Inc. is the most trusted name in e-security, helping organizations build secure, trusted foundations for e-business through its two-factor authentication, access management, encryption and digital signature solutions. RSA Security has the market reach, proven leadership and unrivaled technical and systems experience to address the changing security needs of e-business and bring trust to the new online economy.

A truly global company with more than 8,000 customers, RSA Security is renowned for providing technologies that help organizations conduct e-business with confidence. Headquartered in Bedford, Mass., and with offices around the world, RSA Security is a public company (NASDAQ: RSAS) with 2000 revenues of $282.7 million.

Our Markets and Products

With the proliferation of the Internet and revolutionary new e-business practices, there has never been a more critical need for sophisticated security technologies and solutions. Today, as public and private networks merge and organizations increasingly expand their businesses to the Internet, RSA Security's core offerings are continually evolving to address the critical need for e-security. As the inventor of leading security technologies, RSA Security is focused on three core disciplines of e-security.

Authentication RSA SecurID® systems are a leading solution for two-factor user authentication. RSA SecurID software is designed to protect valuable network resources by helping to ensure that only authorized users are granted access to e-mail, Web servers, intranets, extranets, network operating systems and other resources. The RSA SecurID family offers a wide range of easy-to-use authenticators, from time-synchronous tokens to smart cards, that help to create a strong barrier against unauthorized access, helping to safeguard network resources from potentially devastating accidental or malicious intrusion.

Web Access Management RSA ClearTrust® web access management software is a unified privilege management solution that helps enable secure access to Web-based resources. It is designed to work within intranets, extranets, portals and exchange structures—all while providing users with transparent, single sign-on (SSO) within or across multiple sites and domains. This easy-to-deploy, rules-based solution integrates with existing infrastructures and provides scalability to support growing e-business requirements.

Encryption RSA BSAFE® software is embedded in today's most successful Internet applications, including Web browsers, wireless devices, commerce servers, e-mail systems and virtual private network products. Built to provide implementations of standards, such as SSL, S/MIME, WTLS, IPSec and PKCS, RSA BSAFE products can save developers time and risk in their development schedules and have the security that only comes from a decade of proven, robust performance.

Digital Signatures RSA Keon® solutions are a family of interoperable software modules for managing digital certificates and creating an environment for authenticated, private and legally binding electronic communications and transactions. RSA Keon software is designed to be easy to use and interoperable with other standards-based PKI solutions and to feature enhanced security through its synergy with the RSA SecurID authentication and RSA BSAFE encryption product families.

Commitment to Interoperability

RSA Security's offerings represent a set of open, standards-based products and technologies that integrate easily into organizations' IT environments, with minimal modification to existing applications and network systems. These solutions and technologies are designed to help organizations deploy new applications securely, while maintaining corporate investments in existing infrastructure. In addition, the Company maintains active, strategic partnerships with other leading IT vendors to promote interoperability and enhanced functionality.

Strategic Partnerships

RSA Security has built its business through its commitment to interoperability. Today, through its various partnering programs, the Company has strategic relationships with hundreds of industry-leading companies—including 3COM, AOL/Netscape, BEA, AT&T, Nortel Networks, Cisco Systems, Compaq, IBM, Oracle, Microsoft and Intel—who are delivering integrated, RSA Security technology in more than 1,000 products.

Customers

RSA Security customers span a wide range of industries, including an extensive presence in the e-commerce, banking, government, telecommunications, aerospace, university and healthcare arenas. Today, more than 11 million users across 8,000 organizations—including more than half of the Fortune 100—use RSA SecurID authentication products to protect corporate data. Additionally, more than 500 companies embed RSA BSAFE software in some 1,000 applications, with a combined distribution of approximately one billion units worldwide.

Worldwide Service and Support

RSA Security offers a full complement of world-class service and support offerings to ensure the success of each customer's project or deployment through a range of ongoing customer support and professional services, including assessments, project consulting, implementation, education and training, and developer support. RSA Security's Technical Support organization is known for resolving requests in the shortest possible time, gaining customers' confidence and exceeding expectations.

Distribution

RSA Security has established a multi-channel distribution and sales network to serve the enterprise and data security markets. The Company sells and licenses its products directly to end users through its direct sales force and indirectly through an extensive network of OEMs, VARs and distributors. RSA Security supports its direct and indirect sales effort through strategic marketing relationships and programs.

Global Presence

RSA Security is a truly global e-security provider with major offices in the US, United Kingdom, Singapore and Tokyo, and representation in nearly 50 countries with additional international expansion underway. The RSA SecurWorld™ channel program brings RSA Security's products to value-added resellers and distributors worldwide, including locations in Europe, the Middle East, Africa, the Americas and Asia-Pacific.

For more information about RSA Security, please visit us at: www.rsasecurity.com.

RSA
SECURITY

RSA
SECURED

Mc Graw Hill
OSBORNE

www.ingramcontent.com/pod-product-compliance
Lightning Source LLC
Chambersburg PA
CBHW080119220326
41598CB00032B/4891